Forensics under Fire

Forensics under Fire

Are Bad Science and Dueling Experts Corrupting Criminal Justice?

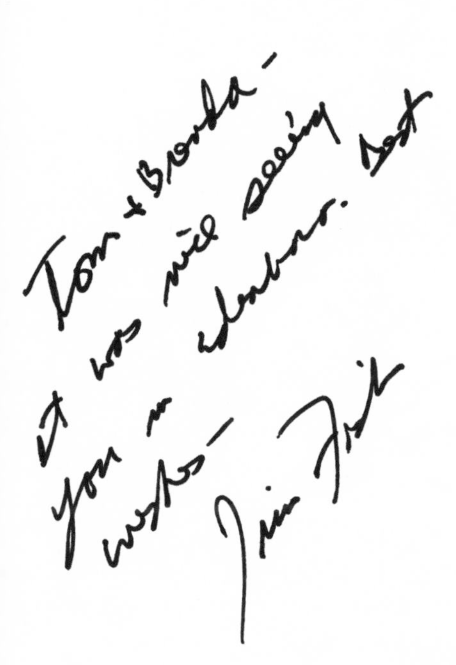

JIM FISHER

RUTGERS UNIVERSITY PRESS
NEW BRUNSWICK, NEW JERSEY, AND LONDON

LIBRARY OF CONGRESS CATALOGING-IN-PUBLICATION DATA

Fisher, Jim, 1939–
Forensics under fire : are bad science and dueling experts corrupting criminal justice? / Jim Fisher.
p. cm.
Includes bibliographical references and index.
ISBN 978–0–8135–4271–3 (hardcover : alk. paper)
1. Criminal investigation—United States. 2. Crime scene searches—United States. 3. Forensic sciences—United States. 4. Evidence, Criminal—United States I. Title.
HV8073.F522 2008
363.25—dc22 2007022050
CIP

A British Cataloging-in-Publication record for this book is available from the British Library.

Visit our Web site: http://rutgerspress.rutgers.edu

Manufactured in the United States of America

It is through clues that we form our opinion about the facts of a case. There is only one alternative: to catch the culprit red-handed.

—Theodore Reik, *The Compulsion to Confess,* 1959

Clues are tangible signs which prove—or seem to prove—that no crime can be committed by thought only and that we live in a world regulated by mechanical laws. The dead man was not killed by a ghostly hand but by a murderer of flesh and blood.

—Theodore Reik, *The Unknown Murderer,* 1945

CONTENTS

PREFACE

I've spent most of my adult life investigating crime, teaching criminal investigation, writing about the subject, and trying to figure out why so many serious crimes in the United States either go unsolved or lead to wrongful convictions. I've concluded that the law enforcement wars on drugs and terrorism have militarized the police and marginalized criminal investigation as a law enforcement function. Since forensic science is mainly in service to criminal investigation, scientific crime detection is not being fully utilized and so hasn't lived up to its full criminal justice potential.

In the 1920s forensic science pioneers and their supporters believed that one day scientific criminal investigation would significantly increase crime solution rates and at the same time reduce the dependence on the unreliable evidence produced by the third degree, eyewitness testimony, and jailhouse informants. This has not happened, at least not to a great enough extent, and to that degree, forensic science is a failed promise.

While the failure of forensic science to fully achieve its potential involves many realities beyond its control, there are problems within the profession that can be fixed by the practitioners themselves. Thinking about the nature and severity of these problems, how they affect the criminal justice system, and how they might be solved led me to write this book.

Jim Fisher
February 13, 2007

ACKNOWLEDGMENTS

I am grateful for the help of Karl Kageff, who read an early draft of the manuscript and made helpful and important suggestions.

At Rutgers University Press, Doreen Valentine's strong editorial hand helped transform a flawed manuscript into a better book. Even experienced writers can learn from a truly outstanding editor.

I am indebted to criminal investigator and author Robert L. Snow, whose review of the manuscript led to significant improvements. I would also like to thank Dr. James D. Fisher for his editorial suggestions. And finally, thanks to Veronica Fisher for her computer expertise.

Forensics under Fire

Introduction

A crime-scene photographer is snapping shots of a dead woman lying face-down in her kitchen. The blood pooled on the floor has been tracked around the house by someone wearing a pair of men's shoes. A detective is sketching the scene while a fingerprint technician examines the .22 caliber pistol lying next to the woman's body. A second criminalist gathers up the shell casing and, from the kitchen table, the handwritten suicide note. Another investigator is placing paper lunch bags over the hands of the victim's husband, who had made the 911 call. At the request of this detective, this man—let's call him Bill Smith—has changed out of his shoes, socks, and trousers. These items have been placed into separate evidence containers. Uniformed officers are sealing off the house and standing guard to keep unauthorized visitors from the site.

At the police station someone from the crime laboratory using a wad of cotton and a special solution swabs Smith's hands for traces of gunshot residue. Then the detectives who were at the crime scene question him. Smith says that his wife, Mary, had been depressed and threatening suicide. Also, she had been drinking heavily and taking sleeping pills. When he left for work that morning at eight she was still in bed. He had called her from the office at two in the afternoon, and other than sounding a little intoxicated, she seemed okay. The firearm, purchased two weeks earlier for home protection, was his. He tells the detectives that he arrived home at five o'clock, entered the house through the front door, took a couple of steps and then spotted his wife's body on the floor. He says he knew at once she was dead and did not approach her body or walk into the kitchen. Instead, he ran out of the house and called 911. A detective asks if he'd be willing to take a polygraph test. Smith says no, he doesn't believe in lie detectors. He does agree to provide samples of his handwriting and to be fingerprinted. He assures the detectives that he has nothing to hide and that he wants to help the police in their investigation.

The autopsy reveals that Mary Smith had been shot once in the chest. The powder-burn pattern on her skin suggests that the shot came from between six and twelve inches away from her. Based on early signs of postmortem lividity—light purple discoloration on the anterior plane (front) of her body caused by the settling of her blood—the forensic pathologist believes that she has been dead much longer than three hours. This means that Bill Smith could not have talked to his wife that afternoon at two o'clock. Early stages of rigor mortis—body stiffening—also suggest a time of death inconsistent with Smith's alibi that he was at work when she killed herself. The pathologist has also found, on the edge of Mary Smith's right hand, a nick made by a bullet, trauma referred to as a defense wound, evidence that belies suicide. As far as the pathologist can determine without a full toxicological analysis, the victim had not been under the influence of alcohol or drugs when she died.

The next day the crime lab issues the gunshot residue report, which reveals that Mary Smith's hands tested negative for traces of gunpowder. But it's a different story for her husband: his right hand tested positive, which means that he, not the victim, had recently fired a handgun. The serological analysis of his shoes, socks, and trousers reveals traces of AB negative blood, which happens to be his wife's blood type. His is A positive. DNA tests are pending. A footwear identification expert from the crime lab has compared the bottoms of Bill Smith's shoes with the bloody impressions in the kitchen. The forensic impression analyst has concluded that the shoes made these murder-scene marks. This means that Bill Smith lied about not approaching his wife's body in the kitchen.

The fingerprint examiner has found, on the .22 caliber pistol, several partial latents left by Bill Smith. There are no fingerprints of his wife's on the firearm. The fingerprint expert has also found several of Bill Smith's latents on the suicide note, but none from his wife. Based upon ejector and firing-pin marks on the shell casing, the crime lab's firearms identification expert has identified the death-scene pistol as the gun that fired the fatal shot.

A forensic document examiner has determined that the handwriting in the suicide note looks more like Bill Smith's than like his wife's. This handwriting expert has also discovered that the signature on a recently purchased life insurance policy for Mary Smith—naming Bill Smith the $500,000 beneficiary—is not his wife's. The document expert is analyzing the signature and Bill Smith's handwriting samples to determine if he is the forger.

Based on the fingerprint, shoe impression, blood, handwriting, gunshot residue, defense wound, range-of-shot, and time-of-death evidence, the forensic pathologist declares Mary Smith's death a criminal homicide. When detectives confront Bill Smith with the forensic science findings, he confesses. His wife had been about to divorce him, which would have left him broke. He had shot her and tried to make the death look like a suicide. He also admits forging his wife's signature on the life insurance document.

The foregoing murder story reflects how police officers, detectives, crime-scene technicians, crime-lab personnel, and forensic pathologists can work together to uncover the truth. Unfortunately, this example of forensic science in action does not represent the way most homicide cases in the United States are handled. Crime scenes are not always protected from contamination; physical evidence is often packaged improperly, lost, or left unaccounted for; forensic experts go unconsulted; and mistakes and omissions on the autopsy table either cut investigations short or send detectives down the wrong investigative path. These and other problems in the practice of forensic science allow offenders to escape justice and can also lead to the imprisonment of innocent people. The proper application of forensic science has the potential to make our criminal justice system work as designed, but until this potential is realized, our process of identifying and prosecuting offenders will remain seriously flawed.

THE PRINCIPAL ROLE of the forensic scientist is to identify physical crime-scene evidence by comparing it to known samples obtained either from a suspect's person or from an object such as a gun, shoe, or burglar tool that this person has possessed, worn, or otherwise been associated with. Forensic science relies on the principle that the criminal leaves part of himself or something that he's associated with at the scene of the crime. Evidence left at the site of a crime might include blood, semen, fingerprints, shoe impressions, bite marks, hair follicles, textile fibers, bullets, and tire tracks. Moreover, the suspect will often inadvertently take something away from the scene. A criminal might, for example, leave the crime site carrying traces of the victim's blood and tissue under his fingernails, or follicles of the victim's hair or fibers from her carpet on his clothing.

The theory that a criminal perpetrator leaves a part of himself at the scene of a crime and takes a piece of the crime site with him was postulated by Edmund Locard, who in 1911 in Lyon, France, established the world's first crime lab. Referred to as the Locard exchange principle, this idea, along with the need to reconstruct what took place at the site of a criminal act, is the basic rationale behind crime-scene investigation. The term "associative evidence" describes traces of things that, pursuant to the Locard principle, connect a suspect to or associate him with the scene of an offense.

Simply put, forensic science involves the application of hard science and technology to the investigation of crime, the proof of guilt or innocence at a criminal trial, and the resolution of factual issues in civil litigation. The most widely used components of forensic science are forensic chemistry, toxicology, biology, mineralogy, serology (bodily fluids), anthropology (bones), pathology (autopsies), and odontology (teeth and bite-mark identification). The process of firearms identification, once called forensic ballistics, incorporates knowledge and science in a variety of fields including gunsmithing, ordnance, ballistics, chemistry, metallurgy, microscopy, photography, and the forensic pathology of gunshot wounds. Latent (crime-scene) fingerprint identification and forensic document examination—the identification of unknown handwriting and the analysis of paper, ink, and printing instruments—are also part of the forensic science field.

Because inferences of guilt or innocence are drawn from the analysis of tangible things or circumstances, physical evidence is, by definition, circumstantial. For example, a suspect's latent fingerprint in safe-insulation powder at the scene of a burglary is direct proof that the suspect was at the site after the safe had been broken into. That the suspect is the safe burglar requires an inference; this requirement makes the crime-scene fingerprint evidence circumstantial. That doesn't necessarily make this evidence weak; on the contrary, unless the suspect can convincingly explain his presence at the burglary scene, he will be convicted. Circumstantial evidence in the form of physical clues and scientific analysis is, at least in theory, more reliable than such direct evidence as eyewitness identifications, confessions, and the testimony of jailhouse informants.

Police detectives today have access to science and technology that criminal investigators prior to the 1980s couldn't have imagined. For example, before 1988, detectives could not identify a crime-scene finger mark by submitting it to a data bank of computerized fingerprint impressions. The Integrated

Automatic Fingerprint Identification System (IAFIS) now allows detectives anywhere in the country to identify a crime-scene print and develop a short list of suspects. In the past, detectives solved crimes, and forensic scientists—if physical evidence had been properly gathered—helped prove the case in court. Today, a computer, not a detective, can solve *and prove* a crime. A rapist who has left his semen at the scene of a crime has, in effect, left a genetic fingerprint. The crime-solving and crime-proving potential of IAFIS and DNA, two of the greatest breakthroughs in the history of forensic science, boggles the mind.

Practitioners of forensic science fall generally into three groups: police officers who arrive at the scene of a crime and whose job it is to secure the physical evidence; crime-scene technicians responsible for finding, photographing, and packaging that physical evidence for crime-lab submission; and forensic scientists working in public and private crime laboratories who analyze the evidence and, if the occasion arises, testify in court as expert witnesses. While uniformed police officers and detectives may be trained in the recognition and handling of physical evidence, they are not scientists and do not work under laboratory conditions. As a result, a lot can—and does—go wrong between the crime-scene investigation and the courtroom.

Television shows like *CSI, Forensic Files,* and *The New Detectives* have fostered public knowledge and interest in forensic science, even ramping up scientific expectations for those involved in real-life criminal investigation and prosecution. Prosecutors call this "the CSI effect," the expectation among jurors that the prosecution will feature physical evidence and expert witnesses. The CSI effect has also caused jurors to expect crime-lab results far beyond the capacity of forensic science. In cases where there is no physical evidence, some prosecutors either eliminate potential jurors who are fans of *CSI* or downplay the necessity and importance of physical evidence as a method of proving a defendant's guilt.

While public expectations of forensic science may be high, even unreasonably so, it is also true that persistent problems within the various forensic fields have kept scientific crime detection from living up to its full potential. Because a serious shortage of qualified personnel has caused DNA-testing logjams, rapists, pedophiles, and serial killers have won time to commit more crimes. The shortage of DNA analysts has also placed a heavy burden on crime-lab personnel, creating problems of quality control. In the past few years, dozens of crime-lab DNA units have been temporarily closed when audits revealed sloppy work, scientific errors, unqualified analysts, weak supervision, poor training, and evidence contamination. Even the highly regarded FBI laboratory has experienced serious problems in DNA and other forensic areas.

Ironically, advances in DNA technology have exposed problems in other fields of forensic science. For example, DNA analysis has revealed that over the years experts have been overstating the identification value of human hair follicles and bite-mark impressions. Hundreds of criminal defendants, if not thousands, have been sent to prison on what many experts now consider unreliable forensic evidence.

A critical shortage of board-certified forensic pathologists has also adversely affected the overall quality of homicide investigation. Overworked forensic pathologists are prone to take shortcuts and make mistakes. The shortage has meant that in many cases of suspicious death, autopsies are not performed.

The field of latent fingerprint identification, while still considered the gold standard of forensic science, has recently come under attack as a result of a handful of high-profile misidentifications. These cases have revealed that not all fingerprint examiners have been properly trained, and that many have either failed or never taken proficiency tests. Questions have also been raised regarding the scientific objectivity of many fingerprint experts. This is particularly true of examiners who, as police officers, see themselves as part of a law enforcement team. Forensic scientists have to be loyal to their science even when it displeases the people who employ them, a stance that takes courage and independence.

There are fakes, incompetents, and charlatans in every profession, but over the years a series of high-profile cases have featured the so-called experts from hell, forensic scientists whose false testimony has helped convict innocent people. Many of these experts from hell are hired guns willing to testify for whoever pays them. The alarming aspect of these expert-from-hell stories is how long such forensic scientists practice before being exposed and defrocked. Just below the expert from hell on the damage scale are well-meaning but incompetent forensic scientists, experts blinded by media attention, and practitioners who bow to prosecutorial pressure. Maintaining the firewall between science and criminal prosecution is a constant challenge—one that is not always met.

Jurors are often called upon to make judgments in trials in which each side presents an expert, each of whose expert testimony contradicts the other's. If forensic science is about *hard* science, how does one explain this dueling-expert phenomenon? If there is a battery of forensic scientists for every prosecutor and for every defense, how can there be credible forensic science for anyone? When jurors are faced with opposing experts, they tend to disregard entirely the physical evidence upon which the experts are testifying.

Most of the problems in forensic science are caused by personnel shortages, poor quality control, the inherent difficulties of crime-scene investigation, the pressures imposed by the adversarial nature of our trial process, the lure of pseudoscience, and the evolving character and complexity of science itself. The criminal cases and forensic science careers featured in this book, while illustrating the various problems in scientific crime detection, do not represent the general state of forensic science in the United States. However, these examples illustrate problems that are not uncommon. But even those who view forensic science in the best possible light agree there is a lot of room for improvement.

PART ONE

Diagnosing Death

Problems in the Science and Practice of Forensic Pathology

> The expert's greatest weakness lies in the fact that he is called as an expert and is most reluctant to admit that he doesn't know all that is knowable about the subject to which he is called to testify.
>
> –Edward Huntington Williams, *The Doctor in Court,* 1929

Forensic pathologists are physicians educated and trained to determine the cause and manner of death in cases involving violent, sudden, or unexplained fatalities. The cause of death is the medical reason the person died. One cause of death is asphyxia–lack of oxygen to the brain. It occurs as a result of drowning, suffocation, manual strangulation, strangulation by ligature (such as a rope, belt, or length of cloth), crushing, or carbon monoxide poisoning. Other causes of death include blunt force trauma, gunshot wound, stabbing, slashing, poisoning, heart attack, stroke, or a sickness such as cancer, pneumonia, or heart disease.

For the forensic pathologist, the most difficult task often involves detecting the *manner* of death–natural, accidental, suicidal, or homicidal. This is because the manner of death isn't always revealed by the physical condition of the body. For example, a death resulting from a drug overdose could be the result of homicide, suicide, or accident. Knowing exactly *how* the fatal drug got into the victim's body requires additional information, data that usually come from a police investigation. A death investigator, for example, will try to find out if the victim had a history of drug abuse, or if there were signs of a struggle at the scene of the death. Had this victim attempted suicide in the past? Did the victim leave a suicide note?

Did someone have a compelling motive to kill this person? Is there evidence of a love triangle, life insurance fraud, hatred, or revenge? And finally, who had access to this person at the time of his or her death? These are basic investigative leads that could help a forensic pathologist determine the manner of death.

When the circumstances of a suspicious death are not ascertained or are sketchy, and the death was not an obvious homicide, the medical examiner (or coroner) might classify the manner of death as "undetermined." Drug overdose cases that are only slightly suspicious and therefore not thoroughly investigated often go into the books as either accident or suicide. This is true of other forms of slightly suspicious death. Because a body is found dead in the water doesn't necessarily mean this person has drowned. This victim could have been murdered and then dumped into the water. Even in a death by drowning, the person could have died after being criminally thrown from a boat or off a pier.

Indeed, there are more sudden, violent, and unexplained deaths in the United States than the nation's four hundred board-certified forensic pathologists can handle. This gruesome workload would require at least a thousand. As a result, not every death that calls for an autopsy gets one. Because there is also a shortage of qualified criminal investigators, not every death that requires an investigation gets the attention it deserves. This means we don't know exactly how many people in this country are murdered every year. And of the cases we know are homicides, about 40 percent go unsolved.

In an obvious homicide—say, when the victim has been shot twice in the back of the head or stabbed many times between the shoulder blades—there will be at least a cursory police investigation. The success of that inquiry will often depend upon what the forensic pathologist is able to determine. An analysis of the fatal wound may, for example, help identify the murder weapon. A time-of-death estimation may break a suspect's alibi or place a suspect at the scene at the estimated time of the murder. An autopsy may reveal that the victim was pregnant, intoxicated, or sexually assaulted. There may be defense wounds on the victim's hands and arms that indicate a struggle. The forensic pathologist may find traces of evidence—hairs, fibers, bodily fluids—that might eventually link the killer to the crime scene. The proper retrieval of a fatal bullet could lead to the identification of the murder gun. The forensic pathologist may also be able to tell if a body has been posed or dragged to where it was found. Perhaps the death occurred miles from where the killing took place. In cases of bodies found at fire scenes, the autopsy will reveal if the victim was killed by some aspect of the fire or died earlier of another cause.

The autopsy, along with the crime-scene investigation, is the starting point, the foundation, of a homicide investigation. If something is missed or mishandled on the autopsy table, if the forensic pathologist draws the wrong conclusion from the evidence, the investigation is doomed.

In cases of homicide where no autopsy is conducted or the medical examiner or coroner has incorrectly designated the death natural, an accident, or a suicide, bringing the killer to justice will be nearly impossible. And medical examiners and coroners are frequently under pressure from politicians, prosecutors, and homicide detectives to make manner-of-death rulings contrary to what the physical evidence tells them. Prominent people who have died under embarrassing circumstances have had their deaths sanitized by cooperating police officers, coroners, and medical examiners–for example, a death by drug overdose could go on the record as a heart attack. Or a detective or prosecutor might ask a forensic pathologist to estimate a time of death that places the main suspect at the crime scene at the moment of death. Medical examiners and coroners who think of themselves as part of a prosecution team are more likely to shade their findings in the name of law enforcement. Forensic pathologists who believe their first duty is to science are less prone to tailor their medicolegal findings to suit a prosecutor's theory of a case.

Chapter 1 features forensic pathologists whose careers crashed and burned as a result of incompetence, bias, and dishonesty. They lost their scientific independence and objectivity by making cause- and manner-of-death findings based on factors outside forensic medicine. These practitioners represent the worst of the worst.

Chapter 2 involves a pair of forensic pathologists who took shortcuts and made mistakes that damaged their professional credibility. After these pathologists rendered controversial opinions in a pair of politically charged deaths, their findings were challenged and their credibility questioned. These cases illustrate the importance of professionalism in the field of forensic pathology and the consequences for practitioners who fail to meet the highest standards of performance.

Chapter 3 deals with the medical science debates over the cause and manner of sudden infant death. The forensic disputes involve how to distinguish between crib deaths caused by fatal illnesses, homicide by suffocation, and truly unknown causes. The chapter chronicles the career of an English child-protection pediatrician, Dr. Roy Meadow, whose theories regarding the relationship between sudden infant death and murder have come under attack.

Chapter 4 further explores the medicolegal problems associated with the diagnosis of sudden infant death. The featured cases involve children whose suspicious deaths were ruled natural on the strength of breathing disorder theory, a hypothesis later discredited.

Chapter 5, which wraps up the sudden infant death discussion, is about the forensic controversy surrounding the medical symptoms–the physical evidence–of the shaken baby syndrome. Researchers have found that many of the signs of this syndrome, such as swollen brains and retinal bleeding, can also be caused by various diseases and birth defects. Until scientists and expert witnesses can agree on what is the result of disease and what of inflicted trauma, innocent defendants will go to prison and guilty people will go free. The second half of the chapter deals with the medicolegal problems associated with distinguishing between children who have been physically battered and those with broken bones that are the result of diseases, vaccines, and other noncriminal causes.

1

Forensic Pathologists from Hell

Bungled Autopsies, Bad Calls, and Blown Cases

Most forensic pathologists are hardworking, well intentioned, and competent. Even the best of them can make honest mistakes. But over the years, there have been several high-profile embarrassments to the profession. These forensic pathologists, because they were careless, incompetent, corrupt, or weak, did great harm to criminal defendants, victims of crime, and forensic science. What follows is a rogue's gallery of run-amok pathologists who represent what can go wrong in the practice of forensic medicine.

Dr. Ralph Erdmann: The Prosecutor's Dream, the Defendant's Nightmare

In 1981, twenty-five years after obtaining a medical degree in Mexico, Dr. Ralph Erdmann moved to Childress in Lubbock County, Texas, and began, on a private contract basis, doing autopsies for five small hospitals in the county. He moved to Amarillo in 1983 and performed autopsies for hire throughout the panhandle region—more than 3,000 autopsies in forty-one west Texas jurisdictions over the next decade. In 1990, at the height of his activity, he performed 480 autopsies. The following year he did 310, most in Lubbock County, from which he received an annual fee of $140,000. In the other, smaller counties, he charged $650 per autopsy. He had a large territory to cover and was constantly on the move, performing autopsies on the run.

Because he covered a rural area, Dr. Erdmann did not always work under ideal conditions. In cases of decomposing bodies, many of the smaller hospitals denied him access to autopsy space because of the stink. As a result, he performed autopsies in funeral home garages, hospital loading docks, parking lots, abandoned houses, and once on a door laid across two fifty-five-gallon drums.

It wasn't just his take-charge work ethic that made Dr. Erdmann so popular with detectives and county prosecutors. What they especially liked about him was his unabashed eagerness to tailor his autopsy findings to their law enforcement needs. If the prosecution needed a victim or suspect to have alcohol in his blood, no problem, whether a blood-alcohol test had been conducted or not. If a certain time of death was necessary to incriminate a defendant, Dr. Erdmann would provide it, even when a precise estimation was not scientifically feasible.

Because Dr. Erdmann made their job so easy, many detectives and prosecutors—according to defense lawyers and other critics of the doctor's work—turned a blind eye to his personal weirdness, sloppy work habits, questionable science, embarrassing omissions, and patent dishonesty. The fact that he was drummed out of the profession indicates that he was, even by the most lenient standards, incompetent and corrupt.

By 1992, as a number of defense attorneys began challenging and exposing his methods and findings, the outlandish nature of Dr. Erdmann's malpractice caught up with him. That year he was forced to surrender his Texas medical license to the State Board of Medical Examiners. He also pleaded guilty to charging several counties for autopsies that he had not conducted. The judge sentenced him to ten years of probation and two hundred hours of community service and ordered him to pay $17,000 in restitution. The following year Dr. Erdmann left Texas for the state of Washington.

A review of Dr. Erdmann's work explains how he had been able to perform so many autopsies. He cut corners. For example, he didn't bother to weigh the internal organs he removed, if, in fact, he removed them at all. He simply estimated their weight. He got caught when the family of a man he had autopsied noticed, in Dr. Erdmann's report, the weight of his spleen noted. This caught the family's attention because years before his death, this man's spleen had been surgically removed.

Even in situations where the cause of death was obviously murder, Erdmann didn't always get it right. In the case of a body found in a dumpster, Dr. Erdmann reported the cause of death as pneumonia. The police later arrested the man who had stolen the dead man's car, shot him in the head, and disposed of his body in that dumpster. Perhaps this man had pneumonia when he was shot, but it was the bullet that killed him. In another body-in-the-dumpster case, Dr. Erdmann lost the dead man's head, the body part containing the bullet that would have connected the suspect to the crime. Without the head, or the bullet, the suspect could not be prosecuted.

In a fatal hit-and-run case, Dr. Erdmann testified that the victim had died instantly of a broken neck. He based this finding on his examination of the fourteen-year-old victim's brain. But when the body was exhumed, another forensic pathologist found that Erdmann had not even bothered to open the boy's skull. In the case of a baby who had died in a bathtub, Dr. Erdmann determined that the victim had been killed by a blow to the stomach. This led to the arrest of the man who was in the house when the baby died. After a second forensic pathologist examined the body, the murder charge was dropped. The baby had drowned accidentally. The cause of death was asphyxia.

As reported in the *ABA Journal*, as a result of Ralph Erdmann's bungled and incomplete autopsies, the defendants in twenty murder cases had grounds to appeal their convictions. The panel of experts who had looked at three hundred of his autopsy reports—a relatively small sampling—found that one-third of the bodies had not even been cut open. When confronted with this evidence, Dr. Erdmann explained it away as clerical error. He never admitted wrongdoing and would continue to insist that he was not dishonest or incompetent—yes, he had made a few mistakes, but he had been forced to work under unfavorable conditions. He accused his detractors of being revenge-minded defense attorneys and characterized the investigation of his work a witch hunt.

Dr. Joan Wood: A Forensic Meltdown

For a criminal justice system to work, its major law enforcement players—the police, prosecutors, and forensic scientists—have to be hardworking, competent, and honest. In Florida's Pinellas and Pasco Counties between 1997 and 2000, the medical examiner's office was not up to par, and the effect on local criminal justice was disastrous. Dr. Joan E. Wood, the head of the Pinellas-Pasco Medical Examiner's Office, the principal source of the problem, resigned under fire in September 2000.

A graduate of the University of South Florida Medical School, Dr. Wood began her career as a forensic pathologist in 1975 as an associate in the

Pinellas-Pasco Medical Examiner's Office. She became the chief medical examiner in 1982, and for six years was the chairperson of Florida's Medical Examiners Commission, the body that regulates the state's forensic pathologists. Her career seemed to be on track until the mid-1990s, when she became involved in a high-profile and controversial homicide case. This case, the 1995 death of a thirty-six-year-old Scientologist named Lisa McPherson, marked the beginning of the end of Dr. Wood's career.

As revealed in court documents and reported in the *St. Petersburg Times,* the sequence of events began at 5:50 in the evening of November 18, 1995, when paramedics responded to a minor traffic accident in downtown Clearwater involving McPherson's sports utility van. She was not injured, but she took off her clothes and walked down the middle of the street, telling paramedics, "I need help. I need to talk to someone." She said she had been doing things that were wrong but didn't know what they were. The paramedics took her to the Morton Plant Hospital for psychiatric evaluation. Following her examination, a group of Scientologists from her church came to the emergency room and escorted her away, promising that she would be cared for by the church, a decision grounded in their distrust of psychiatric medicine. She was taken to the church-owned Fort Harrison Hotel in downtown Clearwater, where troubled Scientologists were taken for rest and relaxation.

On December 5, Lisa's caretakers at the hotel rushed her to a hospital in New Port Richey, a forty-five-minute drive, to see an emergency room physician who was a Scientologist. She had been at the hotel seventeen days, and when she arrived in New Port Richey, the five-foot nine-inch woman weighed 108 pounds. She was covered in bruises, unkempt in appearance, and pale. She was either dead on arrival or pronounced dead shortly thereafter.

At eleven the next morning, Dr. Robert Davis, a forensic pathologist in the Pinellas-Pasco County Medical Examiner's Office, performed the autopsy. Dr. Wood looked on. According to Dr. Davis, Lisa McPherson's death had been caused by an embolism of the left pulmonary artery, which partially obstructed the blood flow that carried oxygen from her heart to her left lung. She had therefore died of asphyxia. A thrombus (blood clot) located behind her left knee had traveled from her leg to her heart and into the lung. At the time of her death, Lisa was severely dehydrated, a factor that contributed to her demise. Her dehydration was so pronounced she would have been unresponsive for more than twenty-four hours before she died, in Dr. Davis's opinion. He believed that the blood clot behind her leg was caused by a combination of dehydration and bed-ridden immobility. Dr. Wood, instead of ruling the manner of death natural or accidental, labeled it undetermined, a manner of death that did not preclude a later finding of homicide.

Because of the condition of Lisa McPherson's body following her stay at the Fort Harrison Hotel, the Clearwater police quietly began looking into the case. Detectives determined that Lisa had been a Scientologist for eighteen years and, during the past two years, had spent about $70,000 on church-related counseling. Before the traffic accident, she had spent relaxation time at the Fort Harrison Hotel. She had worked for a Dallas publishing company that mostly employed Scientologists and had moved to Clearwater when the company relocated there about a year earlier. She had weighed between 140 and 150 pounds when taken to the Fort Harrison Hotel following the traffic accident.

Curious about just what kind of medical care one would get at the Fort Harrison Hotel, the police found that some of Lisa's caretakers had medical training, including one person who had been an anesthesiologist. This caregiver, however, had lost her license because of a drug problem. As far as the police could determine, no one at the hotel had been a licensed physician. The police also learned that during her stay, Lisa had been physically restrained, tied to her bed, and given injections of muscle relaxants and other chemicals.

When word got out that the authorities were looking into Lisa McPherson's death, church officials accused the Clearwater police of religious harassment. In January 1997, the Florida Department of Law Enforcement and the Pinellas-Pasco State Attorney's Office joined the investigation. The following month, Lisa McPherson's family filed a wrongful death suit against the church.

Looking for a second opinion regarding the cause and manner of McPherson's death, in November 1997 Wayne Andrews of the Clearwater Police Department and Agent A. L. Strope of the Florida Department of Law Enforcement traveled to Winston-Salem, North Carolina, to question Dr. George Podgorny, the Forsythe County medical examiner. Dr. Podgorny had reviewed medical records from the Morton Plant and New Port Richey hospitals; pharmacy records of drugs that had been administered to McPherson; and the Pinellas-Pasco autopsy report that chief medical examiner Wood had approved. According to police and court documents, after reviewing this material, Dr. Podgorny opined that the blood clot that had killed Lisa McPherson had been caused by her extreme dehydration and immobility. He told the investigators that if she had been given proper medical treatment and taken to a hospital when she first became ill, she might not have died. What she had needed and did not get was water, salt, vitamins, and extra oxygen. Moreover, her blood-cell count and kidney function should have been closely monitored. When asked if the blood clot in her leg could have been caused by the traffic accident, Dr. Podgorny responded emphatically that such an occurrence would be extremely rare, especially in a thirty-six-year-old woman. He pointed out that people bruise their legs all the time without getting blood clots. In the doctor's opinion, the manner of Lisa McPherson's death boiled down to improper medical care following the traffic accident.

A Pinellas County grand jury on November 13, 1998, returned a two-count indictment charging the Church of Scientology with practicing medicine without a license and abusing or neglecting an adult. The church believed that the McPherson case was being exploited by forces out to destroy the institution and, accustomed to fighting for its survival, hit back hard. One of those on the receiving end was Dr. Joan Wood, the pathologist who had opened the door to the charges with her ruling of an undetermined manner of death. In the months that followed the indictment, defense attorneys representing the church deluged Dr. Wood with subpoenas, demanding all sorts of information. These lawyers wanted her to change the manner of death to accidental on the theory that the blood clot that had killed Lisa McPherson was the result of her traffic mishap. The church also denied practicing medicine at the Fort Harrison Hotel and insisted that Lisa had been properly cared for at the retreat.

In February 2000, more than four years after the autopsy in the McPherson case, Dr. Wood, while insisting that she had not broken under pressure, nevertheless changed the manner-of-death ruling to accidental. Her decision outraged the prosecutor and the police. As far as the prosecutor was concerned, she had folded under pressure. Some of the journalists following the case speculated that pressure and stress had caused the forensic pathologist to come emotionally unglued. Whether she had been bullied into it or not, her manner-of-death reversal destroyed her credibility as an independent forensic scientist and ruined her relationship with the law enforcement community. As reported by the *St. Petersburg Times*, the prosecutor, left with an accidental manner of death, had no choice but to drop the case against the church. Dr. Wood resigned shortly thereafter.

Lisa McPherson's estate in May 2004 settled the wrongful death suit for an undisclosed amount. The case was finally over, but for Dr. Wood, its ramifications would linger long after her resignation.

David Long's Ordeal

For the former medical examiner Dr. Joan Wood, things just kept getting worse. Following her retirement, her rulings in a pair of infant death cases came under scrutiny. The review of these deaths led to further criticism of her work and of the operation of the Pinellas-Pasco Medical Examiner's Office.

In one of these cases, David Long was charged with shaking to death his seven-month-old daughter, Rebecca, in March 1998. Although the child had a history of health problems related to her premature birth at one pound, eleven ounces, Dr. Wood, finding severe hemorrhaging along the baby's spinal cord, established the cause of death as the shaken baby syndrome. This led to David Long's arrest on the charge of murder.

According to the accused father, he had gone into his daughter's room on the morning of March 7 to see why she was crying. His wife had left for work, leaving him at home with the baby, his son, and a twelve-year-old boy who had spent the night as a guest. The twelve-year-old guest in the New Port Richey home told the police that David Long stayed in Rebecca's room about two minutes, and after he had left, the baby had cried for another ten minutes. Long returned to the baby's room after she had stopped crying and found her lying facedown on her bed. She was not breathing. When the child did not respond to CPR, she was rushed to All Children's Hospital in St. Petersburg, where she was pronounced dead.

From the beginning, detectives on the case considered David Long innocent. He had no history of violence and, from all accounts, was a loving father. Moreover, the child had been unhealthy, and after Long had left her room that morning, she was heard crying. Had the father abused the baby by shaking her violently, she would not have been crying when he left her room. The detectives were so sure David Long was innocent, they made an unusual request: they asked for a second expert opinion regarding the cause of the baby's death. This did not please the prosecutor or Dr. Wood.

According to the *St. Petersburg Times*, in September 2000, shortly after she had retired, Dr. Wood received at her home in Clearwater a subpoena requiring her to give deposition testimony in connection with the Long murder case.

She responded by informing the prosecutor's office that she was ill and therefore could not appear for the session. A few months later, the sheriff showed up at her house with another Long case subpoena. He was greeted at the door by a maid who told him that the doctor was ill and not available. After that, Dr. Wood dropped out of sight and remained in hiding almost two years. Meanwhile, David Long sat in jail.

In 2001, Dr. Wood's replacement, Dr. Jon Thogmartin, and four other forensic pathologists performed a second autopsy on Rebecca Long and were astounded by what they found: the baby had died from bronchial pneumonia. They were also shocked by what they didn't find: evidence of brain or retinal bleeding, trauma associated with the shaken baby syndrome. A review of the original autopsy report revealed puzzling contradictions. In one paragraph Dr. Wood had written, "No distinct areas of hemorrhage," and in the next, "The brain . . . has hemorrhage." In her report, Dr. Wood had also referred to the victim as a boy.

As a result of the second autopsy, the prosecutor had no choice but to drop the charges against David Long. He walked out of jail a free man. The ordeal had cost him his job and had driven him into bankruptcy. He had been saved by a detective who had asserted his investigative independence, and by Dr. Thogmartin and the other pathologists who had been willing to contradict the findings of a colleague. Without their intervention, Long might have been convicted and sent to prison for a crime he didn't commit.

Freeing John Peel

The Pinellas County prosecutor in April 2002 asked Dr. Thogmartin and Dr. Stephen Nelson, the medical examiner in Polk County, to evaluate an autopsy performed by Dr. Wood in another shaken baby death. In that 1998 case, after Dr. Wood found brain hemorrhaging consistent with the shaken baby syndrome, eighteen-year-old John Peel had been charged with murdering his eight-week-old son. Although Peel denied shaking his child, he could not overcome this cause- and manner-of-death ruling and was convicted. In 2002, after Thogmartin and Nelson found no evidence supporting this cause and manner of death, John Peel was set free after serving four years behind bars.

A few months after John Peel walked out of prison, Dr. Wood, having been AWOL for more than two years, showed up at a conference of state medical examiners in Gainesville, Florida. A reporter with the *St. Petersburg Times* asked her if her disappearance had anything to do with the McPherson case, and if she planned to get back into forensic science. Dr. Wood denied that her reversal in the McPherson case had anything do with pressure from the Church of Scientology, but she did admit that after twenty-five years as a forensic pathologist, the stress of the job had finally caught up with her. She said she still had panic attacks when she walked into a courtroom. As for her cause-of-death ruling in the Peel case, she was still certain that the baby had been shaken to death. She said she was back in the field as a private consultant, but a few months later, she closed her business and went back into retirement.

In July 2005, Dr. Wood voluntarily relinquished her medical license following a state health department declaration that in the McPherson case, she had become "an advocate for the Church of Scientology," as Susan Martin reported in the *St. Petersburg Times.* Dr. Wood, now sixty, said she had given up medicine due to the accumulated stress of being a forensic pathologist for twenty-five years. She insisted, however, that she had not lost her objectivity in the McPherson case.

Dr. Richard O. Eicher: Too Many Bad Calls

Distracted by her own professional problems, Dr. Joan Wood did not run a tight medical examiner's office. Forensic pathologist Richard O. Eicher, hired by Dr. Wood in August 1999, was forced to resign about a month after the ruling change in the McPherson case. A review of his work over the preceding eight months had revealed serious mistakes in eight of his autopsies.

Dr. Wood had hired the sixty-year-old osteopathic-trained physician without speaking to his former boss at the medical examiner's office in Marathon, Florida. Had she done so, she would have learned that his boss considered Dr. Eicher disorganized and slow with his paperwork. Dr. Wood admitted she had neither conducted a preemployment background check on Dr. Eicher nor closely supervised his work after he was hired. According to her critics, who were not shy about talking to the media, she had been preoccupied with the McPherson case.

One of Dr. Eicher's cases that came under review involved the death of a ninety-two-year-old Tarpon Springs woman. She had been found lying face down on top of boards that had fallen from the broken railing of a deck seven feet above. It was obvious from the scene that the woman had fallen through the deck railing and was killed by the fall. That she had died from blunt force trauma to the chest confirmed that the manner of death was accidental. Dr. Eicher had ruled the cause of death coronary artery disease and the manner of death natural.

Arriving at the correct manner-of-death conclusion, even in a case where no one suspected foul play, is always important. In this case, the condition of the deck railing, and the fact the woman had fallen through it, might have been relevant in a wrongful-death action.

In another of Dr. Eicher's cases, one chronicled in the *St. Petersburg Times,* a thirty-seven-year-old man was found dead in the water next to his burning boat. He had been shot in the head. Dr. Eicher ruled this death a suicide because he believed the man had been shot, point-blank, in the face. In fact, the man had been shot in the back of the head. In addition, he had been recently charged with raping an eleven-year-old girl, another fact that pointed to murder. The person who had killed him may also have set fire to his boat. This death certainly wasn't suicide. Perhaps Dr. Eicher had made a rookie mistake and confused an angry-looking point-blank entrance wound to the head with an exit wound. In any event, his mistake may have allowed someone to get away with murder.

The mistake that may have sent Eicher packing involved the case of a badly burned teenager. As reported by the *St. Petersburg Times,* the Pinellas Park Fire Department, in response to a vehicle fire, found the body of seventeen-year-old Brannon Lee Jones inside the flaming car. Without knowing what had caused the vehicle fire, Dr. Eicher ruled the death a suicide. This confused the police, who considered this a strange way to take one's life, but because the death had been ruled a suicide, they did not conduct a homicide investigation. This failure to investigate infuriated the victim's family. When a second autopsy revealed that Brannon Lee Jones had been beaten to death and then set on fire, the police opened the case as a homicide, but the trail had gone cold. The homicide went into the books as unsolved.

Dr. Eicher had made too many substantive mistakes over too short a period. He had to go, and he did.

Dr. Angelo Ozoa and the Red Sash: Murder or Suicide?

After eighteen years in the Santa Clara County, California, coroner's office, seventy-three-year-old Dr. Angelo Ozoa, a Filipino American, had performed six thousand autopsies since being hired in 1980 as an assistant medical examiner. In 1993 he became chief medical examiner, replacing Dr. John Hauser, who resigned amid criticism that in several cases of highly suspicious death, he had not ruled homicide as the manner of death. But in the summer of 1998, after an extensive review of Dr. Ozoa's work, a civil grand jury in Santa Clara County recommended that Dr. Ozoa resign.

The request for his resignation had more to do with his weaknesses as an administrator than his competence as a forensic pathologist. During Dr. Ozoa's tenure as chief medical examiner, as widely reported by the media, autopsy reports had been falsified and evidence lost, mishandled, and stolen. He had hired forensic pathologists who weren't properly certified and had been criticized for allowing personnel problems such as racial tensions and employee bickering to get out of control, causing low morale in the office. His most serious problem, however, involved a manner-of-death ruling he had made in the 1995 death of a seventy-six-year-old Palo Alto woman, a case that was pending at the time the grand jury recommended that he retire. But unlike Dr. Joan Wood, Dr. Ozoa would not run and hide, nor would he refuse to testify when the case came to trial.

Josephine Galbraith's body was discovered in the guest room of her home by her daughter-in-law on September 18, 1995. The dead woman's husband, eighty-year-old Nelson Galbraith, a retired music school owner and insurance salesman, said he was watching TV in another room when she passed away. A detective from the Palo Alto Police Department and an investigator from the Santa Clara Coroner's Office arrived at the scene to find Josephine Galbraith lying face-up on the bed with three superficial cuts on her left wrist and a red bathrobe sash tied around her neck. Next to her body they found a bloody eight-inch kitchen knife, a razor blade with some blood on it, and a white, five-gallon bucket containing a small quantity of blood.

The bathrobe sash, sixty-two inches in length, had been tightly wrapped around Josephine Galbraith's neck three times. After each wrap, the sash had been tied with a double-knot. The three cuts on her wrists, referred to as "hesitation marks," were typical of the half-hearted attempt of a suicidal person who couldn't bring herself to make the deeper, more painful slashes necessary to cause death by bleeding. There was no suicide note. The coroner's investigator, a man who had been on the job twenty-eight years, recognized the scene as a suicide. The Palo Alto detective, based on the evidence at the scene, agreed with this assessment.

The investigators figured that if Josephine Galbraith had been murdered, the killer would not have made the hesitation cuts. Moreover, the pattern of blood spatter did not suggest a struggle. And the presence of the bucket intended to make the scene less messy was not consistent with a murder scene. Both investigators also knew that while people cannot manually strangle themselves (they pass out before they die), people can strangle themselves to death by ligature—the use of a rope, electrical wire, necktie, or other length of cloth such as a bathrobe sash. They also knew that Josephine Galbraith's death would have been slow enough to allow her to wrap and tie the sash three times. Given the nature of the people involved and the physical evidence at the scene, the investigators had no doubt that she had taken her own life.

Two days after the death, Dr. Ozoa performed the autopsy, a procedure that took him forty-five minutes. He found the cause of death to be "asphyxiation from ligature." This did not surprise anyone. What did shock a lot of people, including the investigators, was his manner-of-death ruling: "strangled by assailant." Dr. Ozoa had based this finding on two assumptions: Josephine Galbraith was too old and frail to have tied the three knots so tightly; and even if she did have the strength, she wouldn't have remained conscious long enough to complete the task. Since Nelson Galbraith was the only other person in the house at the time of his wife's death, if she had been murdered, he must have been the assailant.

Just days before her death, Josephine Galbraith had been diagnosed with Parkinson's disease, an illness that five years earlier had caused the slow and painful death of her sister. Even before the diagnosis, she had told friends and relatives that she wanted to kill herself. She had informed one of her sons that she would like to jump off the Golden Gate Bridge and asked another son, a physician, to provide her with drugs to end her life. He had refused.

Nelson Galbraith, the man implicitly incriminated by Dr. Ozoa's manner-of-death ruling, had severe arthritis of the hands, which would have made it difficult, if not impossible, for him to have tied the knots around his wife's neck. Moreover, there was nothing in Galbraith's background, or in his relationship with his wife, that made him a likely murderer. Investigators, urged on by the district attorney's office, nevertheless pushed forward with the case against him, albeit at a snail's pace. In the meantime, Galbraith's life became a living hell. He told journalist Loren Stein he was being referred to in the media as the "Red Sash Murderer" and was spending thousands of dollars on his defense. Ultimately his defense costs would reach $300,000.

Palo Alto police, guns drawn, stormed Galbraith's house in January 1997 and hauled him away on the charge of first-degree murder. In August 1998, almost three years after his wife's death, Nelson Galbraith, who had been allowed to make bail, went on trial for her murder. The prosecution's key witness, Dr. Ozoa, told the jury that in all his years as a forensic pathologist, he had never heard of a woman killing herself by ligature. Suicide by ligature, however, is a well-recognized method of death that is documented in textbooks and scientific journals.

The defense called forensic pathologists who disagreed with Dr. Ozoa and argued that the defendant could not have physically committed the crime. Following two and a half weeks of testimony, the jury deliberated for one day and returned a verdict of not guilty. Following the acquittal, the district attorney's office sought the opinion of a forensic pathologist in the Santa Clara Coroner's Office regarding the manner of Josephine Galbraith's death. The pathologist agreed with the defense experts: the poor woman had killed herself.

Nelson Galbraith, convinced he had been maliciously prosecuted, sued Doctor Ozoa for $10 million. To bolster his case, Galbraith spent $10,000 to have his wife's body exhumed and sent to Salt Lake City to be examined by Dr. Todd Grey, the medical examiner for the state of Utah. According to Dr. Grey, Dr. Ozoa's incorrect finding of homicide was predicated upon an incomplete autopsy, as Loren Stein notes in a long and detailed article on the Galbraith case. Dr. Ozoa had, among other things, neglected to dissect the dead woman's neck, a procedure that could have helped determine how long it had taken her to die. He had also failed to interpret the swelling in the victim's brain and the broken blood vessels in her face and eyes as evidence of a slow death. Dr. Ozoa, when confronted with Dr. Grey's opinion of his work in the Galbraith case, insisted that he had performed a complete autopsy and that the woman had been murdered.

In July 2002, almost seven years after Dr. Ozoa's autopsy in the Galbraith case, the Medical Board of California, citing Dr. Ozoa's work in that case, voted to suspend his license to practice medicine in the state. The doctor was seventy-seven. Two months later, Nelson Galbraith died. He was eighty-three, and his lawsuit was still pending in the courts.

In January 2003, Dr. Gregory Schmunk, who had replaced Dr. Ozoa as the head of the Santa Clara Coroner's Office, resigned after another civil grand jury issued a blistering report critical of his managerial skills. According to reporter William Dean Hinton, had the officials responsible for his hiring looked into Dr. Schmunk's background, they would have learned there was an outstanding warrant for his arrest in Wisconsin in connection with an alleged theft of textbooks and clothing from his previous employer. The grand jury's investigation also disclosed that in 1999, Dr. Schmunk, in an application to get a concealed gun permit, had failed to mention that in 1994 a Sacramento man had sued him for making threats. The suit was settled for $2,500. During Dr. Schmunk's six-year tenure with the Santa Clara coroner's office, as reported by the *San Jose Mercury News*, he had billed the county $42,000 in travel expenses for conferences and seminars across the country and overseas, an amount many considered excessive.

While problems continued to haunt the Santa Clara County Coroner's Office, Dr. Ozoa was busy organizing medical missions in the Philippines, operations staffed by volunteer Filipino American physicians from northern California. Nelson Galbraith's lawsuit, taken over by one of his sons, was thrown out of federal court in 2004. At present it is pending before the Ninth U.S. Circuit Court of Appeals.

Had Dr. Ozoa not classified Josephine Galbraith's death a homicide, Nelson Galbraith would not have been put through hell trying to stay out of prison and to clear his good name. But Galbraith's ongoing tragedy wasn't all Dr. Ozoa's fault. His homicide ruling did not force the district attorney to prosecute a shaky case, weakened by the debate over the cause and manner of death and the absence of a confession, an eyewitness, and any physical evidence incriminating Galbraith. If that weren't enough, the prosecution's key witness was a forensic pathologist who was under a grand jury recommendation to resign. The prosecutor had nothing to prosecute with, yet the case went forward.

What saved Nelson Galbraith, denied the protection of solid forensic science and sound prosecutorial discretion, was the safety valve of the criminal justice system, a jury of commonsense people. Although Galbraith was lucky in having a good lawyer, he did not come out of the case a winner. Anyone wrongfully accused of a crime loses, regardless of the outcome of the trial.

Dr. Charles Harlan: The Pathologist from Hell

Every so often a forensic pathologist comes along who performs badly in all aspects of the profession. Dr. Charles Harlan, a forensic pathologist who had performed autopsies in dozens of counties throughout the state of Tennessee since 1983, was in April 2005 permanently stripped of his medical license. For three years he had been under investigation by the state board of medical examiners. A three-member panel of his medicolegal colleagues had held hearings in which forensic pathologists who had reviewed his work testified that he had performed incomplete autopsies, mishandled evidence, misidentified bodies, filed false reports, and rendered erroneous cause- and manner-of-death rulings in dozens of cases. In the end, the medical board characterized Dr. Harlan's conduct in office as dishonorable, unprofessional, fraudulent, negligent, and incompetent. For Dr. Harlan, there was not a word of good news in their report.

In 1993, after the mayor of Nashville had publicly accused Dr. Harlan of running a sloppy office, he resigned as head of the Davidson County Medical Examiner's Office. Shortly thereafter, Dr. Harlan was appointed chief medical examiner for the state of Tennessee. When the state refused to extend his contract a year later, he left that job and formed his own company, Forensic Pathology Associates.

The hearings on Dr. Harlan's competency and professional conduct centered on the cause- and manner-of-death rulings in twenty-seven of his cases. In every one, his findings had been challenged by forensic pathologists who had reviewed his work. Almost half of these cases involved infants who had died suddenly in their beds, deaths Dr. Harlan had labeled sudden infant death syndrome (SIDS) cases. Because SIDS is nothing more than a general description of the unexplained deaths of infants who, for whatever reason, have stopped breathing in bed, it is not a medical cause of death.

The review of Dr. Harlan's infant death cases revealed that he had made two kinds of mistakes. He had labeled deaths as SIDS in cases where the causes of death were known, for example, high fever or positional asphyxia (suffocation in the bedcovers). In the other SIDS cases, the causes of death strongly suggested homicide, for example, acute cocaine intoxication and pulmonary edema from blunt force trauma.

Throughout the hearings on his job performance, Dr. Harlan insisted that his work in these and other cases had been professionally correct. He questioned the motives of the doctors who testified against him, many of whom where also in private practice; he accused them of trying to put him out of business to eliminate competition.

Having been so thoroughly rebuked and discredited by his peers, Dr. Harlan could either accept the verdict or question his accusers' motives. He chose the latter.

What's the Solution?

Every year in the United States there are thousands of sudden, violent, and unexplained deaths in the form of homicides, suicides, traffic fatalities, drug overdoses, infant deaths, industrial accidents, and poisonings. Each of these deaths requires an autopsy, and there aren't enough qualified people to do them. This is a serious problem. Too many unqualified people are performing autopsies; among those who are competent, many are stressed, exhausted, or burned out. Too many bodies that should be autopsied are buried without the procedure.

Given the sometimes gruesome nature of the work, and the pressures and responsibility that go with it, it is not surprising that people with medical degrees may not be drawn to forensic pathology. The practitioners profiled in this chapter were able to work as long as they did because there weren't qualified people to replace them. As was the case with Dr. Erdmann, their tenure may be extended because prosecutors find them useful, turning a blind eye to incompetence and dishonesty. Forensic pathologists like Drs. Wood and Eicher showed their fallibility as humans susceptible to stress, exhaustion, and the errors that make forensic pathology, as practiced, something less than an exact science.

When forensic pathologists have problems and make mistakes, people suffer. This eventuality makes checks and balances in the form of independent-minded detectives and prosecutors essential. The criminal justice system requires detectives who are not afraid, and prosecutors who can rise above their political ambitions.

The problem of the forensic pathologist from hell would be solved with at least double the current number of board-certified practitioners in the field. Until that happens, autopsies will be bungled, bad manner-of-death determinations will be made, and the people affected by these bad calls will suffer.

2

A Question of Credibility

Bad Reputations and the Politics of Death

A forensic pathologist is most critical in cases of sudden, unexplained, or violent death where a layperson, or even a physician without specialized training in forensic pathology, would not be able to determine, with a high degree of certainty, if the manner of death was natural, accidental, suicidal, or homicidal. The police might suspect homicide and even have a suspect, but if the forensic pathologist finds that the death is, say, accidental, there will be no criminal investigation. If the forensic pathologist has credibility, it's less likely that people will question his or her call. If the forensic pathologist, for any reason, lacks credibility, the case will remain up in the air. The police and members of the deceased person's family may think that someone has gotten away with murder; the suspected murderer, innocent or not, may live under a cloud of suspicion. No one will think that justice has been done. Credibility and competence are the two critical qualities for an effective forensic pathologist.

To be credible, a forensic pathologist has to be professionally qualified, experienced, and scientifically independent. Once a forensic pathologist has been caught taking shortcuts, making mistakes, or giving in to political pressure, that forensic scientist has lost his or her credibility. This can be especially harmful in close-call, high-profile, or politically charged cases. One would think that forensic pathologists whose reputations have been seriously damaged would be forced out of the field because no one would want to employ their services. But in the United States, a few discredited forensic pathologists are still on the job, tainting the cases in which they are involved. Dr. Michael E. Berkland is one case in point.

The Michael Berkland Debacle

Michael Berkland, without specific training or experience in forensic pathology, entered the field in 1994 when he began working in the Jackson County Medical Examiner's Office in Kansas City, Missouri. Twelve years earlier he had graduated from the University of Health Sciences in Kansas City with a degree in osteopathic medicine. He did his residency in clinical pathology at the Brooke Army Medical Center in San Antonio, Texas, then served six years as chief of pathology at the Munson Community Hospital in Fort Leavenworth, Kansas. Perhaps aware that his résumé was less than impressive in the forensics area, Dr. Berkland claimed he had completed the prestigious forensic pathology fellowship at the Bexar County, Texas, medical examiner's office under the legendary pathologist and gunshot-wound expert Dr. Vincent DiMaio. Dr. Michael Young, the Jackson County medical examiner who hired Doctor Berkland, would later learn from Dr. DiMaio that Dr. Berkland had never been enrolled in that program.

Dr. Berkland's Short Tenure in Missouri

Dr. Berkland had been on the job less than a year when Dr. Young began getting complaints from the Jackson County Prosecutor's Office. According to the prosecutors, Dr. Berkland was not writing reports on all his autopsies, and when he did complete a report, it was often months after he had performed the autopsy. Prosecutors with homicide trials approaching were clamoring for those reports. There were also complaints that Dr. Berkland didn't always make notes within a timely period after doing an autopsy, and when asked how he could remember such detail months later when writing up the report, he assured prosecutors that he had a good memory.

Dr. Berkland's close association with the police caused Dr. Young to be concerned about the pathologist's scientific objectivity. In trials where he had testified for the prosecution, he seemed to treat homicide convictions as personal victories. Doctor Young expected his pathologists to function as independent experts, not as advocates for law enforcement. He cautioned Dr. Berkland against thinking of himself as a member of a law enforcement team.

In 1995, an escapee from a county detention center was found dead in a ditch not far from the fence he had scaled. According to journalists Chris George and Denis Wright, the forensic pathologist who conducted the autopsy, Dr. Brij Mitruka, found that the prisoner had been beaten to death and had ruled the death a homicide. This finding would automatically lead to a criminal investigation to determine who had killed the inmate. The investigation, however, came to a stop after Dr. Berkland took over the case and changed the cause of death to massive blood loss from cuts made by the razor-ribbon fencing. The public defender who had represented the prisoner before his death, suspecting that Dr. Berkland might be part of a corrections facility cover-up, asked Dr. Young to review Mitruka's autopsy and Dr. Berkland's subsequent role in the case.

Dr. Young found that the victim had lost some blood, but not nearly enough to cause his death. He also discovered that Dr. Berkland had not visited the death site, examined the victim's clothing, or looked at the death-scene photographs. Berkland had also failed to question the paramedics who had responded to the scene. Dr. Young ruled that the inmate, as a result of a beating and his attempt to escape, had overtaxed his heart. Notwithstanding this ruling, no charges were filed in connection with this death.

Dr. Berkland had been on the job less than two years when, in February 1996, Dr. Young fired him. Shortly after losing his job in Missouri, Dr. Berkland was hired as an associate medical examiner in Florida's First Judicial District by the acting medical examiner, Dr. Gary Cumberland. Dr. Berkland, who had assured his new boss that the Missouri firing had been politically motivated, would be headquartered in the panhandle city of Pensacola and responsible for performing autopsies in Okaloosa and Walton Counties.

Back in Jackson County, Berkland's replacement, Dr. Sam Gulino, and Dr. Michael Young were making startling discoveries as they reviewed Dr. Berkland's work. They found that 39 percent of Dr. Berkland's autopsies were, in one way or another, incomplete. In 1994, he had failed to follow up twenty-eight autopsies with a report, and in eight autopsies, he had falsely reported that he had sectioned the corpses' brains. The reviewers also suspected that no autopsies had been conducted in several cases in which autopsy reports in fact consisted of details copied from autopsy reports in other cases. As a result of Dr. Berkland's nonreporting and false reporting, nine criminal cases were at risk of being overturned on appeal. It seemed that Michael Berkland had made a forensic mess in Jackson County, Missouri.

Dr. Young filed a formal complaint with the Missouri Administrative Hearing Commission, the state administrative body in charge of oversight in matters involving the state's medical examiners' offices. He and Dr. Gulino testified before the administrative panel in January 1998, and a month later, in reaction to this testimony, a circuit court of Jackson County issued an injunction barring Dr. Berkland from performing autopsies in Missouri. As journalists Chris George and Denis Wright report in "Unwrapped," the court held that Dr. Berkland "poses a substantial probability of serious danger to the health, safety and welfare of his patients, clients and/or the residents of Missouri." The judge characterized the pathologist's job performance as "fraud, misrepresentation, and unprofessional conduct in the practice of medicine."

The court injunction didn't affect Dr. Berkland directly because he was in Florida performing autopsies in and around Fort Walton Beach. Not until the spring of 1999, when the Missouri Administrative Hearing Commission revoked the doctor's license to practice medicine in any capacity in the state did Dr. Stephen Nelson, the chair of the Medical Examiner's Commission for the state of Florida, become aware of Dr. Berkland's status in Missouri. On July 20, 1999, based on Dr. Nelson's recommendation, the Florida Board of Osteopathic Medicine placed Dr. Berkland on temporary suspension for not informing his boss, Dr. Gary Cumberland, of the loss of his license to practice in Missouri.

As George and Wright reported, Dr. Berkland, testifying before the osteopathic panel, referred to the plagiarized autopsy passages as "proofreading errors," called his superiors in Missouri politically motivated, and decried the fact he had not been allowed to defend himself against the charges. To an AP reporter inquiring about Dr. Berkland's future in Okaloosa and Walton Counties, Dr. Gary Cumberland, his boss, said that he still had full confidence in the pathologist and had no plans to replace him. The osteopathic board issued Charles Berkland a "letter of guidance" and lifted his suspension. Dr. Berkland, considered dangerous and dishonest in Missouri, was back in action in Florida.

Dr. Berkland and Political Aide Lori Klausutis

At 8:10 in the morning of July 20, 2001, two years to the day after Dr. Berkland's temporary suspension in Florida, a constituent entered the Fort Walton Beach office of U.S. representative Joseph Scarborough and discovered Lori Klausutis, a twenty-eight-year-old congressional aide, lying dead on the office floor. The victim's car was parked outside, the lights in the office were on, and the front door was unlocked. When Dr. Berkland arrived at the scene two hours later, the area was swarming with police officers, emergency personnel, and TV reporters. A law enforcement spokesperson at the scene informed the media that it appeared as though Lori Klausutis had died sometime the previous evening. According to detectives, there were no signs of a struggle and no evidence of a break-in or robbery. Exactly how she had died would not be known until after the autopsy. But based on appearances, there was no reason to suspect foul play.

Lori and her husband, Timothy Klausutis, a Ph.D. employed at Elgin Air Force Base, had just purchased a house in nearby Niceville. She had been active in Young Republican groups and had worked in the representative's office a little over two years. At the time of her death she was working as a constituent services coordinator in the Fort Walton Beach office.

Dr. Berkland performed the autopsy that afternoon in the presence of a Fort Walton Beach detective, who took notes. The next day, a Saturday, the police told the media that the autopsy revealed no signs of physical trauma. In speaking to reporters himself, Dr. Berkland said that until he received the toxicologist's report, the results of his autopsy would remain "inconclusive." He also said, "Based on physical evidence, I feel comfortable moving the time of death back to the previous day." Although this was not his official call, Dr. Berkland said that Klausutis's death was in all likelihood "accidental due to natural causes." He said, "She has a past medical history that is significant, but it remains to be seen whether that played a role in her death." Dr. Berkland also mentioned that the deceased had been involved in a serious car accident as a teenager and may have been bothered by lingering medical ramifications from this trauma. He said he expected to have the toxicologist's findings by the first of August.

The day after Lori Klausutis was found dead, the story, as shaped by the police, Michael Berkland, and perhaps staffers in Scarborough's office, featured the victim's history of bad health. Tales of her ill health tended to throw water on the idea that she might have been the victim of foul play. And other than suicide or some bizarre office accident, what else could explain the sudden death of a twenty-eight-year-old woman? According to an employee of the local ABC television affiliate, Representative Scarborough called the station three hours after the discovery of the body to inform them that Klausutis had a "complicated" medical history that included "stroke and epilepsy," as journalists Chris George and Denis Wright reported. Scarborough's press secretary told the media, without providing details, that Klausutis had not been in perfect health. The Fort Walton Beach chief of police confirmed that his detectives were investigating the victim's medical history. Members of the Klausutis family, however, were denying that Lori had been unhealthy. She had belonged to a track club and had just posted a good time in an eight-kilometer race.

On August 6, two and a half weeks after Lori Klausutis's death, Dr. Berkland held a press conference to announce his findings. His revelation that the victim had sustained a "scratch and a bruise" on her head evoked questions as to why he had originally denied any signs of physical trauma. Dr. Berkland explained that his original denial had been "designed to prevent undue speculation about the cause of death," reported the *Pensacola News-Journal*'s Derek Pivnick. Apparently Dr. Berkland thought it was his responsibility to prohibit speculative thinking, a role not usually associated with forensic pathology. "The last thing he wanted," Dr. Berkland said, "was forty questions about a head injury." The pathologist went on: "We know for a fact she wasn't whacked in the head because of the nature of the injury." According to Doctor Berkland's analysis, Lori Klausutis, due to a valvular condition in her heart, fainted and fell, bumping her head on the corner of a desk. Even if she hadn't struck her head, she could have died from her heart condition. So her death had been either natural or accidental, or perhaps a combination of both. When asked if she had ever been treated for the heart problem, Dr. Berkland said that to his knowledge, she had not.

Dr. Berkland had intentionally misled the media regarding the condition of the victim's body, once again calling his credibility into question. For that reason, the media wanted to see a copy of the autopsy report. Given Dr. Berkland's history, there was a possibility that he had yet to write it. Three days after Dr. Berkland's press conference, Ralph Routon, the editor of the *Northwest Florida Daily News*, filed a written request for the report. In Florida, autopsy reports are matters of public record, available to anyone, for any reason. In this case, however, the state denied Routon's request. On August 24, one day after Routon wrote an editorial criticizing the authorities for withholding a public document from the press, Dr. Berkland's autopsy report was released. Five days later, the *Northwest Florida Daily News* published the details of his autopsy.

Had Ralph Routon not made an issue of the autopsy report, Dr. Berkland might never have completed it and certainly would not have handed it over to a newspaper, because the report was embarrassing. It revealed that

Dr. Berkland and the police had misled the public from the beginning. It also showed that when the pathologist finally admitted in his August 6 press conference that there had been physical trauma—a scratch and a bruise on the victim's face—he was still manipulating the public. Instead of a scratch and a bruise, Lori Klausutis had suffered a massive head injury that included a seven-and-one-quarter-inch fracture across the top of her skull, an "eggshell" fracture inside her skull above her right eye socket, a scalp contusion on the back of her head, and a subdural hematoma on the left side of her brain that caused it to swell and herniate—break out of—the left side of her skull. Doctor Berkland interpreted the subdural hematoma—called a contracoup injury because it was on the opposite side of the head from the point of impact (causing the brain to slam against the other side of the skull)—as evidence that the victim's moving head had struck a stationary object. He chose this scenario over one more suggestive of foul play: the possibility that a moving object, such as a baseball bat or metal pipe, had struck her head.

According to Dr. Berkland's postmortem analysis, Lori Klausutis had suffered a "cardiac arrhythmia" that had "halted her heart and stopped her breathing." On the way to the ground, her head hit the desk and that "blow to the head had contributed to the death because blood pooled at the point where the fracture occurred." The victim, however, had continued to breathe after falling to the floor, a fact supported by the presence around her mouth and nose of a "foam cone," a bubble of froth made up of mucus and blood. (In drowning victims, this bubble consists of a mixture of breathing passage mucus and water, an indication that the victim had been alive when he or she entered the water.)

Although officially the death had been ruled accidental, it is hard to know from Dr. Berkland's report if the manner of death was accidental or natural. The key factor in determining that she had not been murdered had to do with the source of the head injuries. Because he had sectioned the brain as part of his autopsy, Dr. Berkland knew that the head trauma had been caused by a fall instead of the wielding of a blunt object. Since photographs of the death scene that showed blood and hair on the corner of the desk were never released for public inspection, and there were no death site sketches attached to the autopsy report, one had to take Dr. Berkland at his word. But just how good was his word? He had lied about sectioning eight brains in Jackson County, Missouri, and in the Klausutis case, he had repeatedly misled the public.

By the time the *Northwest Florida Daily News* published the details of Dr. Berkland's autopsy report, the doctor's history of being fired and defrocked in Missouri had been made public. This created, in the local media, a lot of skepticism and rumors of a police cover-up. There is no question that had the truth been told about the victim's injuries on the day of her death, Dr. Berkland's manner-of-death theory notwithstanding, the public would have demanded a full criminal investigation. Moreover, media coverage of the case might have gone national, causing a political scandal for Representative Scarborough. Whether or not an investigation was warranted, it was not Michael Berkland's job to keep the lid on the details until the media had lost interest in the case. A few skeptics urged Dr. Gary Cumberland to review Dr. Berkland's autopsy, but that suggestion died on the vine. The case was over, and the media had moved on to other tragedies and other crimes.

Two years after the Klausutis case, Dr. Berkland's history in Missouri repeated itself in Florida. After six years in Fort Walton Beach, he was fired for failing to complete more than one hundred autopsy reports during 2001 and 2002. The medical examiner's commission suspended his license to perform autopsies and fined him $5,000. Dr. Berkland was also ordered to pay $2,700 in administrative costs. The commission delayed his suspension six months, during which time he was to complete the autopsy reports. It's not unreasonable to conjecture that no one would attach much credibility to these reports, completed on demand, months after the autopsies.

Dr. Berkland's reaction to his second firing was much like his response to his first dismissal. He disputed the number of uncompleted reports and blamed the backlog on being overworked. He claimed that his requests for an assistant had been denied, and that in 2002, he had performed 320 autopsies. This number is well above the 250-a-year limit suggested by the National Association of Medical Examiners. Dr. Berkland may have, in fact, been overworked. But, given his history, who could believe him?

Dr. Berkland's license suspension was lifted again, and in 2005, he was working in Florida as a forensic consultant in private practice. He was also president of the Florida Division of the International Association for Identification, a nonprofit organization made up primarily of fingerprint technicians. In January 2006, he testified as a defense expert in an Escambia County murder trial. The jury found the defendant guilty of murder in the first degree.

Dr. Charles Siebert and the Martin Anderson Boot Camp Case

The death of a fourteen-year-old boy in a Florida panhandle juvenile boot camp run by the Bay County Sheriff's Office would tarnish the career of one medical examiner and cast a bad light on forensic pathology and on the administration of justice in the Sunshine State. In June of 2005, Martin Lee Anderson and his cousins stole his grandmother's Jeep Cherokee from a church parking lot. He was sent to the boot camp, one of six in the state, on January 5, 2006, after he had violated his probation by trespassing on school property. On his first day in camp, following calisthenics in the exercise yard, Martin complained of breathing difficulties and then collapsed. His drill instructors, figuring that he was faking, held ammonia capsules under his nose to rouse him. When that didn't work, they became more aggressive. Thirty minutes later, Martin was in an ambulance on his way to a hospital in Pensacola.

Martin Anderson arrived in the emergency room in bad shape. A cranial monitor indicated that his brain had been seriously damaged. A few hours later, early in the morning of January 6, he died. Since Pensacola is in Escambia County, the medical examiner in that jurisdiction would normally have performed the autopsy. In this case, however, Bay County sheriff Frank McKeithen, the man whose office ran the boot camp, requested that the autopsy be conducted at the medical examiner's office in Panama City by Dr. Charles F. Siebert. In July 2003, the forty-four-year-old forensic pathologist had been appointed interim medical examiner of the Fourteenth Judicial Circuit, a mainly rural region covering the counties of Bay, Calhoun, Gulf, Holmes, and Jackson. A year later, following a review of his work by the Florida medical examiner's commission, Governor Jeb Bush had appointed him to the permanent position.

Following the Anderson autopsy, Dr. Siebert announced that the boy had died a natural death. According to the pathologist, Martin Anderson had the sickle-cell trait, a blood disorder that affects one out of every twelve African Americans, *Lakeland Ledger* journalist Melissa Nelson reported in "Teen's Boot Camp Death 'Natural.'" Dr. Siebert said that the physical stress brought on by the exercise, accompanied by being "restrained" by the drill instructors, had "caused red blood cells to sickle and change shape, causing a whole cascade of events that led to bleeding and hemorrhaging." The bruises and abrasions on the boy's body, according to Dr. Siebert, had been caused by the attempts to resuscitate him.

Martin Anderson's parents, Gina Jones and Robert Anderson, didn't believe their son had died a natural death. They didn't buy the story that the boot-camp drill instructors and guards had merely restrained an uncooperative youth. They believed their son had been beaten to death—murdered—and they said so.

Gina Jones and Robert Anderson weren't the only ones questioning Dr. Siebert's manner-of-death ruling and the story coming out of the boot camp. Statewide, this was the third death of a boot-camp youth in three years. Journalists, normally skeptical, asked the Florida Department of Law Enforcement (FDLE) to release the exercise-yard surveillance camera videotape. The FDLE, headed by Guy Tunnell, the former Bay County sheriff who had started the Panama City boot camp, refused. Reporters also wanted copies of Dr. Seibert's autopsy report, but that wasn't being released either. On February 13, the *Miami Herald* and CNN sued the FDLE for violation of the state's sunshine law. Four days later, bending under pressure from the dead boy's parents, civil rights groups, several state politicians, and the media, the FDLE released the thirty-minute videotape. It became immediately clear why law enforcement officials didn't want the public to see what had happened to Martin Anderson.

The videotape showed seven or eight guards hovering over Martin Anderson as he lay on the ground, his 140-pound body limp. As he lay there, the guards hit him, kneed him, sat on his body, held their hands to his face, and dragged him around. If the boy had been "uncooperative," it was because he was unconscious. A woman dressed in white, presumably some kind of nurse, stood by as the guards worked the boy over. At one point she listened to his heart with a stethoscope. Eventually the boy was lifted onto a gurney, slid into the back of an ambulance, and driven off.

If Martin Anderson had been a suspected terrorist under interrogation, his handling would have been labeled torture. The videotape shocked the nation. People had no idea fourteen-year-old kids in juvenile custody were being treated this way. Dr. Siebert, when asked by reporters if he had seen the tape prior to his autopsy ruling, said he had not. Did seeing the tape change his opinion regarding the manner of Martin Anderson's death? No, it did not.

When Frances Terry saw the boot-camp tape, she too was appalled, but for reasons of her own. On February 20, she drove to Tallahassee from her Bay County home in Blountstown to tell her story to reporters at the *St. Petersburg*

Times. She had a lot to say about Dr. Charles Siebert, and none of it was good. On September 15, 2004, a tornado spawned by hurricane Ivan had hit the family mobile home, killing her husband, James, and their thirty-five-year-old daughter, Donna Faye. Frances, as a way of coping with their deaths, had requested copies of their autopsy reports from the medical examiner's office in Panama City. She received the documents five months later in February 2005, and what she found had shocked and angered her. In Donna's autopsy report, Dr. Seibert had noted that her gallbladder was "not distended," her "uterus is not enlarged," and her ovaries and fallopian tubes were "unremarkable." This was indeed remarkable, since Donna's ovaries and uterus had been surgically removed years earlier. She had also lost her gallbladder and appendix in another operation. In this same autopsy report, Dr. Siebert made the even more startling observation that Donna's prostate gland and testes were also "unremarkable." When Frances Terry pored over his husband's autopsy report, she noted that Dr. Siebert had not mentioned the two prominent scars on his back from a 1999 operation. Terry couldn't understand how a trained forensic pathologist could make such glaring mistakes. Had the doctor actually examined the bodies? Could he be writing up boilerplate reports without actually performing the autopsies? Was this his idea of saving money, cutting costs?

To get answers to these questions, Frances Terry called the medical examiner's office in Panama City. She didn't speak to Dr. Siebert but complained to a junior staffer in the office. Later on, someone from the office called and apologized for the mix-up. Terry next met with Steve Meadows, the state attorney for the Fourteenth Judicial Circuit. Meadows in turn spoke to Dr. Siebert, who assured him that the mistakes were nothing more than "transcription errors," *St. Petersburg Times* journalists Alex Leary and Joni Jones reported. Dr. Siebert said he would correct and revise the two autopsy reports. The matter would have ended there, but when Martin Anderson died at the Bay County Boot Camp and Dr. Siebert classified the death as natural, the issue of his competence and independence made Frances Terry's allegations relevant and newsworthy. Charles Seibert's past had come back to haunt him.

On February 21, the day after the *St. Petersburg Times* broke the Frances Terry story, the *Miami Herald* came out with another revelation that raised more questions about Dr. Siebert's competence and credibility. On January 31, 2006, his state medical license had lapsed because he had not bothered to renew it. Until it was renewed, he was prohibited by law from practicing medicine, and that included performing autopsies. Thus when he autopsied Anderson, he did so illegally, without a valid license.

Two retired medical examiners, when asked to evaluate the seriousness of the mistakes contained in the tornado death autopsies, did not make excuses for Dr. Seibert. Dr. Joseph Davis, the former medical examiner of Miami–Dade County, was quoted in a *Miami Herald* article by Carol Miller and Marc Caputo: "If it's true that he [Dr. Siebert] was coming forth with a female who had male glands, that's not good." Dr. Ron Wright, the former Broward County medical examiner, quoted in the same article, said that mistakes "happen, but not very often. A few people do that sort of thing. They usually find different work."

By early March, Martin Anderson's death had become national news and a political nightmare for Sheriff Frank McKeithen, FDLE commissioner Guy Tunnel, and others associated with the boot camp. A few politicians called for the dismantling of all of Florida's boot camps. Governor Bush, sensing a growing scandal, had to do something, so he appointed a special prosecutor to look into the death. The governor asked Mark Ober, the Hillsborough County state attorney from Tampa, to gather the facts surrounding Martin Anderson's death and to determine if there had been a cover-up. On March 13, pursuant to this inquiry, Ober ordered the exhumation of Martin Anderson's body. The second autopsy would be conducted by Hillsborough County medical examiner Vernard Adams.

Dr. Adams would not be alone in the Tampa autopsy room. In addition to his staff, he would be joined by Dr. Siebert; special prosecutor Mark Ober; Benjamin Crump, the Anderson family attorney; and Dr. Michael Baden, a well-known forensic pathologist from New York featured on the HBO series *Autopsy*, who had been hired as a consultant by the family. At the completion of the autopsy, Dr. Adams said he would withhold his opinion regarding the cause and manner of death until he received the toxicological results of various tissue sample tests. The next day, Dr. Baden and Martin Anderson's parents held a press conference during which Dr. Baden said that in his thirty years of forensic pathology, he had never encountered a case in which the sickle-cell trait figured into the cause of death. "In my opinion, I think he [Martin Anderson] died from what you saw on the videotape. I think he [Dr. Siebert] made a mistake," Baden said, according to Reuters journalist Robert Green. Frank Reddick, president of the Sickle Cell Association of

Florida, agreed with Dr. Baden: "Sickle cell does not cause any type of internal bleeding, does not cause any type of trauma that could lead to this type of death," he told Sara Dorsey of Tampa Bay's Channel 10 News.

Martin Anderson's death was now not only an absolute tragedy for the Anderson family but also an embarrassment to the state's medical examiner system. The videotape and a celebrity forensic expert versus a sloppy pathologist who had let his medical license expire and found testicles on a woman "unremarkable" was no contest. Desperate for some kind of resolution to the case before it got worse, Mark Ober, the governor's special prosecutor, wanted Dr. Siebert to change his manner-of-death ruling to homicide. Dr. Siebert wasn't having any of that, and on March 16, he issued a statement, reported by *New York Times* journalist Abby Goodnough, that contradicted what had become the conventional wisdom in the case:

> My conclusion, based on more than a decade of practice, is that the exertion from exercise triggered Mr. Anderson's sickle cell trait which caused Disseminated Intravascular Coagulation (DIC), resulting in hemorrhaging. [In DIC, small blood clots develop throughout the bloodstream and can cause severe bleeding.]
>
> While I have the utmost respect for the pathologists involved in the second autopsy on Mr. Anderson, I stand behind my findings. My autopsy revealed that Mr. Anderson had sickle cell trait. The evidence revealed that on January 5, 2006, prior to the incident with the drill instructors, Mr. Anderson, after doing push-ups and sit-ups, collapsed to the ground while running laps and complained of having trouble breathing. This clearly indicates that Mr. Anderson was having medical problems prior to his interaction with the drill instructors. It is well documented in the medical literature that sudden idiopathic death after exercise is a risk associated with sickle cell trait.
>
> As a medical professional, I am appalled at the baseless and mean spirited accusations from special interest groups with little or no knowledge of the evidence in this case who are calling for everything from the revocation of my license to criminal charges.

Dr. Joseph Davis, the retired medical examiner of Dade County, who had not taken a position in the dispute, was quoted in a *Miami Herald* article by Carol Miller headlined "Prosecutor Has Options Despite Tests": "I don't recall, in my 40 years, any two autopsies [in the same case] that were so diametrically opposed. This is not a common occurrence." In reality, dueling manner-of-death conclusions in the same case are becoming more common every day. In fact, HBO's Dr. Baden himself had been involved in numerous such battles. For example, in May 2000, he and Dr. Cyril Wecht clashed over whether Las Vegas billionaire Ted Binion had been murdered or had committed suicide. Dr. Wecht testified that his death was a drug-induced suicide; Dr. Baden believed his death was murder by suffocation. The jury went with Dr. Baden, convicting the defendant, Rick Tabish, of the crime. Unfortunately, the kind of medicolegal dispute that occurred in Martin Anderson's boot-camp death was a fairly commonplace occurrence.

The Martin Anderson case had moved from issues of corrections brutality and questionable forensic pathology into the political arena. On March 17, Dr. Baden found himself in the state capital testifying before the Florida House Criminal Justice Committee. As Goodnough reported in the *New York Times,* "There was a time when there were hands over Anderson's mouth," Dr. Baden testified. "There may have been ammonia capsules blocking his nose, hands blocking his mouth that could contribute to asphyxia. He can't breathe, he can't get oxygen. When he leaves on that stretcher he's already mostly brain dead."

In an editorial published on the day of Dr. Baden's testimony in Tallahassee, the *Orlando Sentinel* took the position that Baden was right and Siebert was wrong. The editor wanted Charles Siebert to admit his mistake and rule Martin Anderson's death a homicide. According to the editor's reasoning: "If members of the boot-camp staff get charged with the boy's death, the existence of two conflicting autopsy reports might confuse jurors. That would make it difficult to get convictions." In justifying the view that Dr. Siebert was wrong, the editor pointed out that "numerous respected medical researchers" found the sickle cell explanation off base. The editorial writer also questioned Siebert's expertise, citing the embarrassing business in 2004 related to the Blountstown tornado victims: "In an autopsy report two years ago, he [Siebert] misidentified the sex of a grown woman, and inaccurately described the condition of the woman's father. They were killed in a tornado. Earlier this year Mr. Siebert let his medical license lapse. That reflects poorly on his attention to important details."

Normally, in disputed manner-of-death cases, there are several experts on each side of the issue. So far in the Anderson case, no forensic pathologist had come forward to side with Siebert. Although he was out there by himself, the doctor was not afraid to hit back at his critics. On March 17, in an interview with reporter Bill Kaczor of the Associated Press, Siebert accused Michael Baden of being biased in favor of the people who had employed him, suggesting that he was a hired gun: "I'm charged with coming to a truthful and unbiased manner of death, and I feel that's what I did in this case. I have nothing to gain by conspiring with anybody." Dr. Siebert said he had found no evidence in support of Dr. Baden's theory that the boot-camp guards had smothered the victim and reiterated his opinion that the boy's blood cells, because they had assumed the sickle shape, had not been able to get oxygen to his brain.

The *Miami Herald* on March 28, 2006, ran copies of e-mails FDLE commissioner Guy Tunnell had sent to Bay County sheriff Frank McKeithen and others in which he disparaged critics of his investigation into Martin Anderson's death. In a February 9 e-mail, Tunnell had assured FDLE staffers that the boot-camp video would not be made public. It "ain't gonna happen!" he wrote. Three days after the e-mails were made public, Mark Ober, the special prosecutor, kicked Tunnell and the FDLE off the Martin Anderson investigation, as the *Herald* reported under Carol Miller's byline and the headline "E-mails Lead to FDLE's Ouster in Case." In justifying this move, Ober said that Tunnell had no business sharing information about the case with employees of the agency under investigation.

In a response to an April 25, 2006, letter from Charlie Crist, Florida's attorney general, the state medical examiner's commission agreed to review other autopsies performed by Dr. Siebert. The next day, in a ninety-minute interview with editors and reporters with the *Miami Herald* reported by *Herald* journalists Carol Miller and Jacob Goldstein, Dr. Siebert stood by his findings in the Anderson case, accusing Crist, a Republican running for governor, of launching a politically motivated "witch hunt." Characterizing his Anderson conclusions as "valid" and "backed-by-science," Dr. Siebert claimed that sickle cell–related deaths, while rare, had been documented by forensic pathologists in other cases. Referring to his diagnosis in the Anderson case, Dr. Siebert said, "People just aren't believing it." The pathologist did acknowledge, however, that the boy might have recovered had boot-camp personnel, instead of encouraging him to continue exerting himself, sought medical help. He also declared that while the encounter with the guards might have aggravated Anderson's condition, "there was no trauma significant enough to contribute to his cause of death." He said his "natural" manner-of-death ruling was based on the boy's prior health situation and that manner-of-death rulings such as this were essentially "philosophical" decisions.

On May 5, 2006, a month after FDLE head Guy Tunnell resigned in the midst of the controversy surrounding his handling of the boot-camp death, special prosecutor Mark Ober announced the results of the second autopsy performed by Hillsborough County medical examiner Dr. Vernard Adams. Assisted by a pediatric critical care specialist, a pediatric hematologist, and an expert in cardiac pathology, Dr. Adams found that Martin Anderson had died of asphyxia brought on by ammonia fumes, which caused the boy's vocal cords to spasm, shutting off his airway. Dr. Adams, however, drew no conclusions as to whether the manner of death was a homicide or an accident. Nevertheless, following the release of Dr. Adams's autopsy report, Attorney General Crist issued the following statement, quoted in "New Autopsy Finds Florida Teen Was Suffocated" by the AP's Mitch Stacy: "The medical examiner's report represents the first evidence-based indication that Martin's death was not by natural causes, and makes it even more compelling that unanswered questions be pursued as vigorously as possible."

Dr. Siebert, put on the defensive first by Michael Baden, then by Dr. Adams and his team, said he had ruled out asphyxiation because of the low level of carbon dioxide in Anderson's body. As Abbie Vansickle and Alex Leary reported in the *St. Petersburg Times,* he pointed out that if the boy had suffocated, his carbon dioxide level would have been high because he wouldn't have been able to exhale. "Any second-year medical student would have ruled out suffocation," he said.

On May 25, 2006, Dr. Siebert became the subject of two formal complaints filed with the Florida Medical Examiners Commission by families who accused the pathologist of using his office to cover up police wrongdoing. In a case in which Dr. Siebert had ruled suicide, the mother of the dead boy, Sean Mahoney, alleged that her son had been murdered by a drunken off-duty police officer. The basis of the second complaint involved the 1977 death of Michael Neison, in which Dr. Siebert had ruled the fatality a traffic accident. According to the dead man's family, as reported in the press and supported by court documents, Michael had been beaten to death by police officers. The day after the complaints were filed, Dr. Siebert, again on the defensive, issued the following statement, reported on Orlando's WESH Channel 2: "Filing baseless complaints against me is completely unwarranted. As I have explained numerous times, I am an impartial medical professional who issues unbiased conclusions based on medical and forensic evidence of the case. My medical opinion may not be popular, but it is my job to issue conclusions rooted in scientific fact."

On August 9, 2006, in Jupiter, Florida, an investigative panel of three forensic pathologists who had been reviewing Dr. Siebert's autopsies going back to August 2003 reported their findings to the State Medical Examiners Commission. As *Gainesville Sun* journalist Brian Skoloff reported, the panel found negligent performance in 35 of 698 autopsies, citing a pattern of insufficient description in the autopsy reports. Commission member Dr. Jon Thogmartin noted that all of Dr. Siebert's reports looked the same, suggesting the use of a template. To this, Dr. Siebert replied, "There's only so many ways you can describe what a spleen looks like." Recommending Dr. Siebert's suspension followed by a period of probation, the panel did not render judgment on the forensic pathologist's work in the Anderson case. The full commission rejected the panel's disciplinary recommendations in favor of supervised probation for the remaining ten months of Dr. Siebert's tenure. This would allow the embattled medical examiner to retain his $180,000-a-year position.

Charles Siebert, who was present during the announcement of the panel's findings and recommendation, remained defiant, as Skoloff reported. "I don't think any complaints would have been lodged if not for the Anderson case. Now everybody's jumping on the bandwagon." The forensic pathologist had thirty days to either accept the punishment or appeal the commission's decision. In September 2006, Dr. Seibert notified the commission of his intent to appeal its disciplinary ruling. Pursuant to the procedures available to him, he would take his case to an administrative judge, a process that could take up to a year to unfold. During this time, the commission would not be allowed to impose its disciplinary mandate, the forced professional supervision. Dr. Siebert said he was appealing in an effort to reestablish his professional credibility—to, in essence, clear his name—which would make him eligible for another three-year appointment by the governor when his term ran out.

While the issue of Charles Siebert's work in the Anderson case remained professionally unresolved, the debate over the role of sickle-cell trait in the sudden deaths of young men under physical exertion raged on, as journalist Lisa Greene documented in the *St. Petersburg Times* in October 2006. This cause-of-death issue was at the core of a lawsuit filed by the family of a college football player who collapsed during practice in July 2005. The action was brought when the young man's parents learned that their nineteen-year-old son, upon his collapse, had not been immediately rushed to the hospital. By the time paramedics arrived at the university sports facility, the young man was in full cardiac arrest. He died shortly thereafter. Following the autopsy, the medical examiner in that Missouri county ruled that viral meningitis, an inflammation of brain tissues accompanied by infected cerebral fluid, had caused the death. Another pathologist, citing sickle-cell trait as the cause of death, disagreed with this ruling.

Reporters who had been covering the death of Martin Anderson linked the two cases. If sickle-cell trait had figured in the death of the football player, maybe it had caused the death of Martin Anderson too. Dr. Jon Thogmartin, the medical examiner in charge of Pinellas and Pasco Counties, noted that under rare conditions the trait can cause death. As Greene reported, he then criticized Dr. Baden for saying otherwise in the Anderson case: "Michael Baden saying it [sickle-cell trait] does not harm you, considering the literature, he may as well have walked out and said the world is flat." Greene cited a study in support of sickle-cell trait's causing death, published in the *New England Journal of Medicine* in 1987, which found that among military recruits, those with the trait had a twenty-eight times greater risk of dying during exercise than those without it. Since then, pathologists have attributed dozens of exercise-related deaths to sickle-cell trait. Perhaps Dr. Siebert's analysis was correct after all. But until this cause-of-death issue is resolved within the medicolegal community, how can we know? In the final analysis, it comes down to whose professional opinions we trust and whose we don't.

In the Anderson case, three forensic pathologists came up with three different cause-of-death theories. One pathologist called the death homicide, another natural, and the third had no opinion regarding the manner of death. There were charges of official cover-up and racism. The official handling of the boot-camp death is not a tribute to corrections, law enforcement, or forensic medicine. The case also represents a common problem in forensic science. When a forensic pathologist suffers a loss of credibility in one jurisdiction, he or she can find work in another place because of shortages of practitioners in the field. But when that forensic pathologist continues to run into professional trouble or becomes involved in high-profile or controversial deaths, the past catches up with that individual, and more cases are tarnished by that lack of credibility.

3

The Sudden Infant Death Debate

Dr. Roy Meadow, Munchausen Syndrome by Proxy, and Meadow's Law

Until 1969, whenever a presumably healthy baby died in its bed for no apparent reason, we called it "crib death" or "cot death." These terms described where, not how, the baby had died and didn't sound very scientific. But "sudden infant death syndrome," a purely descriptive term coined by a pediatrician named J. Bruce Beckwith, sounded more technical and more ominous. By describing the suddenness of the death instead of the place it occurs, the term carries an implication of violence and foul play. While breaking new ground rhetorically, the introduction of the letters SIDS into the vocabulary of forensic pathology and criminal investigation added nothing but confusion. The time would come when SIDS, in essence, meant *suspicious* infant death syndrome, a designation that sounded more than vaguely criminal.

Every year in the United States, there are about six hundred reported homicides in which the victims are under five years old. Many who have studied the sudden, violent, or unexplained deaths of children under five believe the homicide rate for that group could be twice that figure. There are, every year, about three thousand unexplained—undetermined—deaths of infants under the age of one. Most SIDS fatalities occur before the infant reaches the age of six months; they happen 80 percent of the time during the fall and winter months. Studies have shown that babies born of mothers who smoke or use opiates have a higher risk of SIDS. And male infants are more likely candidates for sudden death than are female babies. Also at higher risk, statistically, are African American infants, as well as babies who habitually sleep on their stomachs.

For the homicide detective, SIDS cases are difficult to investigate and solve because there aren't many leads to pursue. Unlike most adult murder cases, there aren't many people to interview, activities to trace, enemies to question, and so on. Suspects are usually parents or babysitters, quite often boyfriends of the mother. Unless there is compelling evidence of prior physical abuse, the presumptive manner of death will not be homicide, and detectives will not treat the parents, or anyone else, as murder suspects. Also absent from SIDS cases are some of the more traditional motives for murder—money, sex, and love. When a parent or babysitter murders an infant, the reason is usually either anger or some pathologically vague motive that is difficult to prove. In deaths where forensic pathologists and detectives have ignored evidence of foul play and the cases go uninvestigated, there is no one to speak up for the victim, no friend to kick up a fuss. Detective bureaus in larger cities are made up of units staffed by robbery, arson, burglary, rape, and homicide specialists. SIDS squads made up of investigators specifically trained and suited for cases of suspicious infant death are rare.

An infant who has been criminally suffocated may not exhibit the postmortem signs of trauma normally associated with this cause of death in an adult. Because an infant's life is so fragile and easily snuffed out, there might not be broken capillaries in the eyes or evidence of a struggle. In such cases, the autopsy, the foundation of a homicide investigation, is not always helpful. Because of these realities, there is no way to know how many cases that go on the books as SIDS are really homicides by suffocation. Ideally, forensic pathologists who perform autopsies on infants have training and education beyond standard forensic medicine. Specialists in this field should have knowledge in pediatrics and be trained to recognize subtle indications of abuse and neglect.

The Rise of Dr. Roy Meadow

In 1977, a pediatrician from England published the results of an investigation he had conducted into the cases of eighty-one infants whose deaths had been classified as either SIDS or natural death. The study by Dr. Roy Meadow of St. James University Hospital in Leeds covered a period of eighteen years. His article "Munchausen Syndrome by Proxy: The Hinterlands of Child Abuse," which appeared in the journal the *Lancet*, was shocking in its implications. Dr. Meadow claimed that these eighty-one babies had, in fact, been murdered, and that the forensic pathologists who had performed the autopsies had ignored obvious signs of physical abuse in the form of broken bones, scars, objects lodged in air passages, and toxic substances in their blood and urine. He came close to accusing some of these pathologists of helping parents, mostly mothers, of getting away with murder. In explaining why a mother would kill her baby, Dr. Meadow created another death-related syndrome, the Munchausen by proxy syndrome (MBPS). The Munchausen syndrome, a psychological disorder identified in 1951 by Richard Asher, described patients who injured themselves or made themselves sick to attract sympathy and attention. Richard Asher named the syndrome after Baron von Munchausen, a man known for telling tall tales. Dr. Meadow added "by proxy" because the people gaining sympathy and attention from illness and injuries were not hurting themselves. They were injuring their babies and children or making them sick, therefore getting attention and sympathy by proxy.

In his article, Dr. Meadow profiled some of the pediatric cases that had puzzled him in the early 1970s. For example, he was treating a young boy who had extremely high salt levels in his blood that were adversely affecting his kidneys. Because there was no way the boy could have eaten this much salt, Dr. Meadow came to suspect that the mother, a nurse, was force-feeding salt into the child through a nasal tube. When Dr. Meadow voiced his hypothesis to his colleagues at the hospital, they ridiculed him. In this case, however, the boy's mother confessed to exactly what Dr. Meadow had suspected. Her intent had been not to kill her child, but to use him as a way to make herself a center of attraction at the hospital, an environment she found exciting and romantic.

In MSBP cases, the primary motive is to threaten the victim's health, not kill him. But when an infant or child dies from this form of abuse, the manner of death is homicide and the abuser, in the eyes of the law, is a murderer.

When Dr. Meadow's article appeared, physicians all over the world sent him accounts of cases similar to the ones he had written about. Even Dr. Meadow was shocked by some of these stories—cases that involved punctured eardrums and induced blindness, as well as inflicted respiratory problems, stomach ailments, and allergy attacks. Years later, he would design a controversial experiment in which hidden cameras filmed hospital rooms where suspected MBPS victims were being treated. Of the thirty-nine children under surveillance, the cameras caught parents abusing thirty-three of them, creating breathing problems by putting their hands, bodies, or pillows over the victims' faces. Staff members monitoring nearby television screens quickly entered the hospital rooms, causing the abusers to discontinue their assaults before the children died or were seriously injured. In England and the United States, some of these videotaped episodes were later shown on commercial television. After that, MBPS was no longer an obscure psychological affliction. In 2005, for example, 1,200 cases of MBPS were reported in the United States. Not all these cases, however, involved fatalities.

In the years that followed Dr. Meadow's initial research into these child abuse and infant death cases, he came to believe that the vast majority of MBPS perpetrators were women, and that one-third of them were either nurses or women who worked in some other capacity within the health care industry. His research also suggested that many of these women were married to men who were cold and indifferent, and that at least part of the motive behind making their children ill was an attempt to emotionally energize their spouses. According to Dr. Meadow, many MBPS women also enjoyed being the focus of a hospital physician's attention in a setting they considered glamorous and exciting.

Because of his groundbreaking work on behalf of helpless and endangered children, Roy Meadow received a lot of attention himself. He was in great demand as an MBPS consultant, was asked to give speeches and presentations all over the world, and testified as an expert witness in dozens of high-profile murder trials. In England, he received a knighthood in recognition of his contribution to the field of medicine and forensic science. As a result of his testimony in homicide trials involving multiple SIDS deaths in the same family, his comment that "one [SIDS death] in a family is a tragedy, two is suspicious, and three is murder" became widely known as Meadow's law. In the United States, this would also be referred to as the "rule of three."

In 1987, ten years after Dr. Meadow introduced the startling notion of Munchausen by proxy syndrome, the murder conviction of Marybeth Tinning in Schenectady, New York, gave Meadow's law a face and enhanced its credibility. This unbelievable tale of serial child murder revealed how easy it was for a mother to get away with killing her children. It also made a mockery of forensic pathology in cases of infant and child death. For U.S. criminal justice, the outrageousness of the Tinning case would prove both a blessing and a curse.

Tragedy in the Tinning Family: Meadow's Law in Action

The series of killings began in January 1972 when Marybeth Tinning's eight-day-old daughter, Jennifer, died of acute meningitis. The natural death of her daughter brought Marybeth, a practical nurse married to an unemotional and indifferent husband who worked at the local General Electric plant, a wave of warm attention and sympathy from her coworkers and neighbors. Three weeks after Jennifer's death, Marybeth rushed her two-year-old son, Joseph, to Schenectady's Ellis Hospital, where he was pronounced dead. Without the benefit of an autopsy, the boy's cause of death was diagnosed as a viral infection accompanied by a "seizure disorder." The tragic loss of two children within a span of three weeks brought Marybeth Tinning intense sympathy and attention.

Tragedy hit the Tinning family six weeks later when four-year-old Barbara Tinning died. Following the third death of a Tinning child, the police were notified by hospital personnel, who were becoming suspicious. The pathologist, however, by identifying the cause of death as cardiac arrest, rendered a criminal investigation inappropriate. There was no point in the police looking into a death that had been ruled natural. But death kept coming to the Tinning house. A year and a half later, in December 1973, two-week-old Timothy Tinning died at Ellis Hospital while being treated for a mysterious illness. The pathologist, unable to identify the cause of death, recorded it as SIDS. In September 1975, Nathan Tinning, five months old, died of "pulmonary edema." While hospital personnel were growing increasingly suspicious, the police didn't get involved because all the deaths except one had been classified as natural, and that one had been classified as SIDS.

Three and a half years after Nathan Tinning died, Marybeth lost another child to sudden death. Mary Tinning, age two and a half, had been a healthy child who, for reasons that baffled physicians, simply died. Although she was not an infant, her death was recorded as another SIDS fatality. The sixth death of a Tinning child came more than a year after the publication of Dr. Meadow's article on Munchausen syndrome by proxy. Six deaths, one of which had been recorded as SIDS, in a family in which the mother was a nurse and the father an indifferent husband and parent made this scenario, pursuant to Dr. Meadow's theory, highly suspicious. But in Schenectady, the authorities failed to act.

After the seventh Tinning child died of undetermined causes in March 1980, sympathy for Marybeth was hardening into suspicion of murder. At the hospital, among a dwindling group of Marybeth sympathizers, there was speculation that perhaps the Tinning family had been cursed with a "death gene." On August 2, 1981, emergency personnel rushed three-year-old Michael Tinning, suffering from breathing difficulties, to the hospital, where he died shortly thereafter. The official cause of death—bronchial pneumonia—didn't stem the tide of suspicion, and the fact that Michael had been adopted laid waste to the death-gene theory.

In less than ten years, eight children from the same family had died, and no one in the health care or law enforcement communities had ever asked the person most closely associated with each fatality, Marybeth Tinning, if she had anything to do with the deaths—apparently no one involved was familiar with MSBP or Meadow's law. And a few of Marybeth's friends and neighbors still considered her a tragic, cursed figure deserving of their sympathy. While Marybeth basked in this sympathy and concern, her husband, Joe, seem unfazed by it all.

On the night of December 19, 1985, four years and three months after Michael Tinning's death, Marybeth telephoned her neighbor, also a nurse, and said, "Get over here now!" The neighbor arrived at the house and found three-month-old Tami Lynne, born when Marybeth was forty-two, lying on a changing table. The infant had turned purple and was not breathing. According to Marybeth, the infant had been fussing all evening, and when she couldn't stand it anymore, she got up from watching television to ascertain why the baby was crying. She found the child tangled in blankets and not breathing. Unable to revive her, she had gotten her husband out of bed, called her neighbor, and then telephoned for an ambulance. The emergency personnel reached the house shortly after the neighbor arrived. At St. Clare's Hospital, Tami Lynne was pronounced dead.

The forensic pathologist who performed the autopsy, even though aware of the Tinning family history, and even though the baby's purple coloration suggested death by suffocation, ruled the death SIDS. Hospital personnel, suspecting homicide, showed no sympathy for the mother.

Following the funeral service, Marybeth hosted a brunch at her house where she seemed to enjoy the company and sympathy of her guests, who couldn't help noticing how well she and her husband were coping with the death of their eighth child. Had all this tragedy drained these parents of the ability to feel? If this were the case, why would Marybeth go to the trouble of hosting a postfuneral get-together? Some of Marybeth's friends and neighbors were reluctantly coming to the realization that there was something profoundly wrong with this woman, and whatever that was had led to the deaths of her children. She didn't act insane, wasn't on drugs, and from all appearances, was not a violent person. In fact, she seemed to have been fond of her babies. But if she had murdered her children, wouldn't the police have been on her case?

The local police were suspicious of Marybeth, but because none of the autopsies had resulted in a finding of homicidal death, they had nothing to justify a criminal investigation. All they had was the statistical unlikelihood of so many mysterious deaths occurring under the same roof, to the same mother. But the eighth death was the last straw. After Marybeth's guests left the post-funeral gathering, a detective from the Schenectady Police Department came to the Tinning residence to ask Marybeth about the circumstances of Tami Lynne's death. When he entered the house, Marybeth was reported to have said, "I know what you're here for. You're going to arrest me and take me to jail." The detective did not take her into custody, but the investigation had begun.

Shortly after Tami Lynne's death, detectives from the Schenectady Police Department and investigators from the New York State Police met in Albany to review the eight autopsy reports, hospital records, and statements that had been made by ambulance and emergency room personnel. After the meeting, the consensus was that without physical evidence of child abuse, an eyewitness, or a cooperating spouse, the police needed a confession if they hoped to build a murder case. If Marybeth did not confess to killing her children, they had no case. A little help from the forensic medicine community would have gone a long way in aiding the police, but it just wasn't there. Marybeth would have to confess.

Seven weeks after Tami Lynne's death, a Schenectady detective and a state police investigator came to the Tinning house and asked Marybeth to accompany them to the state police station in Louderville for questioning. Although she knew she was not under arrest, Marybeth agreed to be interrogated about Tami Lynne's death. In the interrogation room Marybeth said she understood her *Miranda* rights and agreed to waive them. She would not, however, sign any papers, including the *Miranda* waiver form. The interrogation, which was not videotaped, proceeded. Initially, according to the police account of the interrogation, Marybeth denied doing anything to her children that would have caused their deaths. The detectives made it clear that they didn't believe this, and not long into the interrogation, they were joined by another officer, a state trooper who had known Marybeth from childhood. "I didn't do it!" she kept saying. But the officers, convinced she was lying, pressed on.

Eventually breaking down under the pressure of the interrogation, Marybeth admitted she had killed three of her children. She said she had used a pillow to smother three-month-old Tami Lynne, five-month-old Nathan, and two-week-old Timothy. She denied foul play in the other five deaths. Detectives brought her husband, Joe, into the interrogation room and she repeated her admission to him. He seemed unfazed by the news that his wife had murdered three of their children. A stenographer typed up the confession in the question-and-answer format, a transcript that ran to thirty-six pages. When describing Tami Lynne's death, Marybeth said, "I got up and went to her crib and tried to do something with her to get her to stop crying. I finally used the pillow from my bed and put it over her head. I held it until she stopped crying."

The bodies of Tami Lynne, Nathan, and Timothy were exhumed, but two of the bodies were badly decomposed, and the third turned out to be a body from a different grave. On December 16, 1986, at the preliminary hearing, Marybeth rescinded her statement, claiming that her interrogators had forced her into confessing falsely. "They were telling me what to say," she said. "A lot of time the police made a statement and then I just repeated it. These gentlemen were telling a story and I just repeated it." Marybeth's attorney asked the judge to exclude the thirty-six-page confession on the grounds it had been obtained in violation of her *Miranda* rights. The judge denied the request. Had the interrogation been videotaped, the constitutionality of Marybeth's confession might not have been an issue.

The prosecutor charged Marybeth Tinning with the murder of Tami Lynne, the last child to die, and the case went to trial. The forensic pathologist who had initially ruled the death a SIDS fatality took the stand and testified that the child had been smothered to death. Five other pathologists for the prosecution agreed with this postmortem diagnosis. The defense countered with six forensic pathologists of their own who attributed Tami Lynne's death to a variety of natural causes. One of the defense experts testified that the child had died from Wernig-Hoffman disease, a genetic disorder that attacks the spinal column.

The jury, confused by the conflicting medical testimony, relied on the defendant's confession to find her guilty. They may also have found it hard to believe that eight children from one family, including an adopted child, could have all died of natural causes. Forensic science, instead of guiding this jury to a rational verdict, had cancelled itself out. Apparently unsure if Marybeth's prime intention had been to kill the child or make her ill for the attention she would get, the jurors found her guilty of murder in the second degree, an offense that carried a maximum sentence of twenty-five years to life. At the sentence hearing, Marybeth, reading from a prepared statement, said: "I just want you to know that I played no part in the death of my daughter, Tami Lynne. . . . I did not commit this crime but will serve the time in prison to the best of my ability. However, I will never stop fighting to prove my innocence." The judge sentenced her to the maximum penalty, twenty-five years to life.

Marybeth's attorney, on the grounds that her confession had been coerced, appealed her conviction, but it was affirmed. Since her conviction, Marybeth has attracted a small group of supporters who believe that most, if not all, of her babies died of natural causes. The proponents of this view believe that multiple SIDS deaths within the same family are caused by a variety of inherited genetic defects. While in general this view would continue to gain support within the medical community, few would apply it in cases where more than four infants in one family had died suddenly.

Although the Tinning case appears to be an extreme manifestation of the Munchausen syndrome by proxy personality disorder, it is not the only case in the history of sudden infant death where an alarming number of babies have died before law enforcement authorities launched criminal investigations. Before MSBP was introduced into the vocabulary of murder, mothers in multiple death cases were less likely to come under suspicion. But even in families where eight children die mysteriously and the mother fits the MSBP profile, suspicion is not enough to sustain a conviction. In the Marybeth Tinning case, the SIDS designation in one of the deaths had been changed to homicide by suffocation, but when the defense presented five forensic pathologists who challenged that cause and manner of death, reasonable doubt kicked in. If Marybeth had not confessed, she would not have been convicted, eight dead babies and MSBP notwithstanding.

The Clark Case and the Repeal of Meadow's Law

In England, the case that started the shift in thinking about Dr. Meadow and his law began in December 1996 in a London hotel where Sally and Stephen Clark, a pair of solicitors from Manchester, were staying with their ten-week-old son, Christopher. The baby, the couple's only child, had been relatively healthy until be began bleeding out of both nostrils and his mouth. With his child's mother out of the hotel, Stephen Clark was in charge of the emergency. On the verge of panic, he called the front desk for help, then tried, without success, to stop the bleeding. With an employee of the hotel looking on, Stephen poured water over the baby's head to clear away the blood and stop his choking. After being treated at a nearby emergency room, the baby was well enough to return home with his parents. Nine days later, while at home with his mother, Christopher died in his crib.

Dr. Alan Williams, a forensic pathologist with the Home Office in London, performed the autopsy. The doctor noted a scratch and a bruise on the infant's upper lip and light bruising on his arms and legs, marks that could have been made during emergency resuscitation. Dr. Williams determined that Christopher had died from a lower respiratory infection and ruled the death natural.

Fourteen months after Christopher's death, the Clark's second baby, Harry, just eight weeks old, died suddenly in his crib. Because Harry had not been ill and was the second child in the Clark family to die in his bed, a cloud of suspicion descended on Stephen and Sally Clark. In performing the autopsy on Harry, the same Dr. Williams, a forensic pathologist without special training in pediatric forensic medicine, found retinal bleeding, swelling of the spinal cord, a healing fracture of a right rib, and a laceration on the baby's brain. The nature of this trauma led Dr. Williams to conclude that someone had shaken this baby to death. This could mean only one thing: baby Harry had been the victim of criminal homicide.

If Harry Clark had been killed, presumably by his mother who was at home with him when he died, what light did this shed on the sudden death of the first infant, Christopher? Meadow's law didn't apply, because according to the first autopsy report, Christopher's death had been caused by an identifiable illness, not SIDS. Now Dr. William's finding that Christopher had died of a respiratory infection seemed illogical. Moreover, if Sally Clark was tried for murder in Harry's death, the finding that Christopher had died naturally would weaken the case. It would make a lot more sense if Christopher had been murdered as well. Perhaps more in service to prosecutorial strategy than science, Dr. Williams changed the cause of Christopher's death to suffocation, and the manner of his death to homicide. By a stroke of Dr. Williams's pen, Christopher had been retroactively murdered. But there was still a problem of consistency. Harry had been shaken to death, not suffocated. Dr. Williams ironed out that wrinkle by changing the cause of Harry's death to suffocation. With the law enforcement tail wagging the forensic dog, the Clarks were maneuvered into a position where they would have to either confess or deny that one or both of them had killed their two children. They had gone from grieving parents to murder defendants. Suddenly, the best they could hope for was not to be convicted of homicide. As for cleansing themselves of the stigma attached to the accusation, they had no hope. They had lost their babies and their good names, and there was a good chance they'd lose their freedom as well.

Following a six-month investigation, detectives from the Cheshire Constabulary arrested Stephen and Sally Clark for murder. Sally was three months pregnant with their third child. In November 1998, Sally gave birth while in custody awaiting her trial. Ten days later, the baby, a boy, was taken from his mother and placed into foster care.

Charges against Stephen Clark were dropped for lack of evidence and he went home. Eighteen months after her arrest, the Crown brought Sally Clark to trial for the murder of Christopher and Harry. The prosecution's case was weak, because Stephen Clark had not implicated his wife and she had not confessed. The case was also weakened by the fact that Dr. Alan Williams had originally ruled Christopher's death natural and then changed the cause of Harry's death from violent shaking to suffocation.

Dr. Meadow Testifies

Dr. Williams testified that his autopsy findings constituted "overwhelming evidence" of criminal homicide. Once homicide had been established, the stage was set for Dr. Roy Meadow. At the time of this trial in November 1999, Dr. Meadow was at the peak of his fame and credibility. It's hard to imagine a more impressive witness. He was Great Britain's version of America's celebrity forensic scientist Dr. Henry Lee. In the witness box he was articulate and confident, a model expert witness. The head of the pediatrics department at St. James Hospital in Leeds, he was the scientist who had exposed to the world a deadly kind of parent, one who would make her babies ill for attention and sympathy, or even intentionally kill them. His Munchausen syndrome by proxy doctrine, as well as the so-called Meadow's law, had been applied to infanticide cases in Great Britain, the United States, and around the world. Just a year earlier, Dr. Meadow had been knighted for his service to child protection. His word on the matter of sudden infant death had indeed become law. Prosecutors loved Dr. Meadow because no expert the defense could put on the stand could be as impressive as Sir Roy. He almost always testified for the prosecution and was a murder defendant's worst nightmare.

The central issue in the Clark trial involved the causes of the children's deaths, something that Dr. Meadow, because he wasn't a forensic pathologist, could not testify on. All he would say was that when two babies in the same family die suddenly in their cribs, call it SIDS or anything else, the deaths are probably not natural. If Dr. Meadow could convince the jury that Christopher and Harry had not died naturally, Sally Clark would be found guilty of killing them both. It all came down to the likelihood of two babies in one family dying naturally in their beds. The issue at hand, therefore, involved genetics and statistics rather than forensic pathology. When Dr. Meadow testified that the odds of two SIDS fatalities in one family was one in seventy-three million, he rendered the quality of Dr. Williams's problematic work in the case irrelevant and sealed Sally Clark's fate.

The Clark defense tried to make forensic science the issue by arguing that Christopher had died of a respiratory infection and that Harry had not been suffocated. If this were true, Dr. Meadow's SIDS statistic had no relevance in this case. In a ten to two vote, the jury found Sally Clark guilty of double murder. Had she been tried in the United States, where guilty verdicts have to be unanimous, the prosecutors would have been forced to retry her or drop the case. The judge sentenced Sally Clark to life in prison.

Now Stephen Clark's wife was in prison and his son, a year old, was in foster care. Having been cleared of any wrongdoing in the death of his children, Stephen regained custody of his son and set out on a mission: to clear his wife's name and get her out of prison.

Dr. David Southall's Report

On April 27, 2000, the airing of a television documentary on the Clark case that included Stephen's account of Christopher's nosebleed at the London hotel set in motion a string of events no one could have foreseen. The ball got rolling when the program fired the imagination of another SIDS researcher, Dr. David Southall, a consultant pediatrician at North Staffordshire Hospital in Stoke-on-Trent. He was under a two-year suspension from the hospital for his covert videotaping of parents suspected of harming their babies for attention and sympathy, a research method once used by Roy Meadow. As widely reported in the UK media, Dr. Southall had also raised eyebrows by declaring that one-third of SIDS cases were criminal homicides. Like Dr. Meadow, his former teacher, Dr. Southall was in demand as a prosecution expert in infanticide trials. He had not, however, testified in the Clark case and had no knowledge of it until he saw the television documentary.

Stephen Clark's account of Christopher's nosebleed and the severity of it led Dr. Southall to conclude that the baby had been murdered by his father, not his mother. He believed that a bilateral nosebleed in a healthy baby was solid evidence that someone had tried to suffocate the infant and was so certain of his long-distance diagnosis that he telephoned the Staffordshire Police Department and reported his theory to a detective who had worked on the Clark case. Detectives don't like to be second-guessed, especially by civilians, and normally dismiss callers like Dr. Southall as either bored mystery readers or cranks. The doctor, however, wasn't a lonely Agatha Christie fan or out of his mind. He was a well-known expert in the field of sudden infant death. Even so, the police were not about to reopen a murder case in which someone had been convicted on the word of some expert who saw something on TV he didn't like. Besides, if Stephen Clark's account of the nosebleed was tantamount to a confession, why hadn't Dr. Meadow picked up on this point earlier in the case? But as a courtesy, and perhaps out of respect for Dr. Southall's status, the police asked him to submit a report detailing his position on the case.

Four months after he had called the police station, Dr. Southall submitted his report. Although he had not looked at any of the documents in the case—medical records, autopsy reports, police files, or court papers—and had not spoken to any of the principals in the investigation, Dr. David Southall characterized Stephen Clark's guilt as "beyond a reasonable doubt."

Merely registering his expert opinion with the police was not, apparently, enough for Dr. Southall. He went public with his shocking and unfounded accusation. If anything was "beyond a reasonable doubt," it was the unprofessional nature of his intrusion into the case. Stephen Clark, in addition to restoring his wife's reputation, now had to clear his own name. Taking his wife's place in prison was not how he had planned to gain her freedom.

Because he couldn't let Dr. Southall's accusation stand unchallenged, Stephen Clark filed a complaint with the General Medical Council, England's equivalent of the American Medical Association. He characterized Dr. Southall's behavior as unprofessional and reckless. He was not alone in this assessment; many of Dr. Southall's colleagues had been appalled by his interference in the Clark case. It was one thing to be colorful and flamboyant, another to be irresponsible. That kind of behavior gave the profession a bad name.

The Doubtful Assumptions of Meadow's Law

It is doubtful that Sally Clark would have been convicted had it not been for Dr. Meadow's testimony that the odds of two siblings dying naturally were one in seventy-three million. That statistic, coming from a world-renowned SIDS expert, had put her in prison for life. At the trial, no one had challenged this aspect of Dr. Meadow's testimony, but after the trial, members of the Royal Statistical Society studied the issue and concluded that the real odds were more like one in a hundred. Because Dr. Meadow's statistic was so far off the mark and had been such an important cog in the prosecution's case, the results of the society's study were forwarded to the judge who had presided over the trial.

When asked how he had arrived at his damning statistic, Dr. Meadow said he had come across it in some government report. While he apologized for the statistic, Dr. Meadow made it clear that he had not changed his mind about Sally Clark. He was no statistician, but he knew infanticide when he saw it.

But did he? In the 1990s, researchers in Great Britain and the United States began questioning the basic assumption upon which Meadow's law was based. They were finding that SIDS could, in fact, run in families. In England, Dr. David Drucker, a research pediatrician at Manchester University, had conducted a series of studies that showed that some cot deaths were associated with a genetic immune system disorder that increased the chances of a second SIDS fatality in the family. Dr. Drucker called Meadow's law "scientifically illiterate."

Pediatricians, forensic psychiatrists, and SIDS researchers in Great Britain and America were also questioning Dr. Meadow's diagnostic protocol in MSBP cases and even raising doubts as to the existence of such a personality disorder. Some of these researchers accused Dr. Meadow of cherry-picking evidence to fit into his MSBP theory. His 1977 article, "Munchausen Syndrome by Proxy: The Hinterlands of Child Abuse," the study that had made him famous and launched a child-protection movement, had never been peer reviewed. Researchers were now requesting access to the raw data behind the MSBP study. Dr. Meadow denied their requests for this information.

The revelation that Dr. Meadow's testimony in the Clark trial had been seriously flawed and misleading caused an appellate court to order a complete review of the prosecution's case. This included a reassessment of Dr. Alan Williams's cause- and manner-of-death findings. Upon review, a pair of consulting forensic pathologists found that Christopher Clark had *not* been suffocated. Finding no evidence that this baby had died of a respiratory infection, which had been Dr. Williams's original diagnosis, the pathologists recommended that the cause of this death be changed to undetermined. Because the autopsy photographs of Christopher were of such poor quality, the forensic consultants were unable to confirm the presence of the scratch on the baby's upper lip or the bruising on his arms and legs.

The consulting pathologists also disagreed with Dr. Williams's finding that Christopher's brother, Harry, had been shaken to death. They found no evidence of retinal bleeding and determined that the brain laceration had been made during the autopsy. The pathologists did confirm the existence of a healing fracture in the rib cage, an injury that could have occurred in a resuscitation attempt. According to the reviewing pathologists, the baby had a staphylococcal infection that had contributed to his natural death. Dr. Williams had found staphylococcus aureus in the baby's cerebrospinal fluid, a fact he had not disclosed at the Clark trial.

The combination of Dr. Meadow's misleading statistic and the findings of the consulting pathologists prompted a panel of appellate judges on January 23, 2003, to quash Sally Clark's homicide convictions. She walked out of prison after having spent three years and eighty-one days behind bars. Her house had been sold and her career destroyed. Her good name was forever damaged. She would live with the possibility that many people still considered her to be the killer of her babies. Sir Roy Meadow continued to make public statements like: "Despite the evidence, people still don't want to believe that mothers will hurt their children." Yes, she had been let out of prison, but on a legal technicality. Yes, forensic science had gotten her out of prison, but it had also put her in. This was not how forensic science was supposed to work. It was supposed to prevent false accusations in the first place. In the Clark case, forensic science had done more harm than good.

Forty-two-year-old Sally Clark died of a heart attack on March 16, 2007, in her Essex, England, home. In poor health and depressed, she never recovered from the ordeal of being accused of murdering her two sons.

The Assault on Meadow's Law Intensifies: The Patel Case

Five months after the appeals court set aside Sally Clark's murder convictions, Dr. Meadow was again back in court testifying as an expert witness for the Crown. The defendant, a thirty-five-year-old pharmacist from Maidenhead in Berkshire named Trupti Patel, stood accused of smothering to death, in a period of four and a half years, three of her babies. At her trial, Dr. Meadow testified that because SIDS did not run in families, it was highly unlikely that these babies had died of natural causes. In the past, Dr. Meadow's testimony had rarely been challenged, but in the wake of the new research and the Clark case, this was no longer true.

To counter Dr. Meadow's assertion, the defense put on the stand Dr. Peter Fleming, a pediatrician at Bristol Royal Children's Hospital and a professor of infant health at the University of Bristol. As journalist David Cohen reported, Dr. Fleming testified that he could "find no convincing evidence that the Patel deaths were caused by applied or imposed injury." In his expert opinion, the babies, ages two to thirteen weeks, had died from a "metabolic disorder."

The defense also called to the stand Trupti Patel's eighty-year-old grandmother, who testified that the sudden deaths of five of her own twelve children as infants had been unexplained as to cause. The intent here was to establish a Patel family history of SIDS, a reality that contradicted the underlying assumption of Meadow's law. The jury weighed the conflicting expert testimony and came down on the side of the defense. They found Trupti Patel not guilty.

Dr. Roy Meadow, the originator of Munchausen syndrome by proxy, the namesake of Meadow's law, and the knighted pediatrician once hailed as the savior of children was losing his credibility as an expert witness. What did this mean? Did it mean that mothers were getting away with murder simply because defense attorneys could now find experts willing to challenge Dr. Meadow's testimony? Or did it mean that Dr. Meadow and his flawed theory had been responsible for dozens and dozens of wrongful convictions? One thing was sure: in the field of SIDS, nothing was certain when it came to distinguishing a tragedy from a murder.

In the United Kingdom, Dr. Meadow's influence over the outcome of infanticide trials continued to erode. Angela Cannings, a mother from Wiltshire, England, found guilty in 2002 of murdering her three babies, a case in which Dr. Meadow had testified for the Crown, was set free in December 2003. That Cannings's grandmother and great-grandmother had lost children to mysterious death caused the appellate court to quash the thirty-eight-year-old's convictions. She had been sentenced to life. Dr. Meadow's word was no longer law. In 2004, he retired with a tarnished reputation and his body of work under attack, an inglorious end to a star-studded career.

The Clark, Patel, and Cannings reversals led to a review of three hundred SIDS/infanticide cases by a special commission of medicolegal experts formed by the British Home Office. As *London Observer* journalist Jamie Doward reported, the SIDS commission in December 2004 found the murder convictions in twenty-eight of these cases "unsafe." The typical unsafe case involved a mother found guilty of murdering two or more of her babies and featured the testimony of Dr. Roy Meadow. In every case, the mother denied any wrongdoing. Some of the trials showcased flawed and misleading testimony by forensic pathologists. As a result of this study, twenty-one murder convictions were set aside, and the mothers, all serving life sentences, were sent home. These mothers had been set free, but, like Sally Clark, they were still shrouded in suspicion.

The Home Office report essentially repealed Meadow's law in Great Britain. It also cast doubt on the validity of Munchausen syndrome by proxy. In July 2005, the General Medical Council, based on Dr. Meadow's testimony in the Clark trial, found him guilty of serious professional misconduct and struck him off the medical register. Dr. Meadow appealed, and in February 2006, the court ruled that expert witnesses were immune from disciplinary action in cases of honest mistakes. The judge, in reinstating Dr. Meadow, called him a first-class pediatrician who had saved many lives. By then, because of the controversy, the term "Munchausen syndrome by proxy" was no longer in use in England. It had been replaced by "fabricated illness."

The Fate of Meadow's Law in the United States

It was unclear how Dr. Meadow's controversial cases in Britain would affect the influence of his work in the United States, particularly the status of MSBP as a recognized pathological motive for child abuse and murder. The case of Marybeth Tinning could hardly be termed "unsafe," and while MSBP may have been overzealously diagnosed in certain cases, Dr. Meadow's syndrome had not been roundly discredited as recently as 2005. For example, in December 2005, a prosecutor in Oklahoma City charged a thirty-year-old mother with the offense of child abuse by Munchausen by proxy syndrome. Sarena Sherrard had been caught on a hospital surveillance tape injecting fecal matter into her two-year-old daughter's catheter. She confessed to injecting her daughter with toxic substances to keep her ill and to gain sympathy and attention for herself. A neurology professor at the University of Oklahoma Health Sciences Center, Dr. Herman Jones, in discussing this case with an Associated Press reporter, said that he encountered on average three MSBP cases a year.

In the United States, the number of SIDS deaths has declined every year since 1994. Deaths that would have earlier been designated SIDS fatalities are now attributed to a variety of natural causes including genetic disorders, rare diseases, infections, and birth defects. According to the National Institute of Child Health and Human Development, pediatricians now believe that babies who sleep on their backs in smoke-free rooms have a reduced risk of SIDS. A study in 2005 conducted by a reproductive epidemiologist in Oakland, California, showed that the use of a pacifier reduces the chances of sudden infant death. A study at Glasgow University in Scotland revealed that adults who sleep in bed with their babies expose these infants to a greater risk of SIDS.

SIDS researchers at the University of Chicago found in 2004 that babies who died suddenly in their cribs appeared to lack a brain chemical called serotonin. As journalist Katy Buchanan reported in January 2006, according to Dr. Ronald Harper of the University of California at Los Angeles, the chemical deficit in the part of the brain that controls breathing leaves babies unable to gasp or otherwise recover from a sudden drop in blood pressure, which causes them to go into a shocklike reaction. The ability to gasp not only helps restore breathing, but also boosts blood pressure and gets oxygen circulating again. Dr. Harper believes "that something has gone wrong with these infants in fetal life."

In 2006, a study conducted at the Harvard Medical School and Boston Children's Hospital revealed that abnormalities in the brainstem—the part of the brain that regulates breathing, blood pressure, body heat, and arousal—may play a major role in cot death. The abnormality involves the way neurons process serotonin, a brain chemical associated with mood and arousal. These defects were particularly obvious in male brains, which may account for the higher risk of SIDS in boys. The results of the study were published in the *Journal of the American Medical Association.*

Munchausen Syndrome by Proxy in Great Britain

In England, Dr. David Southall, the pediatrician and MSBP researcher who, after watching a television documentary, accused Stephen Clark of suffocating his children, was brought before the General Medical Council on charges of professional misconduct. In the fall of 2004, the council found him guilty of this charge and barred him from child protection work for three years. The council could have, but did not, lift his license to practice medicine.

Dr. Alan Williams, the forensic pathologist who had concluded that the Clark babies had been murdered, was censured by the General Medical

Council in 2005 for failing to meet basic medical standards in the case. The council also found that Dr. Williams had withheld evidence that would have shed a different light on the causes of these children's deaths.

Dr. Roy Meadow's status as a SIDS pioneer remains unresolved. In the mid-1990s, the Foundation of the Study of Infant Death sponsored a study to test Dr. Meadow's theory that two SIDS fatalities in one family suggested a serious possibility of foul play. The researchers set out to determine, in families where a second baby had died unexpectedly, what percentage of these deaths had been "unnatural." A leading SIDS expert, Dr. John Emery, professor of pediatric pathology at the University of Sheffield, studied the cases of thirty-five infants who had died unexpectedly following the sudden deaths of their siblings. In 1998, Dr. Emery concluded that in thirteen of these cases, there was insufficient information to support a manner-of-death conclusion. Of the remaining twenty-two cases, Dr. Emery concluded that fourteen of these babies had died "unnaturally." In his professional opinion, between 34.5 and 40 percent of these babies had not died of natural causes. This was not a conclusion his fellow researchers or the sponsor of the study wanted to hear. Dr. Emery died in May 2000, but the research continued. By the end of 2004, eleven more double SIDS cases had been added to the study.

According to investigative journalist Jonathan Gornall, the other six researchers, led by Robert Carpenter, a statistician with the London School of Hygiene and Tropical Medicine, altered Dr. Emery's conclusions by changing six of his findings of unnatural death to natural. Carpenter and the others also reclassified the thirteen cases Dr. Emery had not made rulings on due to insufficient data as naturally caused deaths. As a result of these cause-of-death manipulations, a January 2005 article published in the *Lancet* under Dr. Emery's name did not reflect either his findings or his views regarding Meadow's law. According to the *Lancet* article, only 13 percent of the forty-six infant deaths had been unnaturally caused, a conclusion that ran counter to the underlying assumption of Meadow's law. Published about the time several of Dr. Roy Meadow's murder trials were under appellate review, the article supported the views of his SIDS critics.

In a letter to the editor published in the March 2005 edition of the *Lancet*, Dr. Charles Bacon, the medical advisor to the foundation that had sponsored the SIDS study, expressed his displeasure with its conclusions: "The investigation seems to have taken the benign view that a death should be classified as natural unless there was compelling evidence to the contrary—an approach that is appropriate in the courts but not in scientific debate; . . . this paper will be influential, but those quoting it should be aware that its data do not support such clear-cut conclusions."

Dr. Bacon wasn't alone in his concern that the conclusions of the study might work against the interests of child protection. In November 2005, Dr. Donald Hall, the president of the Royal College of Pediatrics and Child Health, wrote a letter to the *Lancet*'s editor in which he labeled Robert Carpenter's conclusions "seriously misleading."

In December 2006, Jonathan Gornall's article in the *British Medical Journal* cast further doubt on the credibility of the *Lancet* article, whose findings had by then contributed to the repeal of Meadow's law in England.

Dr. Roy Meadow, who had called the results of the foundation's study "wild and amazing," must have felt at least partly vindicated by Gornall's article, entitled "Was Message of Sudden Infant Death Misleading?" which stirred up the Meadow's law debate by energizing the doctor's supporters as well as his detractors.

While Gornall's reportage makes one wonder if Meadow's law is sound after all, the damage done to Dr. Meadow's professional reputation will be difficult to repair. Still, he remains a hero to many, and the day might come when medical science will reinstate this pediatrician to his former status as the pioneering protector of endangered infants. However, his roles in the Clark, Patel, and other "unsafe" murder trials are probably permanent stains on his professional legacy.

NOTWITHSTANDING ADVANCES in pediatric medicine, babies still die suddenly in their cribs of unknown causes. How much suspicion should these deaths arouse? Should two SIDS fatalities in one family be considered, for no other reason than their number, suspicious? Was Dr. Roy Meadow completely off base? And finally, will the day ever come when forensic pathologists will be able to scientifically determine if an infant has been smothered to death? Until this day arrives, if history is a guide, baby killers will get away with their crimes and innocent parents will be wrongfully accused and, in some cases, imprisoned.

4

Infants Who Can't Breathe

Illness or Suffocation?

In the early 1960s, Dr. Alfred Steinschneider, a pediatrician at the Upstate Medical Center in Syracuse, New York, began studying infants and toddlers who had been referred to him with life-threatening breathing problems. Many of these children, after being discharged from his clinic, died at home in their cribs. Occasionally Dr. Steinschneider would treat several children from the same family, and in a few of these cases, after the children left his clinic, they turned blue in their beds and died. Until Dr. Roy Meadow made his Munchausen syndrome by proxy theory public in 1977, multiple crib deaths within a single family, particularly in the absence of physical evidence of abuse, did not arouse the suspicion of the nation's medical examiners. After 1969, these infant deaths by asphyxia were designated SIDS, natural, or undetermined until Dr. Steinschneider formulated his own explanation, a cause-of-death theory that would make him a star in his field.

Dr. Steinschneider's research would convince him that otherwise healthy babies who stopped breathing and died in their cribs suffered from apnea, a genetic sleep disorder that could run in families. According to Dr. Steinschneider, these babies simply "forget to breathe." Infants with siblings who have died this way, pursuant to Dr. Steinschneider's thesis, have a higher risk of apnea and therefore require close monitoring when they sleep. In October 1972, the medical journal *Pediatrics* published his paper "Prolonged Apnea and the Sudden Death Syndrome: Clinical and Lab Observations." The article created a sensation. Within a few years forensic pathologists and pediatricians throughout the world assigned apnea as the cause of death in cases that they would earlier have designated SIDS or fatalities of an undetermined origin. It was the general consensus that Dr. Steinschneider had made sudden infant death a lot less mysterious.

After a period of observation at Dr. Steinschneider's clinic, an infant or toddler was sent home with an expensive electronic breathing monitor that sounded an alarm when the child stopped breathing. The manufacture and sale of these devices would, thanks to Dr. Steinschneider's theory, grow into a multimillion-dollar industry. Once that ball got rolling, a vested, economic interest would for years insulate Dr. Steinschneider's apnea thesis from a growing body of scientific evidence that contradicted it. Throughout the 1970s and 1980s, more than four hundred medical articles cited Dr. Steinschneider's landmark study, and he won a $4.5 million research grant from the National Institutes of Health.

Ironically, the article that brought Dr. Steinschneider so much prominence and prosperity contained within it a ticking bomb that in time would explode, destroying his theory, and along with it, his credibility. The bomb that he had himself planted took the form of a reference in his paper to one of his case studies, which he called "the H. case." The *H* stood for Waneta and Tim Hoyt, a couple from Newark Valley, New York, a small town seventy miles south of Syracuse in rural Tioga County.

The H. Case: The Babies Who "Forgot to Breathe"

In July 1968, emergency-room personnel revived Waneta and Tim Hoyt's newborn baby, Julie, after Waneta found the days-old infant turning blue in her crib. A year and a half earlier, three-month-old Eric Hoyt had stopped breathing and died in his bed. Eric had been born when Waneta was seventeen and still in high school. A year before that, she had given birth to the couple's first child, James. When Waneta gave birth to Julie, she and Tim were living in a trailer home with Julie and James. Because Eric, Julie's sibling, had stopped breathing and died in his crib, Julie was designated a high-risk baby and referred to Dr. Steinschneider's clinic. Her brother James, however, had never been treated for any problems related to his breathing. Other than the normal childhood illnesses, James was in perfect health.

Julie spent several days in Dr. Steinschneider's clinic and, after a period during which she did not experience any further breathing difficulties, was sent home. But shortly after Julie's discharge, on September 5, 1968, Waneta Hoyt found her forty-eight-day-old baby dead in her crib. The little girl had simply stopped breathing. The local medical examiner, finding no indication of illness or disease, ruled the cause of death "undetermined." This finding stunned the nurses at Dr. Steinschneider's clinic. They couldn't understand how this baby could have died. Some of them—the ones who had witnessed Waneta Hoyt's apparent indifference to her baby—thought the death should have been considered a possible homicide.

Two weeks after Julie's death, her two-year-old brother James was dead. According to Waneta, he too had stopped breathing and died in his crib. This was a child who had not been diagnosed with any breathing disorders. His death made the nurses at the Upstate Medical Center even more suspicious. When the medical examiner ruled the death "undetermined," several of them were shocked and dismayed.

A year and a half after James's death, Waneta gave birth to Molly, who, unlike James, began having breathing problems shortly after Waneta brought her home from the hospital. Molly was placed under Dr. Steinschneider's care after Waneta had rushed the unconscious baby to the emergency room, where they had managed to get her breathing again. Instead of notifying the police, Dr. Steinschneider infuriated the nursing staff—some of whom had spoken to him of their suspicions—by apparently discounting their concerns and diagnosing Molly's problem as apnea.

Molly was kept under observation a few weeks in the clinic and then sent home with a breathing monitor. The very next morning, Waneta brought Molly back to the clinic, claiming that during the night her breathing monitor had gone off repeatedly. The baby was placed once again under observation. On July 4, 1970, following a stint in the clinic during which Molly's breathing had returned to normal, she was discharged. That night, at home, she died in her crib. The medical examiner, aware that Molly had been one of Dr. Steinschneider's apnea patients, classified her death as SIDS by apnea.

The fourth Hoyt child to die at home, Molly, who showed no signs of a breathing disorder while being observed at the clinic, was a victim of murder in the minds of many. Dr. Steinschneider, however, interpreted Molly's death as further evidence of his theory that prolonged apnea runs in families.

In August 1971, Noah, the Hoyt's fifth child, in and out of Dr. Steinschneider's clinic throughout his eleven months of life, died in his crib. He was at home, sleeping in a room with his breathing monitor. According to Dr. Steinschneider's records, Noah had suffered twenty-eight attacks of prolonged apnea while under observation at his clinic. Under these circumstances, it is not surprising that his passing was ruled a natural death.

Thirteen months after the fifth Hoyt death, Dr. Steinschneider published the results of his apnea research in *Pediatrics,* the article that would turn him into an internationally known expert in the study of SIDS. Blind to the possibility that the Hoyt case might be the weak link in his thesis, a case that could be his undoing, Dr. Steinschneider featured it in his article.

Dr. Steinschneider's theory, that some babies were born with a defect that made them "forget to breathe" and thus in some cases die, didn't filter down to the nation's homicide investigators. But it did affect the way many forensic pathologists analyzed crib deaths involving children who were being monitored by breathing instruments. Many of these cases weren't ever assigned to homicide detectives because medical examiners, aware of Dr. Steinschneider's work, weren't diagnosing them as homicides. It could be argued that the pediatrician from Syracuse had taken not just the mystery, but also the police, out of crib death.

Dr. Linda Norton, a forensic pathologist from Dallas who specialized in SIDS cases, came across Alfred Steinschneider's article in the late 1970s. Dr. Norton was also familiar with the work of Dr. Roy Meadow in England, and when she read about the H. Case, she thought Dr. Meadow's Munchausen syndrome by proxy theory explained these deaths better than Dr. Steinschneider's apnea thesis did. While consulting on a SIDS case with an assistant prosecutor in the Onondaga County district attorney's office in Syracuse in 1986, Dr. Norton mentioned the H. case to the assistant prosecutor. Doctor Steinschneider's apnea thesis had not aged well since its publication in 1972. Numerous pediatricians in the SIDS field had thoroughly discredited his work and, worse, had accused him of fudging data to support his theory. It was now the conventional wisdom among research pediatricians that apnea was *not* a cause of death in SIDS cases. Nevertheless, many pediatricians and forensic pathologists were still operating under the assumption that apnea was a viable theory, proof that a bad idea can have a life of its own and a very slow death.

The assistant prosecutor to whom Linda Norton spoke, William J. Fitzpatrick, was more than casually interested in Dr. Steinschneider and his theory. The doctor's clinic was located in his home city and the deaths had occurred in a neighboring county. Fitzpatrick knew that the H. case referred to Waneta and Tim Hoyt and he agreed with Dr. Norton that

Dr. Steinschneider's account of the Hoyt children's deaths sounded more like a story of serial murder than a string of tragic, natural deaths. But because the deaths had not occurred in Onondaga County, Fitzpatrick did not have jurisdiction to authorize a criminal investigation. In 1992, when he had moved up to become the Onondaga County district attorney, Fitzpatrick faxed a copy of Dr. Steinschneider's article to Robert J. Simpson, the district attorney of Tioga County, where the Hoyt children had died. Simpson assembled a team to investigate the deaths of Eric, Julie, James, Molly, and Noah Hoyt. It had been twenty-one years since the youngest, Noah, had died. Cold trails in homicide cases usually lead nowhere, and in the Hoyt case, the trail had virtually vanished. District attorney Simpson had no reason to believe that the investigation of Waneta Hoyt, the obvious suspect in the case, would lead to a criminal prosecution.

The only evidence against Waneta Hoyt came from the nurses working at Dr. Steinschneider's clinic at the Upstate Medical Center. In their reports, many had made note of Waneta's apparent indifference to her children. She didn't visit them regularly at the clinic, and when she was there, she didn't interact with them the way other mothers interacted with their ailing babies. Moreover, her babies seemed fine while under observation at the clinic, but when they got home, they developed breathing problems and died. Most of the nurses were convinced that Waneta had killed her babies and were frustrated that Dr. Steinschneider had, in essence, decriminalized the deaths by designating apnea the culprit instead.

District attorney Simpson realized that to successfully prosecute Waneta Hoyt for murdering her five children, he'd need more evidence than the testimony of the breathing clinic nurses. From his perspective, one of the biggest hurdles in the case involved the cause-of-death rulings. He could find forensic pathologists and pediatricians to question Dr. Steinschneider's apnea theory, but besides confusing the jury, what would that accomplish? Muddying the water worked for the defense, not the prosecution. The district attorney had the burden of proving that the babies had been murdered, and attacking apnea as a cause of death, by itself, would not be enough.

Without a confession, there would be no prosecution. The prosecution needed Waneta Hoyt to admit that she had killed her babies. But if, like Marybeth Tinning, she confessed and then quickly took back that admission, claiming police coercion, the jury would have to choose between the rescinded confession of a mother who had lost five babies in seven years and the expert opinion of a highly credentialed pediatric researcher who had been closely associated with the case. The most believable confessions are those supported by forensic science. In this case, because the forensic science was in dispute, the confession would have to hold up on its own. This was like the tent holding up the pole.

Motivated by the need to obtain a confession, Simpson authorized a warrant for Waneta Hoyt's arrest. Officers from the New York State Police took her into custody on March 23, 1994, as she picked up her mail at the local post office. They put the forty-eight-year-old murder suspect into a police cruiser, and on the way to the state police barracks in Oswego, she waived her *Miranda* rights.

In the interrogation room, shortly after detectives began their questioning, Waneta Hoyt broke down and confessed to killing all her children, Frederick Busch would later report in the *New York Times*. Asked why she would do such a thing, she said she did it to stop them from crying. She said they had been driving her crazy. As to how she had killed them, Waneta said she had used a pillow to smother Eric, Molly, and Noah. She had held a towel over the nose and mouth of James, the two-year-old, until he stopped crying. And seven-week-old Julie? Waneta said: "I just picked up Julie and I put her into my arm and my neck like this . . . and I just kept squeezing and squeezing and squeezing."

As Jacques Steinberg would report at the time of her trial about a year later, when asked to sign the written version of her confession, Waneta said she would not sign the statement until she spoke with her husband. Timothy Hoyt was brought into the interrogation room, became agitated, and accused the detectives of forcing Waneta to confess to something she didn't do. He calmed down when Waneta told him, "No, it's true, I did it." She then signed the statement of her guilt.

Between Waneta's confession and her trial, she conferred with an attorney, renounced her confession as coerced and false, and entered a plea of not guilty. Once the pretrial motions were filed and ruled on, with experts on both sides of the apnea debate lined up, nurses subpoenaed, and a jury selected, the trial got under way. District attorney Simpson, aware that the outcome of the case had implications beyond Waneta Hoyt's fate, wanted to make sure the jury understood he didn't believe every baby who had died suddenly in its crib had been murdered. As Steinberg reported, in his opening address he said, "As we are prosecuting Mrs. Hoyt, we are in no way attempting to tell you SIDS doesn't exist and the people who lose children to SIDS are culpable for their deaths." The prosecutor also prepared the jury for the fact that his case rested heavily on the defendant's recanted confession. While the confession did not parallel the cause-of-death rulings of the medical examiner or Dr. Steinschneider's theory of how the babies had died, it was supported by the fact that all five of the babies had died at home with their mother, the only person who had witnessed their deaths. "Each child," Simpson said, "was the subject of her wrath because they were crying and she couldn't put up with their crying."

After all the witnesses had testified and the attorneys had made their closing statements, the jury chose the defendant's confession over Dr. Steinschneider's theory and found Waneta Hoyt guilty of murdering her five children. It was a victory for the prosecution and a defeat for Dr. Steinschneider and the theory that apnea could cause an infant's death. On five counts of second-degree murder, the judge sentenced Waneta to prison for seventy-five years to life.

Six Babies and a Husband: The Gedzius Case

Emboldened by the Hoyt conviction, prosecutors in the state's attorney's office in Chicago decided to investigate Deborah Gedzius, six of whose babies died in their cribs between 1972 and 1987. It wouldn't be an easy case to win. The forensic pathologists in the Cook County medical examiner's office had not classified any of these deaths homicide. The first four fatalities had been designated as SIDS, the last two, as undetermined.

The background was this: when Delos Gedzius married Deborah in 1975, he knew that she was pregnant with her second husband's child, and that her first two babies, fathered by her first husband, had died suddenly in their cribs. Six weeks after the birth of the second husband's child, Barbara Jean, the baby died in her crib. Like Deborah's first two babies, Barbara Jean had been in good health. Deborah explained that a mysterious birth defect had caused the babies to simply stop breathing. Whether Delos should have been suspicious of these deaths depends on whose view of SIDS one subscribed to, Alfred Steinschneider's or Roy Meadow's. But in 1975, Dr. Steinschneider's apnea theory was in vogue and Dr. Meadow's groundbreaking article on MSPB would not be published for two years. Deborah and her characterization of these deaths were, medically speaking, not out of the question.

Delos and Deborah's first child, Jason Gedzius, was born in 1980, four years after the death of Barbara Jean. Shortly after his first birthday, upon waking from a nap they had taken together, Deborah had discovered him dead, she said. This perfectly healthy child had suffocated in his mother's bed. The day after the baby's death, Harry Gedzius, Delos's brother, paid his sister-in-law a visit to see how she was holding up under the tragedy. Delos, working at the family-owned tavern, wasn't home. Harry was shocked by what he encountered at his brother's house: Deborah was laughing and dancing with members of her family and some of her friends. It looked to Harry like a party, a celebration.

Harry Gedzius was so disturbed by what he had witnessed that day, as Lindsay Tanner would later report for the Associated Press, he called the Cook County medical examiner's office. "My nephew just died," he said. "You know, he's the fourth one." Members of the medical examiner's staff knew this was Deborah's fourth infant death, but what appeared to Harry to be a postmortem party did not constitute evidence of murder. The forensic pathologist, following Jason's autopsy, had ruled the case a SIDS fatality, shutting the door, investigatively, on the death. Although not a true cause or manner of death, the designation of SIDS had the effect of eliminating the possibility of even a preliminary criminal inquiry. Deborah Gedzius, who had gone to bed with a healthy child and awakened with a dead one, would not be officially pressed for details.

Jason had been dead two years when Deborah and Delos had their second child, Delos Jr. Early on, this child's life fell into a cycle of home to hospital, hospital to home. At home the baby had trouble breathing, in the hospital he would be fine, at home again he couldn't breathe, and so on. The pattern matched those of the brief lives of babies born to Marybeth Tinning and Waneta Hoyt, patterns that had been preludes to murder. Had Dr. Roy Meadow been in charge of Delos Jr.'s care, one assumes, he would have recommended in the strongest terms that the baby be permanently taken away from his mother, a woman he would have diagnosed as having Munchausen syndrome by proxy. The physicians looking after Delos Jr., however, diagnosed apnea and sent him home with a breathing monitor.

Delos Jr.'s miserable life lasted two years. He died the way his brother, Jason, had died: he took a nap with his mother and didn't wake up. The dead boy's uncle, Harry Gedzius, once again called the Cook County medical examiner's office. How many babies in this family had to die before someone in a position of authority took notice? This time the medical examiner ruled the death "undetermined" forensically, a step closer to homicide than a finding of SIDS, but no one from the Chicago Police Department took the cue. Deborah Gedzius would not be subjected to a homicide investigation.

In 1987, the year the jury in Boston found Marybeth Tinning guilty of murder, Deborah's sixth child, ten-week-old Danny, died in his crib. Three years had passed since Delos Jr. had died between hospitalizations for his breathing problems. Because his doctors believed that Delos Jr. had suffered from apnea, Danny was sleeping with a breathing alarm next to his crib at the time of his death. This time, Deborah denied being home when the baby died. According to her, Danny's breathing alarm was sounding when she came home from work at the Gedzius family tavern. She found Danny dead in his crib and Delos drunk, passed out on the living room sofa. Delos said he was shocked that his wife would try to blame the baby's death on him and claimed the breathing alarm had gone off after Deborah came home. The Cook County medical examiner ruled the death undetermined. No investigation followed, which meant that officially, no one cared who was telling the truth.

Not only had Delos lost a son, but his wife had tried to blame him for the death. It was time, he decided, to seek a divorce from the woman whose six children had died in their cribs. He moved out of the house and into a suburban Chicago apartment. In April 1989, two days after he and Deborah had met to discuss the final details of the divorce settlement, Delos was found dead in his apartment. He had been shot in the head while sitting on his sofa. There were no signs of forced entry and nothing had been taken. Because he had died before the divorce was final, Deborah would come into some life insurance money. This, plus the fact she had access to his apartment and could have been the last person to see him alive, caused the police to consider her a prime suspect in the murder. Harry Gedzius, thinking that Deborah had murdered three or more of her babies, found it easy to believe she had also killed his brother. But believing and proving are two different things, and because the police were unable to take the case to the level of proof, the investigation ground to a halt and died on the vine. Deborah was never charged and the case went into the books as unsolved.

But the murder of Delos Gedzius did prompt Dr. Robert Stern, Cook County's seventy-eight-year-old medical examiner, to ask one of his assistants, Dr. Mary Jumbelic, to review the deaths of Deborah's six children. After poring over medical records and autopsy reports, Dr. Jumbelic came to the conclusion in 1990 that Deborah Gedzius had murdered her children, probably by suffocation. Dr. Jumbelic was not influenced or impressed by the fact that three of the Gedzius babies had been sleeping with breathing monitors at the time of their deaths. The police opened a homicide case, and a pair of Chicago homicide detectives interviewed hospital personnel, Harry Gedzius, and other members of the Gedzius family. In her report, Dr. Jumbelic recommended that the manner of death, in all six cases, be changed to homicide.

Before proceeding officially on the Gedzius case, the state's attorney's office consulted with a number of outside experts in pediatrics and SIDS. Several of the experts agreed with Jumbelic's assessment of the case, but a few did not, and the ones who disagreed cited Dr. Steinschneider's apnea theory—this was five years before the conviction of Waneta Hoyt. Foreseeing a debate among the experts on what had caused the babies to die, the state's attorney decided not to go forward with the case.

In 1995, following Waneta Hoyt's conviction, Dr. Edmund Donohue, the new Cook County medical examiner, changed the manners of death of the Gedzius babies to homicide. A grand jury was convened to investigate the murders. Deborah, now forty years old, had remarried and was living in Las Vegas with her husband, a former Chicago police officer who had been fired after being convicted of theft. Several witnesses appeared before the grand jury, but after a few months, without probable cause to proceed, the investigation came to a halt and the grand jury was disbanded. In July 2002, seven years after the Cook County murder investigation had fizzled out, Deborah was killed in a Las Vegas traffic accident.

Dr. Thomas Truman and the Death of Dr. Steinschneider's Theory

In 1997, Dr. Thomas Truman, a pediatric researcher at Memorial Hospital in Tallahassee, Florida, caused an uproar with his estimation that 10 percent of SIDS fatalities were homicides. If this were true, hundreds of babies a year were victims of undetected murder. In the United States, most SIDS experts believed the homicide percentage to be in the neighborhood of 2 to 5 percent. Pediatricians and forensic pathologists in the U.K., where Dr. Roy Meadow was a child-protection hero, would not have considered Dr. Truman's estimation outlandish. As we have seen, Dr. David Southall, another English researcher in the Munchausen syndrome by proxy field, believed that one-third of SIDS fatalities were, in fact, cases of murder.

Years earlier, Dr. Truman had worked with Dr. Daniel Shannon and Dr. Dorothy Kelly on a thirteen-year SIDS study at Massachusetts General Hospital. The results of that project, influenced by Dr. Steinschneider's apnea theory, were published in the May 1988 issue of *Pediatrics.* Ten years later, in the aftermath of the Hoyt conviction, Dr. Truman criticized that study's findings, claiming that the inappropriate diagnoses of apnea masked numerous cases of child abuse and homicide. Involved in that study were twenty-eight families who each had two or more babies die in their cribs, deaths that were classified as caused by apnea. One family had lost three babies and another, four. Based upon his own studies and the reevaluation of Dr. Steinschneider's work, Dr. Truman guessed that up to one-third of those babies had been murdered. These deaths, in his opinion, should have been investigated by the police instead of being passed over as naturally caused.

When asked by a reporter from the *Boston Globe* why SIDS researchers generally would be so blind to suspicious death, Truman explained that a research pediatrician who reported parents to the police would scare away referrals. Without babies to study, there would be no papers to publish, no government grants, no fame, and no advancement within the field. This, of course, didn't apply to all SIDS researchers, and it was only his hypothesis, Dr. Truman was quick to point out. But the idea that researchers might turn a blind-eye sounded reasonable enough to have a ripple effect that extended to forensic pathologists who, not being pediatricians, had been reluctant to challenge the current thinking on the causes of infant death. The ripples also reached prosecutors, who were not about to contradict their own medical examiners.

Dr. Jerold Lucey, the editor-in-chief of *Pediatrics,* the journal that published Dr. Alfred Steinschneider's apnea study in 1972, apologized for that article twenty-five years later. In the October 1997 issue, he revealed that at the time, none of the editors had been able to make heads or tails out of Steinschneider's data or his methodology. They published the paper anyway. Why? Because the subject was, at the time, such a "hot topic." Dr. Lucey, a pediatrician at the University of Vermont Medical Center in Burlington, made it known that he did not believe that SIDS was a hereditary problem that ran in families. Because of this, he agreed with a growing number of pediatricians and forensic pathologists who believed that cases of multiple SIDS fatalities in a family should be reported to the police for investigation.

The criminalization of sudden infant death, a trend that got off the ground following Waneta Hoyt's conviction, gained momentum in the spring of 1997 with the publication of a nonfiction book called *The Death of Innocents.* Featuring Dr. Steinschneider and his work, the account explains how a flawed cause-of-death theory coupled with economic interests helped some mothers get away with infanticide. The authors, Richard Firstman and Jamie Talon, portrayed Dr. Steinschneider as a researcher so blinded by ambition that he overlooked and, indeed, ran interference for serial killers like Waneta Hoyt.

Forensic pathology had not stopped MaryBeth Tinning and Waneta Hoyt from killing their children, and forensic science had played no role in the convictions of these women. Confessions had been the prosecutions' only hope in both cases. In SIDS cases that would follow in the United States and in England, forensic science would play a role but often would serve to confuse, rather than to clarify, issues of guilt or innocence.

5

Swollen Brains and Broken Bones

Disease or Infanticide?

Shaken baby syndrome (SBS) refers to signs of physical trauma found in children under six who have been violently shaken. When a baby or toddler is shaken too hard, the victim's brain is jarred against the skull, causing it to bleed and swell. The child's proportionally large head and underdeveloped neck causes a whiplash effect. One-quarter to one-third of SBS victims die, and of those who survive, many suffer lifelong disabilities. It is the most common form of child abuse and may be one of the most underreported crimes in the United States.

Most pediatricians and forensic pathologists believe that to diagnose SBS, they must find, at minimum, evidence of subdural hematoma (brain hemorrhaging), retinal bleeding (broken veins in the eyes), and cerebral edema (liquid in the brain that causes it to swell). The conventional wisdom has been that a child with these injuries who has neither been in a car accident nor fallen from a two-story window has been violently shaken. Supportive evidence of SBS might include trauma in the neck and spine, bruises on the arms and torso, and broken ribs. Physicians also look for signs of previous abuse such as healed fractures and scars. Before making a final SBS diagnosis, careful and thorough medical examiners take the time to study the child's medical history and to rule out the possibility of natural causes. Because an SBS ruling in a death means that someone has killed the child, it's a call that should not be made lightly.

Defendants in SBS homicide trials are commonly fathers, mothers, stepfathers, babysitters, or boyfriends of the mother. Defendants may either deny shaking the child or say they panicked in their attempts to revive a child who was choking or turning blue in the crib or bed. Defendants who confess often say they shook the child to stop the youngster's crying. People accused of SBS killings rarely admit that it was their intention to kill the child. As a result, most convictions, particularly those involving plea bargains, involve lesser degrees of criminal homicide. However, convictions of first-degree murder do occur when the victim was a battered child before being shaken to death.

Dr. John Caffrey, a pediatric radiologist, coined the term "shaken baby syndrome" in 1972. The need for such a term suggests that pediatricians and emergency-room physicians saw a lot of children with this type of brain trauma. In the mid-1990s, experts in the field guessed that every year 50,000 children were shaken violently, and that 25 percent, or 12,500, of them died from their injuries, a figure far higher than the FBI's uniform crime reports suggested. The fifty thousand SBS case count is considered a high estimate. Currently, between 2,000 and 3,000 SBS cases are reported every year, of which one-quarter to one-third result in death. Because this is a crime that can easily go undetected, no one really knows how many infants and toddlers are so abused and killed.

In the late 1990s, a handful of pediatric researchers began to question the science behind the standard SBS diagnosis. Could cerebral edema and blood in the eyes and brain have other causes, for example, vitamin deficiency, disease, or reactions to vaccines and drugs? Diseases thought to cause symptoms of SBS included hypophosphatasia, brittle bone disease; Alagilles's sydrome, a liver ailment; Byler's disease, a liver disorder common among the Amish; and glutaric acidura, acid buildup in infants that causes paralysis and retinal bleeding. Some experts theorized that a relatively short fall to a hard surface, say from three feet, could cause damage to the brain similar to that found in SBS victims. If this were the case, then SBS defendants who claimed that the baby had rolled out of bed or had been accidentally dropped might be telling the truth. Expert witnesses on behalf of defendants in homicide trials commonly testified that while violently shaking a child caused injury, it did not cause death. In homicide cases, prosecutors were increasingly under pressure to produce evidence that indicated the victim had been a battered child, so that the cumulative effect of repeated beatings rather than SBS had resulted in the victim's death. An anti-SBS movement took hold around the theme that a lot of innocent people were being persecuted in these cases.

Findings from a newspaper study of the deaths of 2,707 children under the age of six in the state of Washington between 1997 and 2000 seemed to run counter to the idea that flawed medical science and forensic pathology lay behind a number of false accusations of SBS. Indeed, the study suggested that many people got away with murder in SBS cases. The four-month investigation, conducted by the *Seattle Post-Intelligencer,* revealed that of these 2,707 deaths, only fifty-four had been ruled criminal homicides. Of these, thirty-five had led to convictions, five were never solved, five led to acquittal, and three resulted in dropped charges; one suspect was killed by the police; and five defendants committed suicide before a verdict was rendered. According to a computerized model of the predicted ratio of nonhomicide to homicide within this age group, there should have been 78 to 116 rulings of homicide out of these 2,707 deaths. From this study, researchers concluded that in Washington State during the period in question, between twenty-four and sixty-two murders had gone undetected.

Post-Intelligencer journalists, after asking criminal investigators, prosecutors, and forensic pathologists why so many child homicide cases had fallen between the cracks, identified the following problems: (1) In many counties, elected coroners, who are not forensic pathologists, make manner-of-death rulings; (2) many medical examiners are not trained in pediatric pathology; (3) too many homicide detectives in these cases don't bother to visit the death sites, gather medical records, talk to attending physicians, or question parents and others present at the scene; and (4) in cases of suspicious infant death, prosecutors are often reluctant to file homicide charges.

Prosecutors generally aren't interested in pursuing cases they consider lost causes—in this context, any case that isn't a slam dunk. Particularly difficult are cases in which a mother is accused of murdering her child. The presumption of innocence in such cases is so strong that evidence of guilt has to be overwhelming. If there is any doubt whatsoever, a jury is likely to acquit. The testimony of a defense expert in such a homicide trial that the baby's death was natural usually results in the mother's acquittal or a hung jury. Whether or not this is a good thing depends on who was right, the medical examiner or the defense expert. Some accuse prosecutors who don't aggressively pursue cases of suspicious infant death of protecting child abusers. Some accuse prosecutors who do aggressively pursue them of persecuting the innocent, especially when the cause and manner of death are in dispute. Prosecutors are damned if they do, damned if they don't. It's no wonder they tend to avoid cases of this nature.

There are, however, SBS homicide cases in which the forensic science *is* solid and should *not* be in dispute, and there are cases in which people are falsely accused and even wrongfully convicted. Being falsely accused of child abuse and murder is about as bad as it gets; once made, the accusation can't be easily shrugged off. Defendants who are convicted and sent to prison as child abusers can expect to be abused themselves by other prisoners. And even when criminal charges are dropped or dismissed by the court, the stigma remains. People who have lost their children to accidental or natural death, then are walloped again by the criminal justice system, may never recover from the loss of their babies and their reputations. If forensic science has an important role to play in the administration of justice, it is in cases like this. As illustrated by a case in Northumberland County, Pennsylvania, no one is immune from false accusation and governmental bad judgment.

Nightmare in Amish Country: The Glick Family

On December 23, 1999, Liz Glick, the four-month-old daughter of Samuel and Liz Glick, Amish dairy farmers in Dornsife, Pennsylvania, died in the hospital two days after her parents had found her unconscious in her crib. The baby had been ill with a fever and had been vomiting. At the Geisinger Medical Center in Danville, Pennsylvania, pediatricians experienced in treating Amish babies determined that the infant had died of vitamin K deficiency, a genetic and sometimes dietary condition associated with babies born at home and breastfed who have not been given the vitamin through precautionary shots or formula. The symptoms of vitamin K deficiency include bleeding in the brain and eyes as well as the presence of bruises caused by normal handling and movement.

Dr. Michael Kenny, a pathologist at Geisinger, performed the autopsy and, as Kate Rush would later report in "Genomics in Amish Country," concluded that the baby had died of a "closed-head injury" (as opposed to a "penetrating head injury" caused by a bullet, a stabbing instrument, or a blunt object). Since Dr. Kenny was not the medical examiner and it was not his job to make an official manner-of-death ruling, that decision fell to the county coroner, an elected official without a medical degree. Instead of conferring with pediatricians familiar with Amish patients, the coroner took the unusual step of convening an inquest, a jury-empanelled hearing to determine if the death was suspicious enough to warrant a full-scale criminal investigation. The coroner's inquest as a step in the criminal justice process, while still available in most states, is an antiquated way of determining manner of death. Dr. Kenny's "closed-head injury" finding, combined with the bruises and the brain and eye bleeding, led the coroner's jury to rule that Liz Glick might have been the victim of SBS homicide.

The Glick case became national news when a child protection agency speculated that the seven other Glick children, in the wake of the coroner's jury decision, were in danger. For their own protection, the children were placed in foster homes until the Pennsylvania State Police, and perhaps a jury in a homicide trial, determined if their parents had committed criminal homicide. The Glick children were split up and sent to non-Amish foster parents, an action that stunned and terrified the residents of this traditional central Pennsylvania Amish community.

The plight of the Glick family caught the attention of Dr. Holmes Morton, a Harvard-trained pediatrician who in the 1980s had treated Amish patients at Children's Hospital in Philadelphia. He had moved to Strasburg, Pennsylvania, where in 1989 he had founded a nonprofit clinic in the heart of Amish country called the Clinic for Special Children, specializing in the treatment and study of illnesses and disorders affecting the Amish. Supported by donations and fund-raising events, the clinic incorporated a state-of-the-art laboratory for the diagnosis and study of biochemical genetic disorders. Dr. Morton should have been one of the first persons consulted by the authorities in the Glick case. He was well known, had expertise pertinent to the case, and was local. No one, however, sought his opinion on the cause of the Glick baby's death.

Without being asked, Dr. Morton conducted his own inquiry into the Glick baby's medical history. A few days later, he announced that the infant had been born with a genetic liver condition that rendered her body incapable of breaking down vitamin K. The symptoms of vitamin K deficiency—the bleeding in the brain and eyes and the severe bruising—could easily be mistaken for signs of SBS. In Dr. Morton's opinion, the Glick child had not been killed by shaking. There had been a terrible mistake; this baby's death had been natural.

But criminal investigations are like freight trains—once they get rolling they are hard to stop. Even though Dr. Morton had thrown his body across the tracks, the train kept coming. Weeks passed. Finally, in February 2000, the case went before a state medical advisory board of physicians, which heard testimony from several pediatricians who agreed with Dr. Morton's diagnosis. The panel of physicians voted to recommend that the manner of death in the Glick case be changed to natural. The coroner changed his ruling, and shortly afterward, the child protection people gave the Glick children back to their parents. A month later, the district attorney announced that Samuel and Liz Glick were no longer the subject of a homicide investigation. One can only guess how far down the criminal justice track the prosecution train would have rolled had it not been for Dr. Morton's intervention. One or both of these parents could have been sent to prison. Because no one in their own community believed they were killers and the true cause of death was undisputed, their reputations fortunately had been spared.

Dr. Morton's intervention in the Glick case is important because it shows that symptoms of SBS can occur in natural death. Vitamin K deficiency is not a condition limited to Amish babies and it can result from breastfeeding, home birth, liver disorders, premature birth, and diarrhea. Following the Glick case, Doctor Morton and his colleagues, Dr. Kevin Strauss and Dr. Erik Puffenberger, would identify several genetic causes of Amish infant death that previously might have been classified as SIDS cases.

At the dawn of the twenty-first century, medical examiners keeping up with developments in the pathology of child abuse and faced with the traditional triad of SBS symptoms—brain and eye bleeding and cerebral edema—would look for broken bones, scars, bruises, and a medical history that suggested child abuse before ruling a death SBS. They would also have to rule out the various causes of natural death that produced symptoms of SBS.

In forensic pathology and criminal investigation, conventional wisdom has it that a child with a medical history of broken bones is a child who has been battered. But even here medical examiners have to be careful, because some diseases can make a child's bones easily breakable. While brittle bone diseases are in themselves usually not fatal, they can lead a forensic pathologist to a false homicide ruling in cases where other symptoms of SBS are also present. Forensic pathologists and criminal investigators, in cases of potential child abuse, should take nothing for granted.

The Brandy Briggs Case: Never Say You're Sorry

Brandy Briggs, nineteen years old, called 911 on May 5, 1999, after finding her baby, two-month-old Daniel Lemons, limp, barely breathing, and unconscious. The ambulance delivered the baby to Lyndon B. Johnson General Hospital in Houston, where doctors inserted a breathing tube. On May 9, while being cared for at Texas Christian Hospital, Daniel Lemons died.

Dr. Patricia Moore, a forensic pathologist in the Harris County medical examiner's office, performed the autopsy. She had graduated from Southeastern University of the Health Sciences with a doctorate in osteopathy and had received training in pediatric pathology. Dr. Moore had been with the medical examiner's office three years, during which time she had performed some 1,500 autopsies. She was thus well qualified and experienced and, furthermore, was respected by the law enforcement community. In the case of Daniel Lemons, Dr. Moore determined the cause of death to be "craniocerebral trauma with complications" and ruled it an SBS homicide, as the *Houston Chronicle*'s Andrew Tilghman would report in an October 2004 retrospective of the case.

Dr. Moore's ruling led to the arrest of the baby's mother, Brandy Briggs, who faced a homicide trial that could result in a murder conviction with a possible life sentence. Assured by her attorney that a plea bargain had guaranteed her a probated sentence, Briggs pleaded guilty to the charge of child endangerment. In court, standing before the sentencing judge, Briggs admitted that she had shaken the baby to revive him. Obviously reading more into the admission than the defendant had intended, the judge sentenced her to seventeen years in prison. As plea bargains go, this one had not been a very good deal for the defendant.

New lawyers for Brandy Briggs asked two pediatricians to examine the circumstances of her baby's death. Both concluded that the baby had died from complications of a urinary tract infection contracted shortly after birth. In the meantime, as Andrew Tilghman would report in July 2004, Dr. Moore's boss, chief medical examiner Dr. Joyce Carter, was criticizing her for showing bias in favor of prosecutors and police in cases of sudden infant death. The medical examiner accused Dr. Moore of "defective and improper work" and chided her for "not understanding the objectives of neutral medico-legal investigation." Dr. James Bromberg, a physician and defense attorney who had represented a defendant in one of the cases brought to trial on Dr. Moore's ruling, reported that in an eighteen-month period, Dr. Moore had ruled SBS as the cause of death in seven cases. According to Dr. Bromberg, based on data from the U.S. Department of Health and Human Services, a metropolitan area the size of Houston should have only one or two SBS fatalities a year.

In 2002, Dr. Moore left Harris County for a position in the medical examiner's office in Leesburg, Florida. Two years later, she was back in Texas performing autopsies for the Southeast Texas Forensic Center, a private firm operating out of Conroe.

In July 2004, the new Harris County medical examiner, Dr. Louis Sanchez, reviewed Dr. Moore's work in the Briggs case, found no evidence of SBS or any other form of abuse, and changed the manner of Daniel Lemons's death to undetermined. In the meantime, it had been learned that the hospital breathing tube had been mistakenly inserted into the baby's stomach instead of a lung, causing his brain to go without oxygen for forty minutes. Dr. Sanchez then ruled the cause of death asphyxia.

In December 2005, a state appeals court judge set aside Briggs's child-endangerment conviction on the grounds that if she had been tried for murder based on the new cause of death, she probably would have been acquitted. On the day before Christmas, after serving five years in prison, Brandy Briggs was set free. But for her, the ordeal wasn't over. She had been released on $20,000 bond, because the Harris County district attorney's office had not quashed the initial indictment—they wanted to keep the retrial option open.

One could argue that the Texas criminal justice system had not treated Brandy Briggs with a great deal of fairness. One could also make a case for an official apology. Official apologies are rare, however, even in cases where nothing worked the way it should. Officials don't apologize, because to do so requires taking responsibility for what went wrong. And in this case, who could be blamed? When everyone, to some degree, is at fault, singling out one person would be scapegoating. Perhaps that is why, in criminal justice, nothing much gets fixed.

On March 31, 2006, for the fourth time since Brandy Briggs had been released from prison, prosecutors asked for and received more time to decide if she should be tried for murder. In September 2006, the Harris County district attorney dropped the charges against her. It wasn't that he considered her innocent. He said he didn't have enough evidence to go forward with the case.

Hope for more just treatment of such cases in the future comes from a study published in 2006. Researchers at Children's Hospital in Pittsburgh, the University of Pittsburgh School of Medicine, and Yale University School of Medicine found that infants who had been violently shaken might have higher levels of proteins in their blood or cerebrospinal fluid. The doctors had examined the blood and cerebrospinal fluid of ninety-eight children for the presence of certain proteins that are normally associated with brain injury. In this group were fourteen infants suspected by doctors of having been violently shaken. Eleven of these fourteen had elevated protein levels. Doctors hoped that these findings would lead to a screening test for SBS. It will take scientific breakthroughs like this to catch people who have harmed babies and at the same time protect people who have not.

Brittle Bones or Battered Babies?

In 1990, Dr. Colin R. Paterson, a bone specialist (chemical pathologist) from Dundee, Scotland, introduced a form of osteogenesis imperfecta (brittle bone disease) that he called temporary brittle bone disease (TBBD). As described by Dr. Paterson, TBBD causes the bones of babies under the age of six months to break during normal handling and activity. The disorder itself causes no pain and begins to diminish as the infant approaches the age of one. TBBD, in other words, is self-limiting in that it spontaneously cures itself.

Dr. Paterson began diagnosing bone disease at the Perth Royal Infirmary in Perth, Scotland, and later at Ninewells Hospital in Dundee, where in 1969 he became a senior lecturer in the department of medicine. In 1975, he began testifying as an expert witness on behalf of parents in child custody hearings and criminal trials where the defendants stood accused of abusing their babies. Because he was for many years the only TBBD proponent in the world, Dr. Paterson was a busy man, testifying in hundreds of proceedings and trials in the United Kingdom and the United States. He was the only TBBD proponent because, from the beginning, the medical community in Scotland, England, and the United States had viewed TBBD with great skepticism. There seemed to be no scientific research behind the diagnosis, and Dr. Paterson appeared totally blind, in cases of multiple infant fractures, to the possibility of child abuse. He was, however, a persuasive witness in cases in which parents had petitioned the court to have their children returned to them from foster care. In total, seventy-eight babies who had been removed from parents suspected of child abuse would be returned to their families as a result of Dr. Paterson's testimony that the fractures had been caused by TBBD and not by abuse. The family court judges making decisions based on his testimony didn't seem bothered by the fact he was not a pediatrician, a pediatric radiologist, or a forensic pathologist. Indeed, in these cases, Dr. Paterson's testimony was regularly challenged by pediatric radiologists and pediatricians. Nevertheless, judges frequently relied on his opinion and returned the infants to their families.

Dr. Paterson's testimony did not, however, impress every judge familiar with his thesis. In 1990, the year he published his TBBD theory in the *Journal of the Royal Society of Medicine,* Edward Cazalet, a High Court justice in Wales, criticized Dr. Paterson for his testimony at a coroner's hearing involving the death of a baby girl in Nottinghamshire. At the time of the infant's death, she had five fractured ribs, broken bones in both arms and legs, and bleeding in her brain. The local coroner had convened a hearing to determine if the manner of death was accidental, natural, or homicide. In the face of strong physical evidence of child abuse, Dr. Paterson testified that the bone fractures had not been inflicted. The coroner's jury ruled the death natural, prompting Justice Cazalet's criticism that Dr. Paterson had acted as an advocate for his pet theory rather than as a dispassionate expert witness.

Two years later, in a case that would tarnish Dr. Paterson's career, a baby with multiple fractures named Myles Phipps would die. Child protection authorities had taken Myles from his parents after discovering the numerous bone fractures. But as a result of Dr. Paterson's testimony, a family court judge had ordered the return of the baby to his parents. The baby died while in their custody. Dr. Paterson's growing number of critics, presuming that Myles Phipps had been murdered, blamed him for the death. Referring to the fatality as a "cot death," Dr. Paterson insisted that the infant's fractures had been caused by TBBD.

In a case that would strain Dr. Paterson's credibility, twin boys, three months old, upon admission to Royal Hospital for Sick Children in Glasgow, Scotland, were found to have multiple rib fractures. As the *London Daily Telegraph*'s Tara Womersley reported, the boys were taken from their parents in January 1999 after pediatric radiologist Dr. Christine Hall from the Great Ormond Street Hospital for Children in London looked at the X-rays and proclaimed the injuries to be "non-accidental." The parents denied harming the infants. The authorities didn't charge the parents with child abuse but, until the matter was sorted out, placed the boys in the custody of relatives after they couldn't find a suitable foster family. The parents, accompanied by a social worker, were allowed to visit the twins two hours a day. Three months later, the parents obtained permission to move in with the relatives and be closer to their children. They were not, however, allowed to be alone with them.

The parents, in the meantime, had acquired the services of a solicitor and had petitioned the court for full custody of their babies. In October 2000, at a hearing before magistrate John Stewart, Dr. Christine Hall, a board-certified pediatric radiologist, testified that the multiple fractures indicated child abuse. Moreover, after the parents had lost full custody of their twins, neither baby had suffered broken bones.

Dr. Paterson, taking the stand on behalf of the parents, offered this interpretation of the evidence, Womersley reported:

> Temporary Brittle Bone Disease means that a child can suffer multiple fractures and suddenly there are no more fractures, which may be thought to show the presence of child abuse. But it's a medical condition. There needs to be more recognition. It is not at all uncommon and I have come across a number of cases in the past 15 to 20 years. We do not know how common it is as there are cases that you never hear about and you could not examine the x-rays of every three-month-old. In this case, the babies were twins which is one risk factor of the disease, and they were also premature which is another factor.

Aware that his critics considered his advocacy of TBBD dangerous to children, Dr. Paterson stated under oath that he had been involved in 103 custody cases where abuse had been suspected, and that in the 78 of these cases where the babies had been returned to their families, none had suffered further harm at the hands of their parents. According to his analysis, that was because these babies hadn't been abused in the first place. But in his testimony, he failed to mention the 1992 death of Myles Phipps.

The magistrate, John Stewart, was obviously impressed with Dr. Paterson and his testimony, as Womersley reported. "It appears to me," he said, "that those who criticize Dr. Paterson . . . might have cause to reconsider their position." Stewart ordered the babies returned to their parents. In so doing, he had discounted the testimony of Dr. Christine Hall, the board-certified pediatric radiologist who had looked at the twins' original X-rays.

Judge Stewart's ruling shocked pediatric practitioners throughout the United Kingdom and intensified the criticism of Dr. Paterson from judges, child protection advocates, and pediatricians. A month after the decision, the BBC radio show *Frontline Scotland* aired a program called "Brittle Truths" in which Dr. Paterson debated his theory with three pediatric radiologists. The show also featured a parent who said she had been falsely accused of child abuse when in fact her baby's fractures had been caused by TBBD. The most substantive part of the show, however, involved exchanges between Dr. Paterson and Dr. Graham Wilkinson, a pediatric radiologist with Sick Children's Hospital in Edinburgh.

When these two experts looked at the same evidence—multiple infant bone fractures—they drew opposite conclusions. If these fractures had *not* been caused by a traffic accident, according to Dr. Wilkinson, the only reasonable explanation was child abuse. For Dr. Paterson, it was a medical condition. But according to his detractors, Dr. Paterson's theory had originated in his mind, not from a series of extensive, peer-reviewed studies, which explained why he couldn't say what percentage of multiple infant fractures were caused by abuse as opposed to TBBD, other diseases, and accident—he was unable to find support for his theory in the medical literature.

When six pediatric radiologists in Cincinnati, Ohio, and Winnipeg, Canada, conducted a three-year study of infants under the age of one who had multiple bone fractures, they found that 82 percent, thirty-two out of thirty-nine babies, had been victims of child abuse. (The researchers, in classifying a case as child abuse, considered a combination of factors such as a confession by the abuser, the existence of other injuries, no clinical or radiographic evidence of bone fragility, no evidence of a serious accident, and/or an implausible account by a parent or babysitter on how the injuries had been sustained.) Three of the babies had been injured in accidents, one had suffered the fractures at birth, and three had fragile bones caused by osteogenesis imperfecta, rickets, or premature birth. The results of the study were published in the April 4, 2000, issue of *Pediatrics*.

Dr. Paterson's impressive credentials, his record of returning injured babies to their parents, his apparent blindness to the reality of child abuse, and his position as the father and sole proponent of TBBD had made him one of the most successful expert medical witnesses in the United Kingdom. But the more he exposed himself in the courts, the more obvious it became that his testimony was predictable and hardened to a point where it came off as patently unscientific.

For Dr. Paterson, the first sign that his career had peaked came in 2001 following his testimony in a child custody hearing in London before High Court justice Peter Singer. The case began when pediatricians in a London hospital suspected abuse when a twenty-week-old girl came to them with broken ribs, a fractured right ankle, and bruises on her chest, left wrist, and right ankle. The infant's right calf was also badly swollen. Two months earlier, in October 1999, these doctors had treated this baby for bruises on her chest, injuries the parents claimed had occurred accidentally. To spare the child further abuse, the treating physicians now reported the case to child protection officials, who moved the baby into foster care. The parents, pursuant to a petition to regain custody, sought Dr. Paterson's help.

Without reviewing the child's medical file, Dr. Paterson submitted a report to the judge that contained his opinion regarding the source of this infant's fractures. His report, dated January 22, 2001, made no mention of the infant's bruises or of her swollen right calf. Moreover, Dr. Paterson had not spoken to the treating physicians and had not requested documentary evidence of the child's medical history. He nevertheless wrote the following, which appears in the transcript of his General Medical Council (GMC) Professional Conduct Committee Hearing: "Taking all the evidence that I have reviewed to date in this case, I would have thought it more likely than not that [the infant's] fractures were caused by bone disease, probably Temporary Brittle Bone Disease."

A month after Dr. Paterson filed his report with the court, Justice Peter Singer sent him the medical files detailing the extent and nature of the infant's injuries. The judge probably expected that once Dr. Paterson saw all the evidence, he would change or at least modify his TBBD diagnosis. But as the *Guardian*'s Clare Dyer would later report in "Inexpert Witness," Dr. Paterson advised the court that the additional information had not altered his opinion in the case. At the hearing, held in March 2001, Dr. Paterson testified on behalf of the parents. But in this case, the judge was not impressed, and the baby was not returned to his parents. Justice Singer was so disturbed by Dr. Paterson's role in the case, he reported the bone specialist to the president of the High Court's Family Division, referring to the doctor's testimony as "misleading" and "woeful" and accusing him of "tunnel vision." In his letter of complaint, Justice Singer wrote: "Dr. Paterson has, in my view, failed to be objective, omitted factors which do not support his opinion, and lacked proper research in his approach to the case."

This was the second time Dr. Paterson had been criticized by a judge. This time, however, there were serious judicial repercussions. The president of the High Court's Family Division filed a complaint with the GMC, which led to an inquiry into Dr. Paterson's performance as an expert witness in child abuse cases. The GMC's investigation resulted in the issuance of a letter of guidance that required Dr. Paterson to acquire the approval of the judge before he testified in family court. While he could still do research, write papers, and lecture on TBBD, Dr. Colin Paterson was essentially out of business as an expert witness. The GMC's letter of guidance, however, didn't apply in Scotland or in the United States, the world's most lucrative market for experts who testified in court for money.

Dr. Paterson had given lectures in the United States, where—with the exception of a handful of bone specialists—TBBD was not recognized as a medical disorder. In 1996, about six years before receiving his letter of guidance, Dr. Paterson had been contacted by a defense attorney in Tucson, Arizona, whose clients had been charged with abusing their three-month-old daughter. Dr. Paterson planned to be in the States on other business and was willing to testify on behalf of the defendants. But there was a problem: he wouldn't arrive until ten days after the trial was scheduled to begin, and the judge denied him the opportunity to testify. However, in 2002, shortly after receiving his kiss-of-death letter from the GMC, Dr. Paterson did testify on behalf of these defendants after they had been awarded a new trial. It had been the trial court judge's refusal to accommodate Dr. Paterson that had led the Arizona Supreme Court to set aside the child abuse conviction. In England and Wales Dr. Paterson had been discredited and tarnished, but in Arizona, not allowing him to testify had been grounds for a new trial. But Dr. Paterson's involvement in the Arizona case would eventually cause him even more trouble in England and Wales.

The case began when Audia and Martin Talmadge were charged with child abuse after their daughter was hospitalized in January 1995 with fourteen fractured bones. A month earlier she had been treated for extensive bruising. The infant's injuries that led to the arrest of her parents included fractures of her skull, ribs, and legs, bruises on her legs and chest, and bleeding in her brain. When touched or moved, the baby screamed in pain. The parents insisted that they had nothing to do with these injuries, attributing her condition to natural causes. This would be their defense.

At the time, there was only one expert witness in the country who would take the stand and attribute multiple fractures to TBBD, Dr. Marvin Miller. He had looked at the medical evidence in the Talmadge case and concluded that the injuries had been caused by TBBD, not parental abuse. An admirer, if not protégé, of Colin Paterson, Dr. Miller was the director of genetics and professor of pediatrics and obstetrics at Wright State University of Medicine in Dayton, Ohio. He did not want to travel to Tucson and testify in person, but he was willing to be deposed on videotape. The prosecutor didn't want Dr. Miller to testify at all and, in an attempt to disqualify him as an expert witness, filed a six-hundred-page affidavit challenging the scientific and medical basis for TBBD. The trial judge denied the prosecutor's motion to disqualify Dr. Miller but also refused the doctor the luxury of testifying from Dayton. When Dr. Miller informed the defense attorney that he was not coming to Tucson, the Talmadges had no defense. This prompted their attorney to contact Dr. Paterson. Because the defense couldn't afford to pay the doctor's way to the United States, and Dr. Paterson's arrival on other business would make him late for the trial, the trial proceeded without him.

Dr. Marvin Miller's unwillingness to travel to Tucson and Dr. Paterson's late arrival in the country left the Talmadges with Dr. Richard Roberts, other than Dr. Miller the only TBBD believer available in the country. Having diagnosed only one patient with TBBD, Dr. Roberts reviewed the Talmadge baby's X-rays and medical records and agreed to testify for the defense. He was not, however, allowed to base any of his conclusions on Dr. Miller's analysis.

The Talmadge trial began on June 18 and lasted two weeks. The prosecution, through the treating hospital physicians, presented the baby's injuries—the fourteen fractured bones and serious bruising—as its evidence of abuse. The prosecutor also produced a pediatric radiologist who took the stand and ridiculed TBBD and the credentials of Dr. Roberts. When the state rested its case, the defendants were in a hole so deep, only a witness of Dr. Paterson's status could have pulled them out. Dr. Roberts tried, but his testimony was weak and unconvincing. The jury found Audia and Martin Talmadge guilty of child abuse and sentenced them each to seventeen years in prison.

The Talmadges appealed their conviction on two grounds, the denial of Dr. Miller's request to testify by videotape from Dayton and the trial judge's refusal to allow Dr. Paterson to take the stand after the testimonial phase of the trial had ended. The Talmadges argued that these rulings had denied them the only defense witnesses available to them. When the state appeals court rejected this line of reasoning and affirmed the convictions, the defense appealed this decision to the Arizona Supreme Court, which reversed the appeals court ruling and remitted the case to the Prima County Superior Court for retrial. This stunning reversal indirectly gave credence to TBBD as a scientifically viable defense. The justice who wrote the majority opinion in *State v. Talmadge* (one judge out of the nine dissented) summarized his rationale as follows: "Dr. Paterson's testimony was intended to impeach [rebut] Dr. Erickson [the pediatric radiologist who had ridiculed TBBD], something he was capable of doing based in part on his past experience in defining and diagnosing TBBD. Dr. Paterson had diagnoses in excess of 800 cases."

The Prima County prosecutor who would be retrying the Talmadges knew that Dr. Paterson's original TBBD diagnosis in the case had been based on very little information. The doctor had seen a few medical documents and a few X-rays. He had spoken to Mr. and Mrs. Talmadge on the phone for twenty minutes each and had talked with a friend of the family, but he had not seen the police reports, transcripts of the Talmadge interrogations, documents pertinent to the baby's medical history, or transcripts of the trial. Following a phone conversation with a deputy prosecutor in June 2001, Dr. Paterson received 2,784 pages of material relevant to the case. Prosecutors hoped that when Dr. Paterson had access to the full picture, he would reconsider testifying for the defense. The deputy prosecutor spoke to Dr. Peterson again in August and once again in October, and on both occasions Dr. Peterson said he had not looked at the new material. Notwithstanding this admission, the doctor said he had not changed his mind about his TBBD diagnosis.

Before the doctor made the trip to the United States to testify for the defense, the prosecutor wanted Dr. Paterson to know that the bundle of documents he had received included statements from the baby's paternal grandparents that pointed to child abuse. If the doctor came to Arizona, he would be up against a mountain of opposing opinion, and he could also expect to have his credentials, scientific methodology, and medical professionalism challenged on cross-examination. The prosecutors in Arizona knew that Dr. Paterson had just received his letter of guidance from England's GMC. They knew he had been criticized for tunnel vision and subjectivity by Justices Cazalet and Singer. The justices on the Arizona Supreme Court may have been impressed with Dr. Paterson's credentials and resume as an expert witness, but the Talmadge case prosecutors were not.

The Talmadges were tried for the second time in February 2002. Usually, the defendant has a better chance the second time around, but not in this case. The prosecution had put together a stronger case than before and would tear apart the star witness for the defense on cross-examination. Dr. Paterson had been accustomed to impressing juries by claiming that of all the babies he had returned to their parents through his TBBD diagnosis, none had subsequently showed signs of abuse. This implied that none of these children had been abused in the first place. In the United Kingdom, due to his reputation and stature, he hadn't been asked to back this statement up with proof. At the Talmadge trial, Dr. Paterson testified that of the sixty or seventy babies who had been handed back to their parents in the United Kingdom, only two, later on, suffered fractures. In both cases, the injuries had been accidentally caused. Dr. Paterson failed to mention the suspicious death of Myles Phipps in 1992, a fatality he had referred to on an earlier occasion as a "cot death."

On cross-examination, Dr. Paterson was asked how he knew what did or didn't happen to these children after he had finished with their cases. How could he know what went on inside the private lives of these families? Did he follow up on these cases? Dr. Paterson said yes, he did follow up to ensure the well-being of these babies. And what did that follow-up process consist of? Did he make regular visits to the home? Did he regularly check the admission logs of local hospitals? Did he question treating pediatricians? Did he query child protection agencies or police departments? No, he didn't do any of those things. So, what did he do? He called each home once to determine how the baby was progressing. Since TBBD cured itself after the age of one, there was no need to follow up after that. It was obvious to everyone in the courtroom that Dr. Paterson had no idea what had happened to these children after their parents regained custody.

Through their own experts and the blistering cross-examination, the prosecution tried to paint Dr. Paterson as a crank and TBBD as a figment of his imagination. When asked to list physicians who regularly diagnosed TBBD, the doctor named four, misrepresenting, according to the prosecution, the opinions of two. On cross-examination, the prosecutor picked apart dozens of statements Dr. Paterson had made on direct examination, statements not supported by the literature of bone disease. The prosecutor also identified declarations by Dr. Paterson that contradicted the witness's own writings on the subject. Perhaps the most embarrassing aspect of Dr. Paterson's testimony had to do with his general lack of preparation in the case, evidenced by his unfamiliarity with the details of the baby's medical and family history. The prosecutor portrayed him as a Johnny-one-note expert for hire. The trial ended in convictions.

Dr. Paterson's critics in the United Kingdom did not look favorably upon his role in the Talmadge case and considered him an embarrassment to the medical profession. The sixty-six-year-old bone expert's reputation was coming undone, just as the careers of Dr. Roy Meadow and Dr. David Southall had unraveled. In the wake of the Talmadge retrial, Dr. Paterson found himself once again facing charges of misconduct before a panel of five physicians sitting on the GMC's professional conduct committee. The nine-day proceeding in November 2003 featured the testimony of bone disease experts, a review of the documents pertaining to the 2001 child custody hearing before High Court justice Peter Singer, and documents from the Talmadge case.

As the GMC hearing transcript reflects, one of the experts questioning Dr. Paterson's professional qualifications and research, Dr. Betty Spivack, a U.S. researcher in the biomechanics of abusive child injury, had this to say to the panel: "I wish he [Dr. Paterson] would come up with the evidence to convince people like me who think that [TBBD] is a lot of rubbish. He does not have the experience and training in this field." The committee's legal council, Richard Tyson, accused Dr. Paterson of using, in his approach to diagnosing TBBD, the "pick and mix" technique of highlighting symptoms that supported his theory and ignoring evidence that did not. "Being an advocate for a cause," he said, "is not what an expert should be doing. He should be independent in every way." The final blow came when Eileen Shaw, the head of the committee, addressed Dr. Paterson: "You appear to have acted [in the Talmadge case] as an advocate for Temporary Brittle Bone Disease and ignored the significant clinical evidence which was at variance with your published view on the clinical signs of [TBBD]. You risked misleading the court and undermining the confidence which the judiciary is entitled to place in expert medical witnesses." At the conclusion of the hearing, which included a presentation by Doctor Paterson on his own behalf, the committee found the bone specialist guilty of "serious professional misconduct." This was not good, but for Dr. Paterson, the worst was yet to come.

In March 2004, the GMC took the final step against Colin Paterson by striking him off the register, which meant he no longer had a license to practice medicine in England and Wales. Dr. Paterson said he would challenge the ruling, but he dropped his appeal seven months later. David Spicer, the chair of the British Association for the Study and Prevention of Child Abuse and Neglect, called for a complete review of Dr. Paterson's cases in England and Wales, as Clare Dyer reported in the *Guardian* on April 6. In a letter to Margaret Hodge quoted in the *Guardian* article, the children's minister, Spicer expressed his concern: "The innocent explanations for serious injuries advanced by Paterson are now at the very least suspect and probably have no foundation. Local authorities that have allowed or been required to allow children who were seriously injured to remain in households as a result of Paterson's influence must now have reasonable cause to believe these children are suffering or likely to suffer significant harm, and so are under a duty to make inquiries under Section 47 of the Children's Protection Act of 1989 to decide whether they should take any action to safeguard and promote their welfare."

Peter Peacock, the children's minister of Scotland, in May 2004 ordered a review of every Scottish court case that had featured Dr. Paterson's testimony. Just a month later, however, Dr. Paterson was in Gothenburg, Sweden, giving a lecture on TBBD titled "Corruption and Miscarriages of Justice in Child Care Cases." In that talk before a conference of bone pathologists, he said: "[My] work remains very controversial and some have said that the disorder does not exist. However, we have increasingly good evidence that our work is accurate although the cause of the condition is not yet known." In that presentation, Dr. Paterson repeated the claim that had gotten him into so much trouble at the Talmadge trial: "In over seventy patients returned to their parents there has been no subsequent evidence of physical abuse."

In the United Kingdom, TBBD ceased to exist as a justification for returning injured babies to parents suspected of having harmed them or as a defense in child abuse trials. But in the United States and a few other countries, TBBD still had its supporters. Dr. Michael D. Innis, an Australian pediatric neurologist, wrote a letter to the editors of the *British Medical Journal* protesting the action taken by the GMC against Dr. Paterson. In his letter, published in the March 2004 issued of the *BMJ*, Doctor Innis wrote:

> The decision of the General Medical Council to strike Dr. Colin Paterson off the Register is a grave miscarriage of justice. For the record I can state that I have also given evidence similar, in fact almost identical, to that of Dr. Paterson. If Dr. Paterson's accusers wish to challenge me with the same offense I will happily confront them with the evidence to prove the absurdity of their misguided convictions.
>
> Temporary Brittle Bone Disease is a well recognized entity among many pathologists and others with experience in the practicalities of Osteogenesis. Remarks attributed to the chairwoman [of the GMC

> Professional Conduct Committee] that Dr. Paterson was also found to have ignored bruising that was inconsistent with his own published views on the disease are short-sighted and ignore the fact that periosteal reactions found in TBBD, along with bruising, are classical features of vitamin C deficiency which in my experience is associated with some cases of TBBD.
>
> It would appear that Dr. Paterson is the victim of Medical Ignorance as are so many parents deprived of their children because Natural Disorders are being attributed to Non-accidental Injury.

In the United States, Dr. Marvin Miller—the pediatrics and obstetrics professor at Wright State University, the expert in the Talmadge defense who wouldn't travel to Arizona to testify—was still a strong proponent of TBBD. At a 2005 meeting of the Association of American Physicians and Surgeons, Dr. Miller told his audience that over an eleven-year period, he had treated sixty-five babies with TBBD. Unlike Dr. Paterson, Dr. Miller had formed a theory regarding the cause of TBBD. Dr. Miller believed that limited fetal movement—babies who didn't kick, wiggle, and poke around in the womb—were more likely to be born with underdeveloped bones. This explained why twins, premature babies, and infants whose mothers had uterine deformities were prone to TBBD.

Dr. Peter von Kaehne, a general practitioner at the Fernbank Medical Centre in Glasgow, Scotland, was worried that physicians in the child protection field, by dividing into camps and loathing each other, were not working together in the interests of their patients. In a letter to the editor printed in the summer 2004 issue of the *British Medical Journal*, he wrote: "I am a simple General Practitioner and hence neither able nor interested in making judgment on the merits of Dr. Paterson's theory of Temporary Brittle Bone Disease. It may exist or it may not. I would not know. He may also have failed in his role as expert witness and strayed into advocacy—hardly surprising in a field where everyone seems to see him/herself as an advocate or is seen by others as such."

THERE HAVE ALWAYS BEEN disagreements and disputes among scientists. Scientific and medical theories have come and gone. So the real question is: Has science corrupted the trial process, or has the adversarial nature of the criminal justice system corrupted science? If only undisputed science were allowed into the courtroom, forensic science and expert witnesses would cease to exist. But as long as criminal trials are more about winning and losing than about the quest for truth and justice, questionable science will get into evidence. Indeed, if there is a place for bad science to flourish, it is in the courtrooms of the United States.

PART TWO

Crime-Scene Impression Identification

Forensic Science or Subjective Analysis?

> As finger print evidence is comparatively new, difficulty is frequently experienced in convincing a jury that the testimony of finger print experts is competent and positive evidence.
>
> –Frederick Kuhne, *The Finger Print Instructor,* 1916

The chapters in part 2 feature expert analysis and testimony involving the identification of marks and impressions left by fingerprints, bare feet, footwear, and human teeth and ears. In recent years, particularly in the fields of feet, teeth, and ear identification, numerous expert witnesses have been exposed as bogus. Their highly publicized misidentifications produced wrongful convictions and have been an embarrassment to forensic science.

Even fingerprint identification, once the gold standard of forensic science, has come under attack following misidentifications in that field. Until the 1990s, fingerprint examiners were never challenged in court. That has changed, and as a result, serious problems in the fingerprint identification field have been exposed. We now know that many fingerprint examiners are poorly trained, uncertified, and biased in favor of law enforcement.

For decades, forensic bite-mark witnesses connected defendants to murder and rape victims by testifying that teeth impressions in skin are as individual as fingerprints. Following a series of misidentifications uncovered through DNA analysis, it has become clear that bite-mark identification, as a science, is not as precise and reliable as fingerprint identification. While bite-mark witnesses are not supposed to make positive, without-a-doubt identifications but are to qualify their matchups—as in "the crime-scene bite wound is consistent with having come from the defendant's teeth"—bite marks are still powerfully incriminating evidence. Many forensic scientists and defense attorneys, given the unreliability of this evidence, believe that bite-mark analysis should not be the sole basis of a criminal conviction.

With the advent of DNA analysis, the identification of impression evidence is no longer cutting-edge forensic science. Many in the criminal justice system believe that the comparison and identification of fingerprints, teeth, ears, shoes, and vehicle tires is a process too subjective and prone to human error even to be classified as science. Some of these critics think that if forensic science evolves as it should, the day will come when impression identification, particularly in the bite-mark, shoe-impression, and tire-track fields, will be replaced by more scientific and reliable identification techniques.

6

Fingerprint Identification

Trouble in Paradise

The Jennings Case: The First Court to Admit Fingerprint Evidence

At two in the morning a noise coming from her fifteen-year-old daughter's bedroom awoke Mary Hiller. She slipped into her robe and ventured into the hall, where she noticed the gaslight outside her daughter's room was unlit. Fearing that an intruder had entered the house, Mrs. Hiller returned to the master bedroom and shook her husband awake. Clarence Hiller, on the landing en route to his daughter's room, bumped into Thomas Jennings, a thirty-two-year-old paroled burglar in possession of a .38-caliber Smith & Wesson revolver. The men struggled, then tumbled down the stairway. At the foot of the stairs, Jennings, the bigger man, got to his feet, pulled his gun, and fired two shots. The first bullet entered Hiller's right arm, traveled up through his shoulder, and exited the left side of his neck. The second slug slammed into his chest, piercing his heart and lung before coming out his back. Jennings left the house through the front door leaving behind a screaming Mary Hiller, her dead husband, and a terrified fifteen-year-old who had been sexually molested. The crime occurred on September 19, 1910, in Chicago, Illinois.

About a mile from the murder scene, Jennings, walking with a limp and bleeding from cuts on his arm, passed four off-duty police officers waiting for a streetcar. When questioned about his injuries, Jennings said he had fallen off a trolley. One of the officers patted him down and discovered the recently fired handgun. The officers placed Jennings under arrest and took him to the police station.

A few hours later detectives at the murder house found the two .38-caliber bullets that had passed through Clarence Hiller's body. Today a firearms identification expert would be able to match the crime-scene slugs with bullets test-fired through the suspect's gun. But in 1910, this type of forensic identification was fifteen years away. Investigators also determined that the intruder had entered the Hiller house through a kitchen window. A detective who was ahead of his time found four fingerprint impressions on a freshly painted porch rail outside this window. (Paint in those days dried slowly.) A technician with the police department's two-year-old fingerprint bureau photographed the finger marks in the dark gray paint. Mary Hiller, traumatized by the crime, was unable to pick Jennings out of a police lineup. While roughed up and given the third degree, Jennings did not confess.

At Jennings's May 1911 trial, two Chicago fingerprint examiners, one from the police department in Ottawa, Canada, and a private examiner who had studied fingerprinting at Scotland Yard, testified that the impressions on the porch rail matched the ridges on four of the defendant's fingers, placing him at the scene of the murder. While the idea that fingerprints were unique had been around for twenty years, this was the first U.S. jury to be presented with this form of impression evidence. The chance of convicting Jennings was not good, because the prosecution's case—the defendant's arrest one mile from the house, his injuries, his possession of a recently fired gun, and his murder scene fingerprints—was based entirely on circumstantial evidence.

Prior to the testimony of the four fingerprint witnesses, Jennings's attorney objected to the introduction of this evidence on the grounds that this form of forensic identification had not been scientifically tested and was therefore unreliable. The trial judge, in allowing the fingerprint testimony, cited a 1908 arson case, *Carleton v. People,* in which the defendant had been linked to the fire scene by his shoeprints. Jennings took the stand and denied any connection to the murder.

Following a short deliberation, the jury found Thomas Jennings guilty of first-degree murder. To arrive at this verdict, the jurors had placed more weight on physical evidence than on the defendant's claim of innocence. The judge sentenced him to death.

On appeal, Jennings's lawyer argued that there was no scientific proof that fingerprints were unique. By admitting the testimony of so-called fingerprint experts, the trial court had sentenced a man to the gallows on pseudoscience and bogus expertise. The Illinois Supreme Court, on December 21,

1911, ruled that the Jennings trial judge had not made a judicial error by admitting the fingerprint testimony. In spelling out the rationale for this historic decision, Justice C. J. Carter wrote:

> When photography was first introduced it was seriously questioned whether pictures thus created could properly be introduced in evidence, but this method of proof, as well as by means of X-rays and the microscope, is now admitted without question. . . . We are disposed to hold from the evidence of four witnesses who testified, and from the writings we have referred to on this subject, that there is a scientific basis for the system of fingerprint identification, and that the courts cannot refuse to take judicial cognizance of it. Such testimony may or may not be of independent strength, but it is admissible the same as other proof, as tending to make a case. . . .
>
> From the evidence in this record we are disposed to hold that the classification of fingerprint impressions and their method of identification is a science requiring study. While some of the reasons which guide an expert to his conclusions are such as may be weighed by an intelligent person with good eyesight from such exhibits as we have here on record, after being pointed out to him by one versed in the study of fingerprints, the evidence in question does not come within the common experience of all men of common education in the ordinary walks of life and therefore the court and jury were properly aided by witnesses of peculiar and special experience on this subject.

People v. Jennings laid the groundwork for forensic fingerprint identification in this country. By 1925 virtually every court in the United States accepted this form of impression evidence as proof of guilt. For the next sixty-five years the science behind crime-scene fingerprint identification would go unchallenged in U.S. courts. But in the 1990s, with the emergence of a new method of scientific identification, DNA profiling, that would change.

The Collection and Analysis of Fingerprint Evidence

It is what the criminal leaves behind at the scene of the offense that is usually the most incriminating evidence. This kind of associative evidence can be items and impressions left by things the suspect owned or had access to, such as bullets, shell casings, fragments of torn clothing, footwear impressions, and tire tracks. Even more incriminating is evidence of the perpetrator himself in the form of bloodstains, semen traces, bite marks, hair follicles, and prints left by fingers and thumbs. With the exception of DNA analysis, which is costly, scientifically complex, and slow, the identification of a latent fingerprint has been the most reliable, simple, and direct way of placing a suspect at the scene of a crime. Crime-scene fingerprint identification has been the backbone, the gold standard, of forensic science.

A paragraph in criminalist Herbert Leon MacDonell's 1984 book, *The Evidence Never Lies,* while applying to all forms of physical evidence, is especially appropriate when applied to the crime-scene fingerprint:

> You can lead a jury to the truth but you can't make them believe it. Physical evidence cannot be intimidated. It does not forget. It doesn't get excited at the moment something is happening—like people do. It sits there and waits to be detected, preserved, evaluated, and explained. This is what physical evidence is all about. In the course of a trial, defense and prosecuting attorneys may lie, witnesses may lie, the defendant certainly may lie. Even the judge may lie. Only the evidence never lies.

Latent fingerprint identification is predicated on the theory that every ridge configuration on the pad of every finger and thumb is a design that has never been duplicated. It is the uniqueness of the source and the fact that an inked fingerprint impression can be distinguished from or matched to a finger mark or impression left on an object—a gun, piece of paper, beer can, pane of glass—that makes fingerprint identification possible. U.S. courts began admitting latent fingerprint evidence in 1911, only ten years after Scotland Yard became the world's first law enforcement agency to routinely fingerprint arrestees. St. Louis, in 1904, became the first U.S. police department to establish a fingerprint identification bureau.

Inked fingerprint cards, referred to as rolled-on prints, are filed according to a classification system based upon nine fingerprint patterns that fall into one of three basic shapes—loops, whorls, and arches. Named after Scotland Yard's Edward Henry, the founder of the world's first identification bureau, the Henry system of fingerprint card classification and filing is still in use today.

The section of the finger or thumb that contains the ridges that produce the print when something is touched or handled lies between the tip and the first joint. Within this area there are 75 to 175 ridge characteristics that form its signature and comprise one of the nine basic fingerprint patterns. These characteristics, or minutia points, may include ridges that abruptly end, split off, fork, cross over to other ridges, or form enclosures. A ridge configuration is like a city map with streets, intersections, landmarks, rivers, railroad tracks, and buildings. When a crime-scene latent has several characteristics that match its known, inked counterpart, an identification is declared. The latent fingerprint identification process is subjective by nature because it is based upon human observation and judgment. This is all well and good as long as the person making the judgment is a true expert, one who is properly trained and experienced. Most forensic scientists work in crime labs that are independent of law enforcement. In the field of latent fingerprint identification, many of the examiners, sometimes referred to as identification officers or fingerprint technicians, work directly for a law enforcement agency. Many are sworn law enforcement officers.

In large police departments, the people who gather and preserve physical evidence at the scene of a crime are usually not uniformed police officers, detectives, or forensic scientists. They are technical employees specially trained for this specific task. It's a job analogous to a paralegal position in a law firm. Paralegal workers do not argue cases in court, and these crime-scene fingerprint gatherers do not compare crime-scene latents to known prints of suspects and present their findings in court. Ideally, fingerprint experts, people who testify in court as identification experts, should not be police officers. Unfortunately, many are.

Just because someone touches something doesn't necessarily mean they have left behind a latent print. Not all surfaces are receptive to finger marks. For example, it is particularly difficult to retrieve a latent from rough-surface wood, cloth, skin, cardboard, or Styrofoam. Latents can also be incomplete (called partials), smudged, smeared, or on top of each other. Smooth, shiny surfaces are the best sources for clear, complete latents suitable for comparison and identification. At the crime scene, these latents can be brushed with a fine powder and lifted from the surface with strips of transparent tape. It is not a perfect process and can be a hit-and-miss proposition. Ideally, removable objects that may contain the perpetrator's latents—drinking glasses, bottles, cans, guns, cigarette butts, knives, shell casings, and the like—are sent to the crime lab, where special equipment and techniques can make the marks visible for expert processing. The same goes for grocery bags, sheets of paper, envelopes, notebooks, and magazines that might carry latents. Prints on countertops, furniture, doorknobs, telephones, automobiles, and window glass are usually lifted at the scene. Latents that can be seen by the naked eye—for example, prints in safe insulation dust, blood, wet paint, or soot—should be photographed at the scene.

Unless the suspect has an innocent and believable explanation for his or her latent appearing at the scene of the crime, this evidence is highly incriminating. If the suspect has had no reason to be there, the defense's choice, other than to confess and hope for a deal, is to challenge the correctness of the fingerprint identification. This involves finding another fingerprint expert willing to take the stand and testify that the crime-scene latent was not the defendant's, an approach out of the question until recently. Unlike other branches of forensic science, experts in the fingerprint identification field are not accustomed to testifying against each other. Had O. J. Simpson's latent fingerprints been found in blood on the murder knife, the evidence would have gone unchallenged by a fingerprint expert for the defense.

During the first ninety years of fingerprint history, defense attorneys whose clients' fingerprints were found at the scenes of crimes had one option—plead them guilty in return for a lighter sentence. No one considered questioning the credibility or competence of a fingerprint expert, and no one dared challenge the scientific reliability of fingerprint identification. Fingerprints either matched or they didn't. There was nothing to challenge.

The Mitchell Case: The First Challenge to Fingerprint Identification as a Science

The beginning of the end of fingerprint invincibility came in 1991 when FBI agents in Philadelphia arrested Bryon Mitchell and three others on charges that they had held up an armored-car guard carrying $20,000. Mitchell was accused of driving the getaway car. A year later, with two of the defendants dead and a third, a behind-the-scenes accomplice, testifying for the prosecution, Mitchell was the only one left to try. There were no eyewitnesses to the robbery and Mitchell had not confessed. The prosecution, however, had all the evidence they needed. An FBI fingerprint expert testified that a pair of partial latents—one on the gearshift handle and the other on the getaway-car door on the driver's side—belonged to the defendant. This evidence went uncontested, and the jury, quite reasonably, found the defendant guilty.

Mitchell, however, continued to insist that he was innocent. His attorney, an assistant public defender named Robert Epstein, raised an issue unrelated to the fingerprint evidence, filed an appeal, and won. The appellate judge ordered a new trial.

Epstein realized that the same FBI fingerprint expert would testify at the second trial, and the results would be the same. If he had any chance of keeping his client out of prison, the lawyer would have to challenge this evidence. To do that, he would have to file a *Daubert* motion requesting a pretrial hearing before a judge in which he would challenge the admissibility of latent fingerprint evidence as proof of his client's guilt. A *Daubert* motion refers to the 1993 U.S. Supreme Court decision *Daubert v. Merrill Dow Pharmaceutical,* which had established a legal formula that defined the kind of evidence that is inadmissible because it lacks recognition in the scientific community. Epstein would try to convince the judge there was insufficient science underlying the latent fingerprint identification process. If the federal judge granted Epstein his *Daubert* motion, the FBI expert's identification testimony would be excluded from Mitchell's second trial. This first-of-its-kind frontal attack on fingerprint identification was extremely threatening to the fingerprint and law enforcement communities. Of all the various kinds of physical evidence linking suspects to crime scenes, fingerprints impressed jurors the most. If Epstein succeeded in this challenge, it would change the face of law enforcement.

The *Daubert* decision, because it has become the principal legal basis in challenging the court admissibility of fingerprint and other forensic evidence, is considered a threat by many prosecutors and forensic scientists and a vehicle for improvement by defense attorneys and criminal justice reformers. The *Daubert* case began as a civil action in which the parents of two children with birth defects sued Dow Pharmaceutical, the maker of the prescription drug Bendectin. To prove that the mother's use of this drug had caused the birth defects, the plaintiff submitted affidavits from eight experts who had conducted animal experiments and analyzed the chemical components of the product. Dow submitted an affidavit from one expert who had reviewed the scientific literature on the drug and concluded that there were no risk factors for human birth defects. The U.S. district court judge granted a summary judgment in the drug maker's favor, citing a 1923 case, *Frye v. U.S.*, as precedent. In the *Frye* case, the judge had ruled polygraph evidence inadmissible on the grounds that lie-detection technology was not based upon generally accepted science. The judge in the *Daubert* case, in ruling against the plaintiff, had applied this "general acceptance" legal test.

The Supreme Court, in reversing the U.S. district court judge, replaced the general acceptance test with rule 702 under the Federal Rules of Evidence. Pursuant to this statute, a trial judge can determine if the scientific testimony in question will "assist the trier of fact [the jury] to understand the evidence or to determine a fact in issue." The Supreme Court, as a part of this landmark decision, articulated guidelines and criteria to help federal judges make such determinations. Under the *Daubert* rubric, defense attorneys can challenge scientific expert testimony on one or more of the following grounds:

1. The identification theory or forensic technique had not been scientifically tested.
2. The theory of identification had not been subjected to peer review and publication in scientific and professional journals.
3. There was an absence of uniform standards in the application of the theory or technique.
4. There was a lack of general acceptance of this identification process within the scientific community.
5. There was an absence of scientific study documenting the potential error rate associated with this identification process.

At the *Daubert* hearing pertaining to the Mitchell case, held in September 1999, attorney Robert Epstein argued that latent fingerprint identification, as practiced in the United States, failed to meet all the *Daubert* standards. To make his point, he called witnesses to the stand who questioned the scientific validity of fingerprint identification. The government, in support of fingerprint identification, produced three experts of its own, all affiliated with the FBI. But having never faced a challenge like this, the government's experts had no studies showing that fingerprint identification had scientific validity. The government witnesses simply asserted that fingerprint identification had a long history of accuracy. Since fingerprint witnesses were never challenged in court, it was mere speculation, and perhaps wishful thinking, that all these identifications were accurate.

Epstein asserted that the process of matching a latent to an inked impression had never been scientifically studied or tested. His experts didn't question the assumption that fingerprints were unique but raised what seemed to be valid points: there had been no studies, for example, designed to determine how *similar* two ridge configurations could be. If prints shared visual similarities, wasn't it possible, in the case of partials or low-quality latents, for an examiner to make an honest mistake?

Epstein also questioned what constitutes a fingerprint match, noting that in the United States there was no uniform point-of-identity standard, that it varied from state to state. In some jurisdictions, a match could be declared with as little as six points of similarity. In the FBI, which had no point-of-identity minimum, examiners used a nonnumeric standard of identification which consisted of a general, overall assessment of the evidence. In the United Kingdom, this nonnumeric system replaced in 2001 a sixteen-point identification requirement. In Finland, France, Portugal, Sweden, and Germany, the point identification minimum was twelve. With most other nations, the minimum was ten, if it wasn't sixteen. In Los Angeles County, examiners with the sheriff's office could declare a match with ten, but with a supervisor's approval, that number could drop to eight. The only country in the world with an identification minimum as low as eight was Bulgaria.

During the five-day *Daubert* hearing on the Mitchell fingerprint evidence, which produced one thousand pages of transcript, the defense attorney made an assault on the overall qualifications and competence of the nation's fingerprint examiners. This was, perhaps, the strongest and most convincing element of Epstein's argument. Police and sheriff's office examiners were hired without fingerprint experience or backgrounds in science. Police administrators were known to assign fingerprint examination to officers placed on limited duty, or to punish officers who had run afoul of the bureaucracy. Fingerprint training was mostly on-the-job tutelage and self-study and did not involve rigorous and objective proficiency testing. Before becoming eligible to testify in court as experts, fingerprint trainees did not have to pass any kind of certification exam. In 1980, the International Association of Identification (IAI) created a certification program for examiners, but since then, only a fraction of the country's examiners had taken the certification test, and half of those had failed it. The IAI test, however, is not required, and failing it did not disqualify an examiner from testifying in court as an expert witness.

After the fingerprint witnesses on both sides of the issue had testified, the *Daubert* hearing judge in the Mitchell case ruled that fingerprint identification met the standards of scientific reliability enunciated by the *Daubert* decision. The ruling did not surprise anyone, given the cataclysmic consequences had the judge ruled the other way. Three years later a federal court of appeals affirmed the *Daubert* ruling in the Mitchell case.

While defense attorney Epstein had lost a battle, he had started the war. Fingerprint evidence was no longer immune from judicial scrutiny. In the five years following this historic *Daubert* hearing, fingerprint identification would be challenged more than forty times. With the threat of such a challenge in their toolkit, defense attorneys could now negotiate better deals for their clients. The crime-scene latent, still the prosecutor's best friend, had lost some of its power.

For Bryon Mitchell, there was no silver lining in the *Daubert* defeat. At his second trial, the FBI fingerprint expert once again identified the getaway-car latents as his, and that was enough to convict him of robbery. He went to prison proclaiming his innocence.

The New York Scandal: David Harding, Bean Spiller

The power and appeal of fingerprint identification lies in the simplicity of the concept. Either two prints match or they don't. DNA analysis, by comparison, is complex and requires faith in science and expertise. No scientific leap of faith is required to believe in a fingerprint match, because science plays virtually no role in the actual identification. For that reason, convincing a jury that an examiner has misidentified a latent is a hard sell. How could anyone with good eyesight make such a mistake? Even more difficult than making the case that a latent has been misidentified is proving that a fingerprint examiner has lied about the place the defendant's fingerprint was actually found and from which it was retrieved. A latent lifted from, say, a defendant's car and presented in court as coming from the scene of the crime instead is a form of evidence planting and perjury that is difficult to detect. While it would be cynical to believe that this kind of forensic malpractice is commonplace, it's naïve to think that it doesn't happen.

From 1982 to 1992, New York State fingerprint examiners with Troop C in Sidney, sixty-five miles southwest of Syracuse, routinely lied under oath, identifying latents they had lifted from the booking room as coming from burglary scenes. There is no telling how many innocent defendants in these cases went to prison on the strength of this perjury and evidence fabrication. The identification officers might never have been caught had Lieutenant

David Harding not retired and spilled the beans when applying for a job with the CIA. Apparently in an effort to impress the CIA official considering his application, Harding bragged that he had planted phony fingerprint evidence in forty cases. When the CIA employee reported the admission to the New York State Police, the investigation that followed led to the perjury convictions in December 1994 of Harding and four state fingerprint examiners. The five guilty pleas resulted in the closing of all the cases the officers had been working on and caused the reversal of numerous convictions that had been based on their testimony. Fortunately for the New York State Police and the credibility of fingerprint identification in general, the scandal did not make much of a media splash. Perhaps it was viewed as an anomaly, for it triggered no cries for a nationwide audit of fingerprint identification services.

Phony Identification in Los Angeles: Detective William Douglas

In the effort to acquire a confession, detectives are allowed, pursuant to criminal procedural law, to lie to the people they are grilling. Specifically, they can allude to evidence they don't have, and quite often this nonexistent evidence is a crime-scene fingerprint. The theory that permits this interrogation technique holds that an innocent person will not be swayed by the lie into making a false confession. But innocent people do confess, and when a person so inclined realizes that the police are determined enough to fabricate evidence, that understanding alone can push that person into confessing falsely.

Los Angeles detective William Douglas created a fake fingerprint examiner's report in preparation for the interrogation of Jose Corona, the suspect of an attempted robbery of a cab driver. According to the fake document, one of the latents lifted from the taxicab was Corona's. When confronted with the phony report, the suspect refused to take the bait. The case moved forward anyway, and a trial was scheduled. Included in the documents forwarded to the prosecutor's office in advance of the trial was the incriminating faked fingerprint report, which led the prosecutor to believe he had a solid case against the defendant. But the wheels came off the case when Corona's attorney asked to see the report. The police department sent the defense attorney the real fingerprint report, which failed to identify any of the crime-scene latents as belonging to the defendant. In 2000 the judge dismissed the case and issued a report of her own in which she made it clear that she didn't think the sending of the phony report to the prosecutor had been just a mistake. She recommended a criminal investigation into the matter.

The deputy district attorney assigned to the inquiry concluded that Detective Douglas had not committed a crime. In February 2002, the prosecutor, in winding up his investigation, referred to the incident as an "error of judgment and a mistake," according to Jim Crogan in *LA Weekly*. The deputy district attorney also decided that it wasn't necessary to alert defense attorneys representing clients in other cases that involved Douglas and latent fingerprints.

The Rick Jackson Case: The Role of Bias in Fingerprint Misidentification

In 1997, detectives in Upper Darby, Pennsylvania, a community outside Philadelphia, arrested Rick Jackson shortly after Jackson's friend, Alvin Davis, was stabbed to death in Davis's apartment. In the interrogation room detectives showed Jackson a crime-scene photograph of a bloody latent found near the body. According to a pair of fingerprint examiners with the Upper Darby Police Department, one of whom was also a police superintendent, that print had been left by Jackson. The suspect didn't deny that he had been in the apartment, but he denied killing Davis and said he was certain the bloody print wasn't his. Jackson was actually relieved when he realized that the police were basing their case on a misidentified print. He figured that once the police realized their mistake, they would look elsewhere for a suspect.

With Jackson so insistent that the bloody print wasn't his, Michael Malloy, his attorney, took the unique step of having it examined by outside experts. Vernon McCloud and George Wynn, the experts, were retired FBI fingerprint examiners. Between them they had seventy-five years of experience. Both men were also IAI certified. Wynn and McCloud, to their own amazement, found that the bloody crime-scene latent was not Jackson's.

The district attorney, confronted with a defense bolstered by a pair of prominent fingerprint experts who disagreed with the local examiners, pushed forward with the trial anyway. In anticipation of the unheard-of situation of fingerprint examiners squaring off against each other, the district attorney brought in a fingerprint expert from another state, to add quantity if not quality to his side. In 1998, the Jackson case went to trial and the jury, despite the conflicting fingerprint testimony, found Jackson guilty of first-degree murder. The judge sentenced him to life without parole.

Vernon McCloud and George Wynn were so concerned about the fingerprint misidentification, they asked the IAI to empanel a group of experts to review the evidence. When the IAI panel agreed that the crime-scene latent was not Jackson's, the district attorney began to doubt his own experts and sent a photograph of the bloody latent to the FBI lab for analysis. The examiners in Quantico, Virginia, agreed with McCloud, Wynn, and the IAI panel: Rick Jackson had been sent to prison on a misidentified crime-scene fingerprint.

In December 1999, after Jackson had spent two years behind bars, his conviction was set aside and he was set free. The out-of-state fingerprint examiner was fired, but the Upper Darby examiners were not disciplined or prohibited from future fingerprint work. Moreover, they would continue to insist that they were right and all the other experts were wrong. Jackson sued these local examiners in 2001 but lost the case.

Judge Louis H. Pollak: Raising the Issue of Examiner Proficiency

The second *Daubert* challenge of fingerprint evidence arose out of a federal drug case involving four Philadelphia homicides in the Eastern District of Pennsylvania. Held in January 2002 before federal district judge Louis H. Pollak, the hearing featured many of the same documents and much of the same testimony presented in the first *Daubert* challenge, brought by attorney Robert Epstein in 1999. The hearing also included the testimony of David Stoney, the director of the McCrone Research Institute in Chicago. According to Dr. Stoney, fingerprint identification was unscientific and subjective and therefore prone to human error. To back up his position, Dr. Stoney cited the high error rate in fingerprint certification examinations. Judge Pollak, the former dean of Yale University and the University of Pennsylvania Law School, stipulated that fingerprints were unique but found the certification test results worrisome. This led him to a historic ruling—the first of its kind in the United States: in the Eastern District of Pennsylvania, fingerprint experts could state their *belief* in a fingerprint match but could not declare an identification beyond all doubt. The judge's decision didn't bring down the house of fingerprints, but it did weaken the foundation and sent shock waves through the forensic identification community.

A month later, pursuant to the same case, Judge Pollak presided over a follow-up *Daubert* challenge to the admissibility of fingerprint evidence in federal court. The three-day hearing featured the testimony of Stephen B. Meagher, an FBI lab examiner, and Allan Bayle, a retired Scotland Yard fingerprint expert brought into the case by the defense. Meagher described the bureau's nonnumeric identification process, a system called ACE-V, which stood for Analysis, Comparison, Evaluation, and Verification. According to Meagher, this system had never resulted in the misidentification of a latent print. He pointed out that the high certification test error rates did not involve FBI examiners. The FBI had its own proficiency tests and the scores were high.

Allan Bayle argued that fingerprint identification was not as foolproof as its practitioners claimed. He said the FBI proficiency test used only inked fingerprint impressions for comparison and so did not reflect the reality or difficulty of latent fingerprint identification. According to the former Scotland Yard examiner, the FBI tests were so easy, anyone with six weeks of training could pass them, Steve Berry reported in the *Los Angeles Times*. "If I gave my experts these tests, they would fall about laughing," he said. Bayle argued that the claim that the FBI had never misidentified a latent print might simply reflect the fact that until recently, no one had ever challenged an FBI fingerprint identification.

As a result of this follow-up hearing, Judge Pollak modified his ruling by finding that fingerprint examination as practiced by the FBI met *Daubert* standards. He based this ruling on the fact that there had never been, among FBI examiners, a case where a fingerprint had been misidentified. But a year after Bayle's testimony, FBI fingerprint examiners would make a misidentification in a high-profile case.

The Cowans Case: The Police Fingerprint Bureau as a Personnel Dumping Ground

The identification values of DNA and fingerprint analysis clashed several years after the 1998 wrongful conviction of Stephen Cowans. In 1997 a burglar shot and wounded a Boston police officer, then fled to a nearby house where he hid until it was safe to leave the scene. Detectives suspected Stephen Cowans, a twenty-seven-year-old burglar and petty thief, of the shooting. They showed his photograph to the wounded police officer and to the homeowner the shooter had taken hostage. Both men identified Cowans as the perpetrator of the crimes. Fingerprint examiners from the Boston Police Department, Dennis LeBlanc and Rosemary McLaughlin, identified as Cowans's a latent lifted off a coffee mug at the hostage's house.

A year after the shooting and home invasion, a jury, obviously impressed with the fingerprint evidence, found Cowans guilty. The Suffolk County judge sentenced him to thirty-five to fifty years in prison. Cowans claimed that he had been framed.

In the fall of 2003, Cowans's attorney, James Dilday, requested a DNA analysis of saliva traces on the coffee mug from which the incriminating print had been lifted. The DNA test revealed that the saliva wasn't Cowans's. The district attorney, concerned about the reliability of the fingerprint identification, sent the latent to the FBI lab, where examiners identified it as belonging not to Cowans but to the hostage. A judge set aside Cowans's conviction, and on January 23, 2004, after serving six and a half years in prison, he was let out.

The officers who had identified the coffee mug latent as Cowans's were placed on suspension pending an investigation by the Massachusetts attorney general's office. Three months later, the state attorney general announced there was not enough evidence to support charges of perjury. According to the attorney general's investigation, the identification officers had made an honest mistake. In September 2004, the Boston identification unit was shut down and its duties taken over by the Massachusetts State Police.

For decades, the Boston Identification Bureau had been used as a dumping ground for officers who had been caught drinking on the job, using excess force, and stealing. Not every identification officer had such a record, but just one at any given time would have been bad enough. Further, the examiners were not adequately trained, and few, if any, had been certified by the IAI or any other certifying organization. Most of them were, in essence, amateurs. No one will ever know how many misidentifications had come out of this fingerprint bureau or how many innocent people had been sent to prison.

The Patterson Case, Act 1: Group Identification of Latents

Following oral arguments for and against the scientific reliability of fingerprint identification in a case involving an officer of the notorious Boston Police Department Identification Bureau, six judges on the Massachusetts

Supreme Court would decide on September 7, 2005, first, if fingerprint evidence should have been excluded at a 1995 trial in which Terry L. Patterson had been convicted of murdering an off-duty Boston police officer; and second, if all Massachusetts courts should exclude fingerprint evidence in general.

The history of the case was this. At three in the morning of September 26, 1993, Detective John Milligan's body was found in his truck outside a Walgreen's drugstore in the Roslindale section of Boston. He had been killed while moonlighting as a security officer, shot five times at close range, and his sidearm was missing. The police arrested Patterson, a man from nearby Dorchester, and charged him with the detective's murder. At Patterson's trial, held in January 1995, Sergeant Robert Foilb of the Boston police latent fingerprint section identified as Patterson's four latents lifted from the window of the driver's side door of Detective Milligan's truck. According to Foilb, adjacent prints from the same hand had been pressed onto the glass at the same time. In describing the four latents, the examiner referred to them as "simultaneous prints." Officer Foilb concluded that the four prints corresponded to the little finger, ring finger, middle finger, and index finger of Patterson's left hand. Foilb had found thirteen points of identity from the four prints—one latent had two points of similarity, one had five, another had six, and one had none.

On cross-examination, Foilb conceded that none of the individual latents met the eight-point minimum standard that was the identification norm in Massachusetts. He said, however, that in analyzing the four latents as a single unit, he applied the nonnumeric ACE-V method used by the FBI and fingerprint examiners in the United Kingdom. Based upon the thirteen points of identity in the four latents, Sergeant Foilb was confident the defendant had touched the dead officer's car. He assured the court that similarities in simultaneous impressions "can be counted as a total number because there is no other way to have those fingerprints put on an object." In other words, they all had to have come from the same hand.

On the strength of Sergeant Foilb's fingerprint testimony, as well as testimony from Stephen Meagher of the fingerprint section of the FBI lab, Terry Patterson was found guilty and sentenced to life. An appeals court, however, reversed the conviction on unrelated procedural grounds on December 6, 2000. Fingerprint identification was under attack by then, and Patterson's attorney, John H. Cunha Jr., decided to join the battle with a *Daubert* challenge of his own.

In May 2004, the *Daubert* hearing, featured several fingerprint experts who for five days argued the merits of the Patterson identification and fingerprint science in general before a superior court judge. Fingerprint expert Stephen Meagher, Dr. William J. Babler, a biological anthropologist, and David R. Ashbaugh, a fingerprint examiner with the Royal Canadian Mounted Police went up against Dr. James E. Starrs, a professor of forensic science at Washington University, Dr. David Stoney, now with the University of California at Berkeley, and Simon A. Cole, a criminologist at the University of California at Irvine. (Babler, Ashbaugh, Stoney, and Cole didn't actually take the stand. Their testimony in *Mitchell* was entered into the record.) On November 29, 2004, the judge ruled that the fingerprint identification methodology used in the Patterson case met the *Daubert* criteria. The decision was a defeat for Terry Patterson and a victory for fingerprint identification—which in 2004 was otherwise having a bad year.

Misidentification in the Madrid Bombing Case

On March 11, 2004, terrorists in Madrid bombed a passenger train, killing 191 people. The Spanish National Police sent the FBI digital images of eight latents found at the bombsite. These images were fed into the FBI's Integrated Automatic Fingerprint Identification System (IAFIS), a $640 million supercomputer housed in Clarksburg, West Virginia. The computer selected from its collection of 48 million fingerprint sets fifteen digital latent images as possible matches. Three FBI examiners matched one of the fifteen possibilities to a latent from Spain that had been left on a plastic bag containing bomb detonators. The examiners believed this print belonged to a thirty-seven-year-old lawyer from Portland, Oregon, named Brandon Mayfield. If the FBI examiners were correct, Brandon Mayfield had been at the scene of the Madrid bombing. The fact that Mayfield, a former army lieutenant, had converted to Islam heightened the FBI's suspicion that he had been involved in the bombing.

Fingerprint examiners in Spain agreed that Mayfield's print and the crime-scene latent shared eight points of similarity, but the numerous dissimilarities kept them from declaring a match. The FBI responded by having a fourth examiner look at the evidence, and he too, declared a match. Shortly thereafter, FBI agents arrested Mayfield. In the meantime, the police in Spain announced that the crime-scene latent belonged to an Algerian suspect,

Ouhnane Daoud. FBI examiners traveled to Madrid and realized their mistake when they compared Mayfield's print to the actual latent. Blaming the low-resolution image of the digital latent, the FBI apologized to Mayfield. However, as Jennifer Mnookin reported in the *Washington Post*, when a panel of fingerprint experts reviewed the evidence, they found that the misidentification had nothing to do with the quality of the digitized latent. The four FBI examiners had simply overlooked "easily observed" dissimilarities between the two prints.

Mayfield filed a lawsuit, and on November 29, 2006, the federal government agreed to pay him $2 million in damages. The Justice Department augmented the settlement with an official apology, stating that misidentifications of this nature were rare. University of California at Irvine professor Simon Cole disagreed. Responding to news of the settlement, he told *Los Angeles Times* journalist Sam Verhovek that "this is a tip-of-the-iceberg phenomenon. The argument has always been that no two people have fingerprints exactly alike. But that's not what you need to have an error. What you need is for two people to have very similar fingerprints and that's what happened here."

The FBI could no longer claim that its examiners had never misidentified a latent fingerprint. Moreover, in the wake of the Madrid embarrassment, the FBI proficiency test came under attack once again as too easy and unrealistic. Even worse, according to an FBI whistleblower, the test never changed and cheating was commonplace.

The Patterson Case, Act 2

On September 5, 2005, Terry Patterson's attorney, John Cunha, again stood before the Massachusetts Supreme Court, a panel of six judges led by Chief Justice Margaret H. Marshall. He based his hope that the lower court's *Daubert* decision would be overturned on, first, the FBI's misidentification of the Madrid bombsite latent and, second, the shutdown of the Boston Police Department's fingerprint operation after the Stephen Cowans misidentification.

In anticipation of Terry Patterson's second trial, Cunha's immediate goal was to have the fingerprint evidence excluded on the grounds that Boston police sergeant Robert Foilb's grouping of four "simultaneous" latents was an unreliable method of fingerprint identification. During the past twelve years, these latents, lifted from the driver's side window of the murdered detective's truck, had been analyzed by several examiners with the Boston Police

Department, five with the Massachusetts state crime-scene services, and three with a private fingerprint firm, Ron Smith and Associates. All these examiners concurred with Foilb's simultaneous latent identification. But through the presentation of two amicus briefs—friend-of-the-court reports filed by legal scholars and others not parties to the actual case—Cunha would argue that the FBI's ACE-V methodology used by Foilb in the identification of the four adjacent latents was not scientifically reliable. The amicus argument went something like this: Assuming that a single latent has up to 175 ridge characteristics that can be identified, matching six, ten, or even twelve in the four latents isn't enough to individualize the set.

The second prong of Cunha's challenge, more abstract and far-reaching, encompassed the entire forensic field of fingerprint identification. Cunha wanted the Massachusetts Supreme Court to declare *all* fingerprint identification evidence inadmissible. If the top court so ruled, no criminal defendant in the commonwealth would ever again be convicted on the basis of a crime-scene latent.

Suffolk County assistant district attorney Donna Patalano was in charge of protecting the fingerprint evidence in the Patterson case and defending the reliability of fingerprint identification generally. She was up against a well-prepared battery of experts who thought they were reforming fingerprint identification as a forensic science in Massachusetts. There was a lot at stake, and if she lost, the defeat would be felt throughout the country and perhaps the world.

In one of the two amicus briefs on behalf of the defense, professors from three Boston area law schools were bolstered by the opinions of seven experts, including Simon Cole. Citing twenty-two cases of fingerprint misidentification, the experts argued, among other things:

1. Simultaneous fingerprint identification was not supported by scientific study.
2. The ACE-V method of identification was too subjective to be scientific.
3. Fingerprint examiners are merely technicians and self-proclaimed experts.

The second amicus brief, prepared by attorney Lisa J. Steele for the National Association of Criminal Defense Lawyers, alleged:

1. The number of known fingerprint misidentifications is just the tip of the iceberg.

2. When one fingerprint examiner identifies a print, the second examiner is biased in favor of that conclusion.
3. The ACE-V method was subjective and judgmental.
4. All fingerprints, latents and inked impressions, are by nature distorted, making the identification process flawed and unreliable.

A pair of amicus briefs, one by Norfolk County, Massachusetts, and the other by the state attorney general's office, were submitted on behalf of the commonwealth and in support of fingerprint identification. At the oral presentations before the panel of justices, Donna Patalano and John Cunha, as well as their expert witnesses, were given fifteen minutes to make their points. The justices frequently interrupted these presenters with pointed questions.

On December 28, 2005, the Massachusetts Supreme Court handed down its decision, a narrow ruling and perhaps a compromise. Fingerprint identification based upon the ACE-V methodology, as applied to simultaneous latents, would no longer be admissible in Massachusetts courts. Fingerprint identification in general met the *Daubert* test and therefore would be admissible in courts throughout the commonwealth. While a small victory for the fingerprint reform movement, the decision was a big win for Terry Patterson. If the commonwealth wanted to retry him for the detective's murder, they would have to do it without the fingerprint evidence. A second trial became moot when Patterson pleaded guilty in February 2006 to reduced charges of manslaughter. Having served twelve years of his original sentence of twenty-five to thirty years, that plea meant he could look forward to release as early as the upcoming summer.

Professor Simon Cole, the University of California criminologist and one of the more active crusaders for fingerprint identification reform, reacted to the Massachusetts Supreme Court decision this way: "It's a little chink in fingerprint's armor. It would have taken a great deal of courage [to reject fingerprint evidence entirely] but we hoped that we had a court that had that courage." It seems possible that the justices on the Massachusetts Supreme Court considered their decision more a matter of law than courage.

One thing is certain: the end of the war against fingerprints is still nowhere in sight.

7

Fingerprints Never Lie

Except in Scotland

In January 1997, Scottish officers from the Strathclyde Police Department responded to the scene of a murder in nearby Kilmarnock. Marion Ross, a fifty-one-year-old former bank clerk, had been stabbed to death in her bathroom. Her ribs were crushed and she had been stabbed in the eye and throat with a pair of scissors that had been left stuck in her neck. There was no sign of forced entry. Police officers theorized that Marian Ross had been killed by one of the men who recently had been doing remodeling work in her home.

Shortly after the crime the police arrested twenty-three-year-old David Asbury, a construction worker from Kilbirnie in Ayshire. Although no latent fingerprints belonging to Asbury had been found at the scene, examiners with the Scottish Criminal Records Office (SCRO) identified a print on a container, a biscuit tin, found in the suspect's apartment as being the murder victim's. The tin contained £1,800, money the police believed the killer had stolen. Asbury claimed that the money and the tin were his.

The all-important latent on the biscuit tin had been lifted at Asbury's apartment by Shirley McKie, a thirty-four-year-old detective constable with the Strathclyde Police Department. McKie, whose father, Iain, had retired after thirty years with the same department, was a respected officer who loved her job. Her feeling of accomplishment in discovering the key piece of evidence in a homicide case ended abruptly when she was called on the carpet for leaving her own print at the scene of the murder. SCRO examiners had identified a bloody left thumbprint on the bathroom doorframe as hers. Leaving one's own latent at the scene of a crime was an embarrassing mistake for a detective. Had she forgotten to bring her latex gloves? According to

McKie, she had gone to the scene three times but had never gotten beyond the front porch. That latent, therefore, could not have been hers. The SCRO examiners must have made a mistake.

McKie's denial in the face of incontrovertible evidence infuriated her supervisors. It was one thing to make an honest mistake, another to lie about it. As far as they were concerned, she had sneaked a look at the gory bathroom tableau and had gotten caught. If she didn't 'fess up, she could lose her job. Several of McKie's fellow officers went to her house and begged her to admit the truth. Even her father, Iain, didn't believe her. Fingerprints don't lie, people do. Too depressed to work, McKie went on sick leave for two months.

In Scotland, justice can be quick. In May 1997, just three months after his arrest, David Asbury was brought to trial in Glasgow, still maintaining his innocence. McKie took the stand. She would be followed by the SCRO fingerprint examiners who had connected Asbury to the victim through the biscuit-tin latent McKie had lifted. On cross-examination, Asbury's solicitor, George More, in an effort to muddy the water on the latent fingerprint identification, asked McKie if she had helped process the murder scene for evidence. She testified that she had not been inside the house. But didn't examiners from the SCRO identify one of the latents in the murdered woman's bathroom as hers? Yes, they had. But didn't you say you weren't in the house? Yes. So the SCRO examiners had made an incorrect fingerprint identification? That latent was not mine, replied the witness.

Following the thirteen-day trial, the jury chose to believe that the SCRO had correctly identified the biscuit-tin latent as Asbury's and convicted him of murder. By implication, the jury believed that detective McKie had been at the murder scene as well and had committed perjury. McKie's police superiors were angry that, to save herself embarrassment, the detective had committed perjury and in so doing had put a murder conviction at risk. She had also challenged the reputation of the SCRO.

Shirley McKie's Arrest and Quest for Vindication

As bad as it was for Shirley McKie, things would get a lot worse. In March 1998, police came to her house and arrested her for committing perjury in the Asbury trial. If convicted, she could go to prison for up to eight years. Taken to the police station in Ayr, she was booked, strip-searched, and locked in a cell. She made bail the next day and learned that she had been suspended from the force pending the outcome of her trial. Unless she could prove that the print at the Ross murder scene wasn't hers, she would be convicted of perjury and possibly sent to prison.

Through an Internet search, McKie learned of fingerprint analyst Patrick A. Wertheim with the Arizona Department of Public Safety. A member of the IAI, the Fingerprint Society of Great Britain, and the Canadian Identification Society, Wertheim worked in the state crime lab in Tucson. He had been trained in Texas with the Department of Public Safety in Austin and had taken the advanced fingerprint-comparison course at the FBI Academy in Quantico. He had been analyzing fingerprint evidence for twenty years. At no cost to McKie, Wertheim agreed to fly to Glasgow to compare McKie's known prints to the murder scene latent, referred to by the SCRO examiners as Print Y7. To his utter shock, Wertheim found that the two prints did not match. This was the first time in his career that he had not been in agreement with another fingerprint expert. For Shirley McKie, it was an exhilarating moment of exoneration. Maybe now people would believe her. Wertheim agreed to testify on her behalf at her upcoming perjury trial.

After reviewing the evidence, David L. Grieve, a fingerprint examiner with the Illinois state crime lab, and New Scotland Yard's Allan Bayle, the United Kingdom's foremost fingerprint expert, agreed to join Wertheim in testifying against the SCRO experts at McKie's trial. (Bayle had gained prominence after identifying a latent fingerprint in the 1988 bombing of Pan Am Flight 103 over Lockerbie.)

When the trial commenced in May 1999, the stakes were high for everyone concerned. Shirley McKie was fighting for her freedom and her good name. The four SCRO examiners, on suspension themselves pending the results of the trial, were defending their credibility and the honor of the Scottish Identification Bureau. The eyes of the fingerprint world were focused on this historic fingerprint confrontation. The SCRO examiners testified that there were sixteen points of similarity between Print Y7 and McKie's left thumb impression. Pat Wertheim, David Grieve, and Allan Bayle testified that there were only five points of identity and numerous points of dissimilarity that proved that the two prints were not the same. In their professional opinions, someone other that Detective McKie had touched the doorframe near the murdered woman's body.

In his closing argument the Crown advocate, Sean Murphy, argued that latent fingerprint identification was a matter of interpretation and that jurors, when making their decision, should remember that the SCRO experts between them had one hundred years' experience in the field. He urged the jury to select the Scottish Identification Bureau over New Scotland Yard and the Americans. The jury, deliberating less than an hour, came back with a verdict in favor of Shirley McKie. Judge Lord Johnston responded to the verdict by saying that he personally respected Shirley McKie for her "dignity and courage." The acquittal was an embarrassing defeat for the SCRO and a victory, albeit a costly one, for Shirley McKie. But this was not the end of the story, just the start of a new phase in what would develop into an international fingerprint identification scandal.

The SCRO, having committed itself to the claim that Print Y7 had not been misidentified, had no intention of admitting its mistake and apologizing. It was too late for that. Instead, the SCRO went on the attack in a ridiculous attempt to control the damage to the image of the bureau. This included the circulation of a memo to law enforcement agencies across Scotland that trumpeted the bureau's "integrity." There was legitimate concern within the Scottish criminal justice system that if the SCRO lost its credibility, dozens, if not hundreds, of fingerprint identification cases could unravel. By raising questions regarding Pat Wertheim's qualifications and by accusing Shirley McKie of leaving her latent behind at yet another murder scene (an accusation that turned out to be false), SCRO officials tried to discredit the acquittal in the perjury trial.

In the closed and insulated criminal justice community, McKie was forever tainted and unwelcome, having been forced into becoming a law enforcement whistleblower, a role no police officer in his or her right mind would wish for. She was worse than a mere civilian—she was a traitor. The same held true for Allan Bayle, who was perceived by many of his fingerprint colleagues as a turncoat. He went into early retirement, ending a brilliant career in public service. The four SCRO examiners, on the other hand, were still on the job.

In December 1999, despite her perjury acquittal, Shirley McKie was dismissed from the Strathclyde Police Department on medical grounds. On suspension since March 1998, the dismissal made her ineligible for a pension. A few weeks after her dismissal, her supporting fingerprint expert from Arizona, Pat Wertheim, posted the following Internet message on a forensic science blog:

> For an "expert" to claim an identification between the crime scene mark in this case and Shirley McKie's fingerprint is sheer incompetence.

> To testify knowingly to this erroneous identification, as some believed happened at Ms. McKie's [perjury] trial, is perjury. For the police to cover up the mistake is despicable. For a whole police administration to blindly support their "experts" without seeking a competent outside review is foolhardy.

The BBC television documentary series *Frontline Scotland* aired a show on January 18, 2000, called "Finger of Suspicion" that featured host Shelley Jofre's interviews of Shirley McKie, Pat Wertheim, and five other fingerprint experts who disagreed with the SCRO identification of the Ross murder-scene latent. Among these experts were Ray Broadstock, the chair of the National Fingerprint Society; Ron Cook, a British examiner with thirty-three years' experience; and a former SCRO examiner who asked not to be identified. What follows are excerpts that reflect the theme of the program:

MCKIE: I was persecuted for two and a half years. I had to give up my job, and, as far as I can see, absolutely nothing is happening to the people responsible for this. . . . It just seemed like every time it got bad, you think, right, that's as far as they're going to take this. It was like, they just kick you again and it got worse, and it got worse, and it got worse, and I really felt that they were trying to kill me. . . . I was so scared, because I thought if David Asbury gets found not guilty, then everybody is going to blame me. . . . I opened the door [to my house] and there was a detective chief superintendent and two female detectives outside. I was being arrested for perjury, and I was just in an absolute daze. . . . I'd actually thought about what I'd do if I got found guilty, what would be the easiest way to make it go away. It was really deciding what was the quickest way to kill myself, really. I'm not a brave person at all, and I couldn't have coped. If I'd been found guilty for something I hadn't done—no—I couldn't have lived with that.

PAT WERTHEIM: I had a sinking feeling in my stomach because it was clear to me that a mistake had been made. This was not an identification. There was something horrible going on here. I've never been put in a position of calling another fingerprint expert wrong, and to see a print so obviously not the same. . . . Quite frankly, my stomach just knotted up. . . . I think it's disinformation to try to minimize the damage that might be done to other fingerprint cases. Those four experts I know must have numerous fingerprint cases. . . . And the damage that might accrue to some of those, I suspect, could be considerable. So I suspect what's happening is just damage control. . . . I suspect that if I hadn't caught this, Shirley McKie would be sitting in jail right now. You see, if this can happen to Shirley, it can happen to anyone who has been fingerprinted for any reason.

FORMER SCRO EXAMINER: The system of fingerprint identification is infallible. The expert individually is not. Any expert can make a mistake, and should be seen to admit making a mistake.

HOST: How did a solitary fingerprint end up as the cause of so much controversy? Could it date back to a simple mistake made by the first SCRO expert?

PAT WERTHEIM: I believe he carelessly applied a lower standard to an officer's fingerprints than he would have applied to a suspect's fingerprints. And I believe the situation escalated.

SHIRLEY MCKIE: I've totally lost most, not all, of my confidence. I can't see a police car or a police officer without feeling totally nauseous and sick. When the phone rings, or someone comes to the door, there's still the fear in me that it's going to be the police again, they're going to take me out of the house again. I've lost a job that was to be a career for me. I'm thirty-seven years old and I'm terrified of my future.

The *Frontline Scotland* program prompted the Association of Chief Police Officers in Scotland to ask William Taylor, Her Majesty's chief inspector of constabulary, to investigate the fingerprint identification procedures used by the SCRO. Taylor agreed to look into the matter. Shirley McKie, accompanied by a fingerprint expert from the Lothian Police Department and an examiner from the police station in Borders, visited the Scottish Parliament in March. The purpose of her appearance was to urge criminal justice minister Jim Wallace to take an interest in the problem.

The SCRO's apparent misidentification of Print Y7 had great relevance to David Asbury, who had gone to prison on the strength of the SCRO's identification of the murder victim's latent on the biscuit tin found in his house. If the fingerprint bureau had misidentified one print in the case, perhaps they had made a mistake on the other one as well. Yet George More, the legal aid solicitor who had defended Asbury at trial, had made no effort to have the biscuit-tin latent reexamined by outside experts. BBC Scotland, having taken up Shirley McKie's cause and the crusade to expose the SCRO, flew Pat Wertheim back to Scotland in May. The Arizona fingerprint examiner had agreed to determine if the latent on the biscuit tin did in fact belong to Marion Ross. Allan Bayle, formerly of New Scotland Yard, would also review the evidence for the BBC.

At Asbury's trial, the SCRO experts, in identifying the biscuit-tin latent as Marion Ross's, had used the standard fingerprint comparison courtroom exhibit consisting of photographs of the two prints side-by-side, connected by a series of red lines depicting the points of similarity between the latent and the known, inked impression. According to the exhibit and the testimony of the SCRO examiners, the biscuit-tin latent met the UK standard of sixteen points of identity with the known impression. (This occurred before the United Kingdom adopted the nonnumeric system in 2001.) When Wertheim and Bayle looked at the SCRO exhibit, they were surprised to find that the two prints were not the same. Even more shocking, the exhibit itself didn't reveal sixteen points of identity between the two prints. Many of the connecting red lines failed to depict similarity points. As Wertheim put it, they simply ended "out in the middle of nowhere." The exhibit that sent David Asbury to prison was phony, and so was the testimony that went with it. Honest fingerprint identification mistakes were bad enough—this was perjury.

On May 16, 2000, Allan Bayle and Pat Wertheim appeared on a second *Frontline Scotland* show to discuss the growing fingerprint scandal. Called "False Impression," this segment included an interview with Amelia Crisp, David Asbury's mother, who said that the biscuit tin had been in her son's bedroom for three years. The money inside it was his as well, a sum that had taken him years to save. He had drawn it out of the bank to buy a car. Mrs. Crisp didn't know anything about fingerprint identification, but she did know that Marion Ross had not touched her son's biscuit tin. As far as she was concerned, her son had been framed.

Turning the Tables on the SCRO

In June, following a three-month inquiry into the SCRO and its role in the McKie case, Chief Inspector of Constabulary William Taylor called for an overhaul of the Scottish fingerprint bureau. He recommended improvements in training and proficiency testing, and the imposition of a program whereby SCRO fingerprint identifications would be double-checked by outside examiners. These suggestions outraged Harry Bell, the head of the SCRO, who had not conceded that his examiners had made mistakes in the McKie and Asbury cases. William Taylor did not, however, call for the termination or criminal prosecution of the SCRO examiners. Jim Wallace, the Scottish minister of justice, publicly apologized to Shirley McKie and, in so doing, upset everyone in the SCRO. Wallace also offended everyone in the fingerprint identification community when he said that fingerprinting was "not an exact science."

Two months later, Lord Advocate Colin Boyd (the Scottish version of the U.S. attorney general) appointed a special prosecutor to determine if crimes had been committed by the SCRO examiners. The special prosecutor, Bill Gilchrist, would determine if the fingerprint scandal had included a cover-up conspiracy, perjury, or both. The four SCRO fingerprint examiners were placed on suspension pending the outcome of the investigation. Once again, SCRO head Harry Bell defended his experts against what he called a witch hunt.

At a conference on August 15, 2000, at the Tulliallan Police College in England, two of the SCRO examiners who had identified the murder-scene latent as McKie's and the biscuit-tin latent as Marion Ross's presented their interpretation of the evidence to a room full of fingerprint examiners from around the world. The suspended examiners, desperate to salvage the reputation of the SCRO, had prepared a new fingerprint comparison exhibit that illustrated forty-five points of identity between Shirley McKie's inked thumb impression and Print Y7. Arie Zeelenberg, head of the National Fingerprint Service of the Netherlands, was in attendance and had determined in his own mind that the two prints did not match by the time the SCRO presentation was over. The Dutch expert also suspected that the exhibit had been doctored. Zeelenberg would become one of many in his field to publicly stand up for Shirley McKie.

Two days after the Tulliallan conference, Lord Gill of the Court of Criminal Appeal in Edinburgh granted David Asbury "interim liberation" pending the appeal of his murder conviction. In response, the Ross case prosecutor, advocate Gerry Hanretty, acknowledged problems with the Crown's fingerprint evidence. Asbury had been in custody more than three years.

In September 2001, when it seemed that some measure of justice would arise out of the McKie and Asbury ordeals, Bill Gilchrist, the special prosecutor who had been investigating the SCRO, announced that he had not found enough "admissible evidence" to prove the commission of any crimes. He said there was no evidence of intent and he couldn't prosecute on the basis of differing opinions regarding the identification of a pair of fingerprints. This finding allowed the four SCRO examiners to go back on the job. Gilchrist made the announcement after meeting with Shirley McKie and her father, Iain. The McKies were understandably disappointed and bitter. Having received no satisfaction from the SCRO, or even an apology from the chief constable of the Strathclyde Police Department, Shirley McKie decided to file a civil action against her former employer. Three months later, Lord Emslie of the Court of Session dismissed the suit, calling it "fundamentally irrelevant." According to the judge, McKie had failed to prove that her ex-colleagues and bosses had acted with malice.

On August 14, 2002, six months after the Court of Session in Edinburgh threw out McKie's lawsuit, Lord Gill of the Court of Criminal Appeal quashed the Asbury murder conviction due to "considerable concern" over the SCRO fingerprint identification. David Asbury would not be retried and the Marion Ross murder case would revert to the status of unsolved. Four weeks later, U.S. fingerprint experts Pat Wertheim and David Grieve along with England's Allan Bayle and the Dutch examiner Arie Zeelenberg, in a petition to the Scottish Parliament, asked for another government investigation of the SCRO. Aware of the effect the scandal was having on fingerprint identification generally, Allan Bayle said: "The SCRO has to put its house in order. They are still denying and saying it [fingerprint identification] is only opinion. It is not opinion. It's a fact in black and white. Either it is or it isn't or it's inconclusive, nothing else." The petition included the names of 160 fingerprint experts from around the world who had reviewed the evidence and concluded that Print Y7 was not Shirley McKie's.

Having lost confidence that the Scottish government would aggressively pursue justice on her behalf, Shirley McKie, on October 9, 2003, sued the Scottish executive for £850,000. To bolster her claim of cover-up, conspiracy, and perjury, McKie made a startling accusation. She alleged that from the beginning there had been disagreement within the SCRO over the identity of Print Y7. Five examiners had opined that Shirley McKie had *not* left the bloody latent at the Ross murder scene. Had the agency not suppressed this information, she would not have been dismissed from the force or charged with perjury. In discussing her suit on BBC Scotland radio, McKie said: "Quite frankly, they should be in jail for what they've done to me, but they're pulling the wool over the public's eyes and they've been allowed to get away with it."

Defending the government's refusal to prosecute, fire, or even discipline the SCRO examiners, Cathy Jamieson, the Scottish justice minister, stuck to the "let's not criminalize dissenting opinions" rationale. On a later BBC Scotland radio show, Jamieson threw fingerprint identification down the well in articulating the government's position when she said, "Fingerprinting has never been an exact science." This may have been the awful truth, but fingerprint identification, compared to the fields of bullet, tool-mark, hair, fiber, bite-mark, and handwriting identification, has long been considered a more exact and objective science. The SCRO scandal was having the effect of grouping fingerprint identification with these less scientifically respected fields. The controversy was spinning out of control, escalating into something far beyond the misidentification of a couple of latent fingerprints.

In a move that shows just how vindictive the police can be against former colleagues who have challenged their integrity, the Strathclyde Police Department sued McKie for the legal costs incurred in the defense of her suit against them. To pay the £1,300 legal bill, McKie would have to sell her house. On the BBC radio program *Good Morning Scotland,* Iain McKie, who had spent thirty years in law enforcement, expressed his outrage over how the authorities had "destroyed" his daughter: "It's just yet another blow after several years of blows." Two months later, an anonymous donor paid the legal bill to get the police department off McKie's back.

The Case Mounts against the SCRO

More than seven years after Marion Ross's murder, the fingerprint scandal was still alive and unresolved. The strain had taken its toll on Shirley McKie's health. She had been consumed by the case, under enormous stress, and unable to get on with her life. She was worn down, depressed, and angry, and there was no end in sight.

Forensic science does not recognize national borders, so the public disgrace of any expert makes itself felt around the world. Allan Bayle, the English expert who had been involved with the McKie case since the beginning, was becoming increasingly worried that the SCRO's refusal to deal with the problem and the continuing scandal was damaging the entire profession. On September 20, 2004, in a radio interview broadcast on BBC's *Good Morning Scotland,* Bayle expressed his concern and frustration:

> There's parts of the latent [Print Y7] you can compare without being an expert, and it just shows quite blatantly that the SCRO [examiners] are not telling the truth. They've had all these inquiries but they still haven't changed, they've got a long way to go, they're still making mistakes and that's alarming. I'm getting phone calls from all over the world from experts saying they're being asked about the McKie case.

In the summer of 2005, three Scottish fingerprint experts from the Grampian Police Department, John Dingwall, John McGregor, and Gary Dempster, with fifty-four years experience between them, happened to come across the SCRO's original Print Y7 comparison exhibit. Upon reviewing the photographs, the three examiners experienced the shock of recognition that had hit Allan Bayle and the two U.S. experts. As the BBC reported in "Fingerprint Battle Takes New Turn," the Scottish examiners were so disturbed by what they saw, they sent a letter to the lord advocate that concluded: "We are satisfied beyond any reasonable doubt that the mark disclosed on the crime-scene photograph was not made by the left thumb of Shirley McKie." In response to this letter, the SCRO offered the following response: "[The] SCRO rejects any suggestion that it is failing to play its part in delivering fair and effective justice. We believe that the organization continues to provide an effective and professional fingerprint service."

That fall, solicitors representing the SCRO against Shirley McKie's £850, 000 lawsuit were moving for a settlement, an outcome that didn't sit well with the Print Y7 examiners. In a letter to the judge who would hear the case if it went to trial, the examiners made it clear they were sticking to their guns, as Liam McDougall would report in the United Kingdom's *Sunday Herald* on February 12, 2006: "As we, along with other independent experts, have stated it [Print Y7] is a valid identification, we wonder why [our lawyers] are making this unnecessary admission, as this makes not only our position untenable, but that of the fingerprint service in Scotland." The SCRO examiners at this point were not the only problem. The issue was evolving into a political scandal. There was no longer serious debate over the identification of Print Y7—virtually everyone in the profession knew it wasn't McKie's. If the civil action went to trial, dozens of the world's most prominent fingerprint experts would settle the issue on a highly public stage. The McKie case was already known around the world as the "Scotch botch" and the "Highland Hiccup." The trial had to be avoided at all costs. The suit had to be settled.

On February 7, 2006, the day before McKie's fingerprint experts were scheduled to start testifying, the Scottish ministry announced the settlement. Shirley McKie would be awarded £750,000, just £100,000 less than she had asked for. By agreeing to this settlement, McKie had let the government off the hook, but not cheaply. In an absurd attempt to save face, the ministers released the following statement: "We want to make it clear . . . that this settlement has been made on the basis that the identification was an honest mistake made in good faith." It was too late, however, to convince anyone that the settlement wasn't an admission of perjury and cover-up. Notwithstanding the opinions of two hundred fingerprint experts from around the world, Fiona McBride, the only SCRO examiner to be publicly identified, insisted that Print Y7 was McKie's. (The other SCRO examiners were Hugh MacPherson, Charles Stewart, and Anthony McKenna.) Having been denied her day in court, McBride said she was outraged by the payoff. She felt betrayed and characterized Shirley McKie's allegations against the SCRO as "malicious and unfounded."

The next day, Iain McKie said on BBC's *Good Morning Scotland:*

> If you can get £750,000 for a so-called honest mistake in Scotland, I wonder what you would get if liability was admitted—the fact remains that they are liable and the fact is that they were not honest mistakes. There could well be people languishing in prison in Scotland who should not be there because of similar mistakes. The Scottish Executive, the Justice Minister, and the Crown Office have to sort out this horrendous mess. They have been running about like headless chickens for nine years and it is time for them to stop running. We want a totally open public inquiry so people can see what happened. The public in Scotland are paying for my daughter's compensation, they should be given the answers why.

Pat Wertheim, the fingerprint expert from Arizona who had been the first to spot the misidentification, said on the same radio broadcast: "I can't express my disgust with the SCRO. They have done more damage to the science of fingerprinting than I ever thought possible." The Dutch expert, Arie Zeelenberg, one of the examiners who would have testified had the civil suit gone to trial, echoed this concern: "The SCRO has had nine years to study this print—and today we hear it's an honest mistake. So I don't think we made much progress. The international fingerprint community is waiting for an admission of this mistake because it is haunting the profession all over the world."

Did Patrick Docherty Kill Marion Ross?

On February 25, 2006, Scotland's national newspaper, the *Sunday Herald,* broke a story that explained why Stephen Heath, the senior investigating officer in the Marion Ross murder case, wanted the McKie suit settled out of court and the scandal put to rest. A few days after the January 8, 1997, murder, Heath and his men learned, from a confidential informant, that a thirty-three-year-old career criminal named Patrick Docherty had claimed responsibility for Marion Ross's murder. Heath was about to have Docherty arrested when, on January 31, 1997, the SCRO reported that they had found Marion Ross's latent on David Asbury's biscuit tin. Instead of Docherty, the police arrested Asbury. Five and a half years later, an intruder murdered ninety-one-year-old Margaret Irvine in Galston, Ayshire. She was found in her apartment lying on her bed beaten and bound. A duster had been shoved into her mouth and a pillowcase pulled over her head. In March 2005, Patrick Docherty and another man, Brendon Dixon, were found guilty, in the High Court in Kilmarnock, of the Irvine murder. Docherty, with a history of fifty-four previous felony convictions, was sentenced to life in prison.

If Patrick Docherty had in fact killed Marion Ross, if Print Y7 had been his instead of Shirley McKie's, the "honest mistake" that steered detectives away from Patrick Docherty had led to the violent death of another woman. Iain McKie, in responding to these revelations in the *Sunday Herald* on February 5, 2006, told reporter Liam McDougall, "It seems now that the behavior of the officers in this investigation, led by Mr. Heath, must be opened up and examined as a matter of urgency."

Cathy Jamieson, the justice minister, and John McConnell, the first minister of the Scottish Parliament, made it clear that they had no plans to initiate another investigation of the SCRO. Frustrated by the Scottish government's unwillingness to pursue the matter further, Iain McKie established a Fighting Fund to finance a private prosecution of the SCRO examiners. Two anonymous donors had contributed £25,000 each to the cause. SCRO examiner Fiona McBride said that she and her colleagues welcomed the prospect of a private inquiry. McDougall reported that an SCRO spokesperson responded to the Fighting Fund by saying: "We continue to have every confidence in our staff who deliver a sound professional service."

Whenever a forensic scientist's credibility is challenged, prosecutors who have benefited from that person's testimony, or who may need it in upcoming trials, will usually defend that expert. It's more a matter of self-interest than loyalty. In Scotland, an entire identification bureau was under attack, and the criminal justice establishment had come to its aid in an effort to prevent an investigation that could expose malpractice in hundreds of cases, setting off an avalanche of appeals. There was legitimate concern that the Shirley McKie and David Asbury cases represented the tip of an injustice iceberg, and that fear wasn't confined to Scotland.

The McKie–FBI Connection

In February 2006, the SCRO scandal spread to the FBI and threatened the credibility of the fingerprint evidence in the Pan Am Flight 103 bombing that had helped convict the Libyans. According to Juval Aviv, Pan Am's senior Lockerbie investigator, before the McKie perjury case went to trial in 1999, FBI agents, worried that a disagreement among SCRO examiners over the identify of Print Y7 might discredit and weaken the latent fingerprint evidence in the upcoming Pan Am trial in the Hague, flew to Scotland to urge the SCRO to unify behind the McKie identification. Aviv said he had received phone calls from SCRO examiners telling him of the FBI's mission to have the Scottish examiners present a united front against McKie. SCRO officials did not deny having meetings with the FBI over the Lockerbie case, but they insisted that the talks had nothing to do with Shirley McKie's perjury trial. This denial, however, did not ring true. The incriminating latent in the Lockerbie case, found on a travel document in Malta, had not been identified by SCRO examiners. Among others, Allan Bayle, then of New Scotland Yard, had analyzed this print. If the SCRO hadn't been involved in the Pan Am case, what exactly were they discussing with the FBI if not the McKie case? Because there were only twelve matching points of identity between the Libyan's print and the incriminating latent, the Lockerbie prosecutors may have felt vulnerable to a fingerprint-identification challenge. This had not been a good time for a public fingerprint controversy.

Pan Am's Juval Aviv was not the only one accusing the FBI of aiding and abetting a SCRO cover-up in the McKie case. David Grieve and Pat Wertheim, the U.S. fingerprint experts, reported that the FBI had approached them as well. FBI agents had asked both examiners not to testify for the defense at McKie's perjury trial. The FBI didn't want the fingerprint men rocking the boat in the bigger, more important Lockerbie trial.

The possible cross-fertilization of the McKie and Lockerbie cases caused several members of the Scottish Parliament to call for another public inquiry. For the SCRO, the FBI, and the fingerprint community, the McKie case had escalated into an international scandal that threatened the credibility of forensic science. If you couldn't trust fingerprint evidence, what could you trust?

In March 2006, fingerprint experts Pat Wertheim, Gary Dempster of the Grampian Police Department, and Arie Zeelenberg, now with Interpol, appeared on another *Frontline Scotland* television program dealing with the scandal. As the *Scotsman*'s Eben Harrell reported on March 9, the experts accused the SCRO examiners of evidence tampering in the McKie perjury trial. According to this allegation, someone connected to the SCRO, in an effort to lead the McKie jury to a guilty verdict, had cropped and blurred photographs of Print Y7 and the impression of McKie's left thumb to hide the points of dissimilarity between the two prints. Gary Dempster, the first Scottish fingerprint expert to criticize the SCRO, said, "From the information that we have seen, I don't believe it was an honest mistake." Pat Wertheim, in his remarks, left nothing to the imagination: "The SCRO knew before the trial that it was an erroneous identification and removed the differences in an attempt to mislead the court."

If all this were true, the SCRO, to insure the prosecution of the Libyans accused of the Lockerbie bombing, tampered with evidence and committed perjury to convict Shirley McKie of the same crime. The ploy didn't work because the FBI couldn't talk Pat Wertheim and David Grieve out of testifying for the McKie defense. McKie was acquitted, the SCRO disgraced, the FBI embarrassed, and the science of fingerprint identification knocked off its pedestal.

The McKie Case Legacy: A Damaged Form of Forensic Identification

The McKie case is an example of how much strength, resilience, and resolve it takes to fight the criminal justice system and win, if you call what happened to Shirley McKie winning. It also requires a lot of help from the right people and a little luck. It seems that government officials, no matter where, for reasons that are not always apparent, seldom take responsibility for their mistakes and their crimes, even after they have been caught red-handed.

The SCRO's refusal to take responsibility for what they did to Shirley McKie and David Asbury outraged fingerprint experts everywhere. Robert MacKenzie, the SCRO's deputy head, had arranged to have the Fingerprint Society's 2006 conference in Scotland. As a result of the more recent revelations surrounding the McKie case, examiners from around the world were boycotting the three-day conference, Liam McDougall reported in the

United Kingdom's *Herald* on March 12, 2006. In declaring his intentions, Allan Bayle said: "I won't be going this year because the Fingerprint Society has sat on the fence as this case devastates fingerprinting. They've done nothing to resolve it. How can I go to a conference organized in Scotland by the very people who are damaging fingerprinting?"

The FBI, tarnished by Pan Am Lockerbie investigator Juval Aviv's allegations that two agents in 1999 had pressured two members of the SCRO who had misgivings over the McKie fingerprint identification to fall in line with the evidence against her, announced an internal investigation. FBI spokesperson Paul Bresson said, "We have been fact-finding and interviewing relevant people to find out what happened." The FBI inquiry was being carried out by the bureau's Public Affairs Team. Retired FBI agent Richard Marquise, the bureau's head Lockerbie investigator from 1988 to 1992, said he had no knowledge of such interference by the FBI. He found the allegation "highly unlikely." The FBI/Lockerbie/McKie story made headlines in Scotland, but, except for a 2001 article in the *New Yorker*, was virtually ignored by the media in the United States, a break for the American fingerprint community. As of 2007 the FBI inquiry into the McKie/Lockerbie fingerprint connection has not been resolved.

8

Shoe-Print Identification and Foot Morphology

The Lay Witness and the Cinderella Analysis

Comparing a crime-scene shoe print left on a hard surface, or an impression in dirt, mud, or snow, to the bottom of a specific shoe is not unlike the process of latent fingerprint identification. In many crime laboratories the latent fingerprint people also handle footwear and tire-track evidence and occasionally the identification of tool marks. William Bodziak, for example, the world's leading footwear identification expert, who was employed in the FBI lab from 1973 to 1998, also identified tire tracks and worked in the questioned document section of the laboratory. Compared to DNA analysis, toxicology, and various aspects of forensic pathology, the identification of shoe marks, latent fingerprints, crime-scene bullets, tool marks, and handwriting involves less science than it does informed human observation. Does the shoe print look like it has been made by the bottom of the suspect's shoe?

A crime-scene shoe print or impression can be identified as part of a footwear group according to size, brand, and model or identified as coming from one shoe to the exclusion of all other shoes. Every year 1.5 billion pairs of shoes are sold in the United States. At any given time there could be as many as 100,000 pairs of size ten Nike sneakers of a certain model and tread design. There could be 5,000 pairs of these shoes in circulation in the Chicago area alone. The criminalistic or incriminating value of a group identification depends upon the size of the group. These group, or class, identifications occur when the crime-scene print or impression is not detailed enough for a match to a specific shoe or when the shoe that made the mark is not available for comparison. Probably the most famous group identification of shoe prints came at O. J. Simpson's double murder trial when Bodziak identified several crime-scene prints in blood as having been made by a pair of size twelve Bruno Magli Lorenzo shoes, luxury footwear made in Italy. Bodziak's testimony tended to incriminate Simpson in two ways: the identification involved a relatively small footwear group, and Simpson, after denying that he owned Bruno Magli shoes, was seen in a television clip wearing a pair. The shoes that made the bloody prints were never located.

An individual shoe, boot, or sandal can be linked to a crime-scene print or impression the way a latent fingerprint can be matched to its inked, rolled-on counterpart. Instead of comparing ridge configurations, the footwear examiner looks at a shoe's sole and heel for unique signs of wear that show up in the print or impression. Every shoe that has been worn awhile is as unique as a fingerprint. The more wear, the more potential for identification. Footwear identification, unlike fingerprint matching in most states, does not require a minimum number of similarity points. The credibility of a shoe identification depends upon the training, experience, and objectivity of the examiner as well as the quality, clarity, and uniqueness of the characteristics being compared. New methods and techniques are constantly being developed, for example, to lift footwear impressions from dust and even preserve shoeprints made in snow.

Shoe prints left in dust, blood, or soot or in heat insulation powder from a safe that has been broken into is photographed (next to a reference ruler), then peeled off the surface the way a latent fingerprint is lifted. Footwear impressions are usually preserved with plaster-of-paris casts of the depressions. Shoes and their crime-scene prints and impressions can be compared side-by-side or through the use of transparent overlays. Footwear comparison photographs made for courtroom presentation resemble exhibits that show the similarities between a latent fingerprint and its inked counterpart.

To connect a suspect to a crime scene through footwear evidence, detectives need three things: a good print or impression, the shoe that made it, and a way to link the suspect to the footwear. In the Simpson case the detectives had shoe prints in blood, but none of the footwear in Simpson's possession matched the murder-scene evidence. The prosecution had to settle for a group identification, that Simpson had once worn a brand, model and size consistent with the crime-scene prints. As evidence goes, this was relatively weak.

Dr. Louise Robbins: The Analyst Who Could See Things Other Experts Couldn't

Perhaps fortunately for O. J. Simpson, the world's only footwear identification expert who might have identified the crime-scene prints as having been made by shoes worn by him without having access to the actual footwear had died eight years before his trial. Dr. Louise Robbins, an anthropology professor at the University of North Carolina at Greensboro, wasn't interested in matching the bottoms of shoes to corresponding crime-scene latents. She would have claimed she could identify the crime-scene prints in the Simpson case by examining other shoes in Simpson's possession. Robbins's method of identification, a process she called "wear pattern analysis," was based on her theory that no two people have the same shaped feet or walk in exactly the same way. According to her, this unique feature reveals itself inside the shoes people wear *and* in the prints or impressions they leave behind.

Dr. Robbins claimed she could look at a crime-scene shoe print and determine that it had been made by the wearer of shoes other than the one that had actually left the crime-scene print. Her critics, and there were many, called this the "Cinderella analysis." If a defense attorney had a client in a case in which Dr. Robbins was testifying for the prosecution, that defendant's foot always seemed to end up fitting the shoe that had made the crime-scene print or impression. The jury, without access to the actual source of the print or impression, simply had to take her word for it. It's not surprising that prosecutors with insufficient evidence and weak cases loved this woman. Defense attorneys called her the expert from hell.

In her work as an anthropologist, Dr. Robbins had frequently exhibited the ability to see things that her colleagues could not. When working in Africa she garnered worldwide publicity after identifying a 3.5 million-year-old fossilized footprint as made by a woman who was five and a half months pregnant. She claimed generally that from a footprint she could tell a person's height within an inch, plus their weight, gender, and race. Dr. Timothy White, a professor of anthropology at the University of California at Berkeley, who had worked with Dr. Robbins in Africa, characterized her conclusions as pure nonsense.

If Dr. Robbins had confined her ideas to the classroom, she would have been harmless and no one would have been greatly bothered by her patently ridiculous theories. But in 1976, when she took her nonsense into the courtroom as a forensic footwear identification expert, people not only started to worry, defendants started going to prison. Between 1976 and 1986 Dr. Robbins testified, for fees up to $9,000 a case, in ten states and Canada. During this period at least twelve defendants went to prison on the strength of her crime-scene shoe-print or impression identifications. Her career as an expert witness came to an end in 1987 when she died of brain cancer at the age of fifty-eight.

That year, the American Academy of Forensic Sciences sponsored a panel of 135 anthropologists, forensic scientists, lawyers, and legal scholars to review her cases and her work. The panel concluded that her identification methodology had no basis in science, as Marc Hanson reported in the *ABA Journal*. Melvin Lewis, a law professor at John Marshall, called her work "complete hogwash." Lewis, who operated an expert witness referral service, was dismayed that so many judges had qualified Robbins as an expert witness. Russell Tuttle, a professor of physical anthropology at the University of Chicago, was also not a big fan of Dr. Robbins's work: "It's just cockamamie stuff," he said. "Why do we allow this kind of rot, this pseudoscience, into our courts?" William Bodziak, who had testified against Dr. Robbins in several trials, agreed: "Nobody else has ever dreamed of saying the kinds of things she said."

Dr. Robbins had not only wormed her way into courtrooms and the hearts of desperate prosecutors, she had impressed juries. She had a Ph.D., taught at a major university, and had been written up in *Time* magazine. In 1985, she published a book, *Footprints: Collection, Analysis, and Interpretation*. As a self-validating expert who used scientific terminology to advance an absurd theory, she came off as extremely confident and sure of her conclusions. Moreover, some prosecutors portrayed her as a pioneer in a new field of scientific identification. One prosecutor, in defending Dr. Robbins against her critics, reminded the jury that it had taken four hundred years for Galileo's theories to gain acceptance in the scientific community.

The Dennis Ferguson Case: Dr. Robbins and Her Wear Pattern Nonsense

In 1984, Dr. Robbins was in Wheaton, Illinois, helping the state's attorney for DuPage County put Dennis Ferguson away for life. Ferguson was being tried for the murder of fifty-year-old Andrea Young, who had been raped and stabbed by a man who had broken into her Elmhurst home shortly after midnight. The victim had crawled to a neighbor's house but died a few days later. The police made casts of shoe impressions outside the house, impressions believed to have been made by the intruder. The victim did not, before her death, identify or describe her attacker.

Suspecting Ferguson, detectives went to the twenty-eight-year-old's house looking for the murder weapon, bloody clothing, and the footwear that had made the crime-scene impressions. They seized three pairs of Ferguson's shoes but did not find a knife or any other evidence linking

Ferguson to the murder. Faced with no confession, no fingerprint evidence, no murder weapon, and three pairs of shoes that didn't come close to matching the crime-scene impressions, the state's attorney turned to Dr. Robbins—the only expert in the world who could connect a suspect to a shoe impression without the shoe that had made it. To no one's surprise, after analyzing Ferguson's shoes and the crime-scene plaster casts, Dr. Robbins concluded that Ferguson's feet had been in the unidentified shoes that had made the crime-site impressions. The jury, with Dr. Robbins's wear pattern analysis and the testimony of a jailhouse informant ringing in their ears, found Ferguson guilty. The judge sent him away for life.

In 1988, four years after Ferguson began serving his sentence, an Illinois appellate court ruled that wear pattern analysis was nonsense and set aside the conviction. But Ferguson's future was still up in the air, because state's attorney John Kinsella decided to retry him. The prosecutor didn't have Dr. Robbins as a star witness, but he still had the jailhouse snitch, and he had another way to connect Ferguson to the shoe impressions. This time the prosecutor would try to place the defendant at the scene of the crime by connecting him to the size, brand, and style of the shoe that had left the general tread pattern at the scene. The police had found a shoebox in Ferguson's closet indicating that he had once owned such shoes, which he had purchased from a store in Melrose Park. Richard Nosek, Ferguson's attorney, pointed out that any number of people could have owned this kind of shoe and argued that "the evidence of the state is at best weak, at worst it is fraudulent and incredible," Ray Gibson reported in the *Chicago Tribune.* The judge agreed, directing a verdict of acquittal. Without Dr. Robbins and her wear pattern testimony, the prosecutor was left without a case.

Vonnie Ray Bullard and the Bloody Footprint

In 1984, the year of Dr. Robbins's testimony in the Dennis Ferguson case, the supreme court in her home state of North Carolina upheld a 1982 murder conviction that had been based entirely on her identification of a crime-scene footprint made in blood. Cumberland County sheriff's deputies had arrested Vonnie Ray Bullard shortly after the 1981 shooting and stabbing death of Pedro Hales. The key evidence in the case featured a barefoot print in blood on asphalt near the murder scene. Because Bullard's pickup had been seen in the area about the time of the murder, inked impressions of his feet were taken for comparison with the bloody foot latent. But there was a problem. The crime-scene print lacked the ridge detail necessary for an identification. The prosecutor solved this problem by calling upon Dr. Robbins, who examined the blurred latent and concluded that it had been made by one of the defendant's feet. Having a world-renowned expert place the defendant's foot in blood at the scene of the crime was too much for the defense to overcome. The jury found Bullard guilty and the judge sentenced him to life. Vonnie Ray insisted that he was innocent.

Three years after the North Carolina Supreme Court affirmed the Bullard conviction, the American Academy of Forensic Sciences panel concluded that Dr. Robbins's identifications had nothing to do with forensic science. On the strength of this finding, Vonnie Ray appealed his case to the U.S. Supreme Court. In 2002, the nation's highest court refused to hear his plea. With his legal remedies exhausted, Vonnie Ray Bullard will remain, for the rest of his life, a prisoner of Dr. Robbins's nonsensical footprint identification.

The DuPage Three

In 1985 Dr. Robbins was back in DuPage County, Illinois, helping another prosecutor with a weak case. Her expert testimony, combined with that of a jailhouse snitch and the coerced confession of a man with borderline intelligence, sent a pair of murder defendants to prison. The case involved the February 25, 1983, kidnapping, rape, and bludgeoning death of Jeanine Nicarico, a ten-year-old whose body was found, two days after the abduction, along a road six miles from her home in Naperville. Police arrested a petty thief with a low I.Q. named Rolando Cruz. Under intense grilling, Cruz said he had held the victim down while Stephen Buckley, an itinerant laborer from Aurora, and his friend Alex Hernandez raped her. Buckley and Hernandez were arrested. Neither man confessed. The only physical evidence in the case consisted of a boot print left by the abductor when he kicked in the front door of the house. The police had not found the boot that had made this crime-scene mark.

Cruz, Hernandez, and Buckley were tried together, and the case against them was flimsy. Cruz had confessed to an accomplice role but denied raping or killing the victim. He implicated Hernandez and Buckley, who had pleaded not guilty. Buckley was connected to the crime scene by Dr. Robbins, who testified that his foot had worn the boot, wherever it was, that had made the mark on the door. Buckley was also implicated by Eric Hook, the jailhouse snitch. That was the prosecution's case. Given the quality of the evidence—the coerced confession of a slow-witted and frightened man, the jailhouse informant, and Dr. Robbins's Cinderella analysis—there shouldn't have been enough proof to sustain an indictment, let alone a conviction. Detective John Sam, one of the DuPage investigators who had worked on the case, believed that the defendants were innocent, resigned his job in protest, and testified for the defense. Relying heavily on Rolando Cruz's confession, the jury found Cruz and Hernandez guilty but split on the issue of Buckley's guilt. Half of them were apparently unimpressed with the jailhouse snitch and Dr. Robbins's wear pattern analysis. The state's attorney, Jim Ryon, vowed to bring Buckley back for retrial.

Early in 1986, six months after the double conviction and hung jury in the Nicarico case, Brian Dugan, a twenty-nine-year-old burglar and sex offender from Aurora whose MO included kicking in doors, confessed to a pair of sexual murders in the northern Illinois communities of Somonvak and Geneva. The high school dropout also admitted raping and killing Jeanine Nicarico, a crime he said he had committed alone. Dugan offered to plead guilty to all three murders to avoid the death penalty. The prosecutor agreed to the deal in the northern Illinois cases but refused to acknowledge the Nicarico confession. Two men had already been convicted of that crime, and a third, Stephen Buckley, was about to be retried for that murder. To accept Dugan's confession in the Nicarico case would amount to an admission that Jim Ryon and his lead DuPage County prosecutor, Thomas Knight, had prosecuted two innocent men and were going after a third party whose trial had ended in a hung jury.

When Buckley's attorney learned that Brian Dugan had confessed to the murder his client was being retried for, he attempted to subpoena Dugan as a witness. The state's attorney objected, and the judge, ruling against Buckley, refused to admit Dugan's testimony, justifying his decision on the grounds that Dugan's confession was "untrustworthy." Nevertheless, on March 5, 1987, the DuPage County state's attorney dropped the murder charges against Stephen Buckley—not out of a sense of justice or compassion for an innocent man who had spent three years in jail, in my opinion, but because the prosecutor no longer had a case. Dr. Robbins was too sick to testify and the jailhouse snitch, having received his reward for helping out the good guys, was long gone. In the meantime, Rolando Cruz and Alex Hernandez were rotting away on death row.

The Cruz and Hernandez convictions were set aside by the Illinois Supreme Court in January 1988 on procedural grounds. The court held that the two defendants should have been tried separately. Both men, however, were quickly retried. Without the benefit of Brian Dugan's confession, they were again convicted. Six years later, both convictions were set aside for the second time. The appeals court found that the trial judge had erred in excluding evidence of the Dugan confession.

In 1995, thanks to the publicity given DNA by the O. J. Simpson trial, men who had been convicted of rape and murder based on the testimony of victims and jailhouse informants began requesting DNA testing of the evidence in their cases. For Hernandez and Cruz, convicted twice in legally flawed trials and still vulnerable to being tried again, and for Stephen Buckley, who had escaped conviction but had not been exonerated, a DNA analysis of the rape-scene evidence was the only hope of proving their innocence. The tests were made and the results were stunning. The DNA analysis revealed that none of the three—Cruz, Hernandez, or Buckley—had raped Jeanine Nicarico and that she had been assaulted by Brian Dugan, the man who had confessed to the rape and murder.

In October 1995, notwithstanding the DNA results, Jim Ryon and Thomas Knight put Rolando Cruz on trial for the third time, which meant that Brian Dugan would not be charged in connection with the Nicarico murder. The case came to an abrupt halt when Judge Ronald Mehling directed a verdict of acquittal, outraged by the apparent disregard for evidence and lack of prosecutorial discretion shown by the DuPage County State's Attorney's Office. His concern led to the appointment of a special prosecutor to conduct an investigation into the handling of the Nicarico case and the operation of the DuPage County criminal justice system.

In December 1996, the special prosecutor's investigation resulted in a forty-seven-count indictment charging Jim Ryon, Thomas Knight, another DuPage County prosecutor, and four detectives with perjury, obstruction of justice, and conspiracy in the criminal prosecution of Rolando Cruz, Alex Hernandez, and Stephen Buckley. The law enforcement defendants in the case became known as the DuPage Seven. By 1999, following a series of trials, the seven defendants were acquitted on all charges. A year later the taxpayers of DuPage County paid Cruz, Hernandez, and Buckley, pursuant to their civil suit settlements, $3.5 million. Former DuPage County state's attorney Jim Ryon ran for governor in 2002 and was defeated.

The Jeanine Nicarico case jumped back into the news in November 2005 when Brian Dugan, now forty-nine and serving two life sentences and 155 years for the Sononauk and Geneva murders, was indicted for the 1983 rape and killing of the Naperville girl. Joseph Birkett, the DuPage County state's attorney pushing the case, said he would ask for the death penalty, which, in Dugan's case, would be mainly symbolic. In trying Dugan, Birkett risked exposing the bad judgment associated with his predecessors' prosecutions of Cruz, Hernandez, and Buckley. Given the passage of so much time, the boot that had left the print on the crime-scene door, like Dr. Robbins, was no longer around. Had more attention been paid to Brian Dugan in 1986 when he confessed, the boot that had made the impression on the crime-scene door might have been identified.

Two Roommates, One Boot, and Bad Science: Sergeant Kennedy's Barefoot Morphology

Dr. Robbins was dead, but her method of fitting feet into shoes and connecting suspects to crime-scene prints or impressions without the shoes that made the marks had not died with her. In 1998, ten years after the American Academy of Forensic Sciences discredited this form of footwear identification, it reappeared in a double murder case in Lexington County, South Carolina. The killer, or someone else present at the scene, had walked blood into the kitchen of the murder house, leaving behind the distinct print of a boot. The police interrogated a local man named James Brown, who confessed to his involvement in the murders. Promised that he wouldn't receive the death penalty, Brown agreed to testify against his roommate, Jeffrey Jones.

In the house Brown and Jones shared, the police found a boot with traces of blood and a tread pattern that matched the crime-scene print. James Brown said the boot belonged to Jones. Jones, who denied any role in the killings, said the boot was Brown's. To make a case against Jones, the prosecutor, Dayton Riddle, would have to link the boot to the murder scene, then connect Jones to the boot. Establishing the first link involved traditional crime-scene forensics; the second required something else. Whatever that was, it would have to overcome a serious problem. The boot that left the bloody print was a size 9½, but all Jones's shoes were either size 11 or 11½. Since James Brown wore 9½-sized shoes, common sense would suggest that the bloody crime-scene mark had been left by one of his boots. But the prosecutor already had Brown; he needed Jones, and this is how bad science found its way into the murder trial.

Prosecutor Riddle, in linking Jeffrey Jones to the murder scene, put two experts on the stand who figuratively slipped the bloody boot onto the defendant's foot. Steven Derrick, from the state crime lab, testified that the shape of the defendant's foot and the distance between his toes led him to the conclusion that the crime-scene boot was the defendant's. Because this was Derrick's first Cinderella analysis, the prosecutor backed up his testimony with an expert who had come all the way from Ottawa, Canada. If an expert is, as they saying goes, a guy wearing a necktie who's from another state, a sergeant with the Royal Canadian Mounted Police was an expert with a capital *E*. Sergeant Robert B. Kennedy, a veteran fingerprint examiner from the RCMP crime lab in Ottawa, had picked up Louise Robbins's torch in the early 1990s. Since then, he had been visiting military bases and other sites collecting inked impressions of bare feet for his foot identification data bank. Sergeant Kennedy, although a latent fingerprint analyst, was not interested in the ridge configuration on the bottom of a person's heel and sole. He was cataloguing, from the weight-bearing areas of the foot, various anatomical features that made each foot unique. Through what he called "forensic barefoot morphology," Kennedy claimed the ability to match an individual foot to a particular shoe. If that shoe were the source of a crime-scene print or impression, Sergeant Kennedy could link the wearer to that shoe and by extension to the crime scene.

Sergeant Kennedy took the stand in the Jones trial and assured the jury that because he had helped Steven Derrick with his analysis of the defendant's foot and the bloody print, there was no doubt in his mind that when that boot made the print at the crime scene, the defendant was the one wearing it. The jury found Jones guilty, not because he had been linked to the scene through his fingerprints, a hair follicle, or blood, but through a boot that was too small to fit him. Dr. Louise Robbins would have been proud. The judge sentenced Jeffrey Jones to death.

The Supreme Court of South Carolina, three years later, set aside Jones's conviction, holding that barefoot morphology was not a true forensic science. Riddle promised to retry Jones using the same evidence that had led to the reversal. Referring to Sergeant Kennedy's barefoot morphology, Riddle said: "That's good evidence, despite the fact [the conviction] got reversed. I think what happened . . . is that I was a little bit ahead of the curve." If by this the prosecutor meant finding a witness who could see things true scientists could not, he was indeed ahead of the curve. Dr. Robbins had been ahead of the curve as well, and because of it, innocent people had gone to prison. Wear pattern analysis and barefoot morphology amount to cutting-edge hogwash and have no place in a court of law. Notwithstanding Dayton Riddle's promise, Jeffrey Jones was not retried for murder.

Rex Penland's Cowboy Boots

In the context of a criminal trial, which is worse, a pseudoscientist like Dr. Robbins or impression identification witness who's not a scientist? For Rex Penland, it was six of one and a half dozen of the other.

Vernice Alford, a twenty-nine-year-old waitress and part-time prostitute, should not have gotten into Rex Penland's pickup on the night of November 30, 1992. From Patterson Avenue in Winston-Salem, North Carolina, Rex and Vernice, accompanied by Rex's eighteen-year-old nephews, Larry and Gary Sapp, drove north into Stokes County. Three days later, a pair of surveyors discovered Vernice in a ditch at the end of a logging road. Stokes County sheriff's detectives found a bloodstained tree trunk and lengths of black and yellow nylon rope that had been used to tie the victim to the tree. Stabbed eighteen times and raped, she had managed to crawl to the ditch where the surveyors had stumbled across her body. A deputy made a plaster cast of a section of tire track and did the same with what looked like an impression made by a cowboy boot. There were also impressions of a tennis shoe with "pony" embossed on the sole, tracks the deputies believed had been made by the victim. The officers also recovered an empty Monarch cigarette pack from the crime scene.

Acting on an anonymous tip, deputies rounded up the Sapp brothers and hauled them in for questioning. The twins admitted being at the scene but accused their uncle of the murder. According to their version of the events, Penland had made them tie the victim to the tree; while they drank beer in his Chevy, he repeatedly stabbed her. When Penland returned to the truck, according to his nephews, he was carrying the bloody knife and bleeding himself, explaining that he had accidentally cut his leg while stabbing the woman.

Deputies arrested Penland at his home in Germantown. They searched the trailer and found a knife (later determined to contain traces of dried blood), a pack of Monarch cigarettes, and a pair of muddy cowboy boots. Officers also found a pair of trousers that had been cut or torn and noticed that Penland had recently cut his leg. Penland said he had injured himself on a barbed-wire fence while hunting deer. Hanging from the gun rack in the cab of his truck, deputies found and seized a pair of handcuffs.

Rex Penland did not deny being at the scene of the rape and murder. According to his version of the crime, he was passed out in the pickup when the Sapp brothers tied the victim to the tree and murdered her. He said he never stepped out of the truck and that his knife had nothing to do with Vernice's death. If it contained traces of blood, the blood was his.

Joyce Petzka, an analyst with the State Bureau of Investigation crime lab in Raleigh, compared Penland's cowboy boots to the crime-scene footwear impression. If she matched one of the boots to the cast, Penland's story that he hadn't been out of the truck would be broken. In her report to the Stokes County district attorney's office, Petzka advised that because of the lack of detail on the plaster footwear cast, she was unable to identify either of Penland's boots as the source of the crime-scene impression. This was not good news for James Yeatts, the Stokes County assistant district attorney prosecuting the case. Yeatts also received bad news from the crime-lab serologist, who reported an insufficient amount of crime-scene sperm cells to allow a DNA profile. (In 1992 DNA was an emerging and callow science.) There was, however, some positive news from the crime lab. Fibers removed from the handcuffs found in Penland's truck were consistent with fibers from the victim's denim jacket.

Penland was the first to go on trial, in January 1994. The Sapp brothers were witnesses for the prosecution and both testified that when the victim asked for money shortly after getting into the truck, Penland told her to get into the handcuffs. After she had done so, according to the brothers, Penland struck her in the face. At the scene, after having sex with the victim behind the truck, Penland reportedly said, "I am going to ice this bitch." Acting on Penland's orders, the Sapp brothers wrapped the rope around Vernice and the tree three or four times. According to their testimony, they purposely didn't knot the rope so she could escape. The brothers returned to the truck and were drinking beer when Penland went to the tree and stabbed the victim. The Sapps admitted having sex with Vernice but claimed it was consensual. Back at Penland's trailer, Penland told his nephews that he had "iced that bitch."

Footwear examiner Joyce Petzka connected the crime-scene tire tracks to Penland's pickup but wasn't asked about her boot impression analysis.

Prosecutor Yeatts didn't want a footwear expert contradicting the testimony of two lay witnesses. William Lemons, an agent with the North Carolina Bureau of Investigation and Lieutenant Albert Tuttle of the sheriff's office testified that the sole and heel pattern on the plaster cast was "consistent" with one of Penland's cowboy boots.

The prosecutor knew how important it was to prove that Penland had lied about never leaving his truck that night and stressed the point in his address to the jury: "If Rex Dean Penland was passed out drunk in the trunk, how come his cowboy boot print was out there in the sand that night? What does that tell you about Larry and Gary Sapp? Tells you he wasn't passed out. It was just like they said." After deliberating forty-six minutes, the jury found Rex Penland guilty of kidnapping, rape, and murder. He was sentenced to death. Two weeks later, the Sapp brothers cut a deal which led to guilty pleas to second-degree murder. Although sentenced to fifteen years, they were released in 1998 after serving four years. Gary Sapp would later be convicted of assaulting two women, and his brother would be sent back to prison on a firearms violation.

Penland, in February 1996, appealed his death sentence to the North Carolina Supreme Court. His attorney argued that the victim had gotten into the truck voluntarily and had consented to the sexual intercourse, which would have voided the kidnap and rape charges and taken the death penalty off the table. The court rejected this argument and the death sentence stood.

On July 31, 2005, Judge John O Craig set aside Penland's 1994 conviction and ordered a new trial. By hiding the fact that a crime-lab expert had been unable to identify Penland's boot as the source of the crime-scene impression, the prosecutor had denied the defendant a fair trial. The harmfulness of this omission had been made worse by the testimony of the two lay witnesses contradicting the expert's opinion. The judge held that under the circumstances, these police witnesses should not have been allowed to testify as experts.

Since the Penland conviction, DNA science had advanced greatly, and a DNA analysis revealed that Penland had not had sex with the victim. DNA tests also showed that the blood on his knife was his. Undeterred by the scientific and legal developments in Penland's favor, the district attorney pledged a retrial, assuring the citizens of Stokes County that there was still enough evidence to convict him. As of June 2007, Rex Penland has not been retried.

In the Penland case, a judge eventually excluded the footwear identification testimony of two lay witnesses because their opinions contradicted the report of a shoe identification expert. However, in general, courts do not exclude the shoe identification testimony of ordinary police officers. Since most judges don't consider footwear identification a science, experts aren't needed. Jurors can see for themselves if a crime-scene shoe print or impression matches the bottom of a particular shoe. In *Ratliff v. Alaska,* an examiner with the state crime lab in 2002 identified the defendant's shoes as having made prints left at the scene of a burglary in Juneau. The defendant's attorney filed a *Daubert* motion challenging the science underlying footwear identification. The judge ruled that the *Daubert* decision didn't apply to the case because footwear identification wasn't a forensic science. The crime-lab examiner was not an expert, but merely a technician who could point out similarities and dissimilarities in the comparison process. Footwear identification, unlike DNA, toxicology, and forensic medicine, does not require science. Jurors don't have to believe in anything but their eyes. An appellate court affirmed the trial judge's denial of the *Daubert* motion.

One might argue that if footwear comparison is not a science, then neither is fingerprint identification. In fighting *Daubert* challenges to fingerprint science, prosecutors might consider taking this approach. If latent fingerprint analysis is not a matter of science, there is no problem with lightly trained police officers making these identifications in court. There would be a price to pay, however, if the criminalistic fields of footwear and fingerprint identification were treated as nonscientific. Witnesses in these fields would lose their authority and much of their power to sway a jury. An expert's credibility comes from more than training, experience, and knowledge. Jurors believe experts they perceive as objective, independent, and committed to their science.

That's why police officers, because they are law enforcement officials, should not identify footwear evidence in court. Like the police witnesses in the Penland case, who were convinced of Penland's guilt, they may see things in footwear evidence that an expert doesn't see. This lack of objectivity should also preclude the use of police officers in the field of fingerprint identification. The McKie and Penland cases reveal that impression misidentifications can occur when the so-called experts are more loyal to law enforcement than to forensic science.

9

Bite-Mark Identification

Do Teeth Leave Prints?

After putting out a house fire set by an arsonist, firefighters in Upstate New York discovered the body of a Cayuga County social worker named Sabina Kulakowski. The victim, found naked on a dirt road three hundred yards from the torched farmhouse in Aurelius, had been beaten, stabbed, strangled, and bitten. Three days after the May 23, 1991, discovery the police questioned Roy A. Brown, a heavy drinker with a criminal record. A week before the murder Brown had been released from jail after serving eight months for threatening another social worker. Brown denied any involvement in the arson or the killing. He insisted that he didn't even know the victim.

With no evidence connecting Brown to the scene of the fire or the murder sites, detectives asked Dr. Edward Mofson, a local dentist, to connect the suspect to the victim through the bite marks on her body. Although he had no forensic training, Dr. Mofson matched one of the wounds on Kulkaowski's body to Brown's upper front teeth. Brown was charged with murder and taken into custody.

In 1992, when he took the stand as an expert witness at Brown's trial, Dr. Mofson testified that he had identified one of the four bite marks on the victim's body as having been made by the defendant. On cross-examination, the dentist was asked to explain how a man missing two upper front teeth could have left a bite mark comprised of six upper tooth impressions. Dr. Mofson speculated that while being bitten, the victim's skin must have twisted in a way that filled the gap that would otherwise have shown up in the bite-mark impression. Of the three bite marks that were not identified, none of them reflected such a space in the upper dentition.

The defense countered Dr. Mofson's testimony with that of Dr. Homer Campbell, a forensic dentist from New Mexico. Pointing out that a man missing his two front teeth would leave a bite mark vastly different than the ones on the victim's body, Dr. Campbell made it clear in no uncertain terms that the defendant was not the person who had bitten Sabina Kulakowski.

Absent any other evidence connecting Roy Brown to the murder, the jury had to choose between the prosecution's unqualified bite-mark witness and the expert for the defense, who was a bite-mark identification specialist. The jury chose Dr. Mofson, finding the defendant guilty of second-degree murder. The judge sentenced him to the maximum sentence of twenty-five years to life.

In January 2007, after serving fifteen years at the Elmira Correctional Facility, Roy Brown was vindicated. A DNA analysis of the saliva on the victim's nightshirt fit the profile of another man. A forensic dentist also identified this man as the one who had bitten Sabina Kulakowski. The man incriminated by DNA and bite-mark analysis was dead. He had killed himself in 2003 by putting himself in the path of an oncoming train. While this new evidence did not immediately open the prison door for Roy Brown, it was grounds for a new trial. He was set free in March 2007 after the prosecutor decided not to retry him for the Kulakowski murder.

The Brown case illustrates the devastating effect an unqualified expert witness can have on the criminal justice system. It also shows that judges, not jurors, should be responsible for weeding out unqualified expert witnesses.

Ted Bundy and the Popularization of Bite-Mark Identification

The identification of a series of bruises or abrasions, usually in the shape of two semi-circles or brackets, as a human bite mark made by a particular set of teeth is a function of forensic dentistry referred to as bite-mark identification. This form of impression identification, also called forensic odontology, is based on the assumption that no two people in the world have front teeth that are identical in thickness, shape, relationship to each other, and patterns of wear. The process of comparing a bite mark to a known set of teeth is not unlike the identification of latent fingerprints, footwear, and tire-track impressions. Bite-mark wounds are found on victims of murder, rape, and child molestation. This type of crime-scene evidence is preserved by life-size photography, tooth-mark tracings onto transparent sheets, and dental casts of the impressions themselves. A suspect might be asked to bite down on a pliable surface for an impression sample, have a cast made of his teeth, or both. Usually, connecting a suspect to a victim through expert bite-mark testimony will be enough evidence, by itself, to sustain a criminal conviction.

The field of bite-mark identification exploded in the 1980s, and hundreds, if not thousands, of defendants have gone to prison since then on the strength of bite-mark testimony. Although bite-mark identification had been a recognized branch of forensic science since 1970, it was the 1979 trial of serial killer Ted Bundy in South Florida that put this form of identification on the map the way the O. J. Simpson case, in the mid-1990s, popularized DNA profiling. After killing thirty to one hundred college-aged women in the states of Washington, Utah, and Colorado, Bundy entered a Florida State University sorority house in Tallahassee, where he sexually assaulted and murdered two Chi Omega sorority members. He was arrested shortly thereafter in a stolen car, then put on trial for murder in Dade County.

The Dade County prosecutor had plenty of nonphysical evidence against Bundy, but it was circumstantial and didn't place him in the Chi Omega murder room. The killer had left a hair follicle on a crime-scene stocking mask that looked similar to samples from Bundy's head. But hair identification wouldn't be enough, by itself, to carry a conviction. The rapist had left traces of semen, but these murders were committed more than ten years before the dawn of DNA analysis. If the prosecutor had any chance of convicting Bundy, he would have to connect him to the murder scene through a bite wound on one of the victims. To do that, the Dade County prosecutor called on Dr. Richard Souviron, the chief odontologist for the Miami–Dade County medical examiner's office; Dr. Lowell J. Levine, a forensic dentistry consultant to the New York City medical examiner's office; and Dr. Norman Sperber, with the San Diego and Imperial County medical examiner's office. These three experts were the nation's top bite-mark identification specialists.

Dr. Souviron took the stand and testified that the sorority house attacker had left a postmortem double-bite mark on one victim's left buttock. He had bitten once, turned sideways, then clamped down again. The killer's top bicuspids and his lateral and central incisors had remained in the same position, but he had made two wound brackets with his lower front teeth. When he photographed Bundy's teeth, Dr. Souviron noticed they were of uneven length, chipped, and oddly aligned, factors that helped individualize his bite-mark pattern. To show that the teeth matched the murder scene wound, Dr. Souviron laid a photograph of the defendant's teeth, depicted on a transparent sheet, over an enlarged photograph of the bite mark. The bottom edges of Bundy's teeth lined up perfectly with the crime-scene mark. Dr. Sperber and Dr. Levine took the stand and lent their expertise to the identification. Bundy, representing himself, brought in his own dental expert, an odontologist from Maryland who testified that "the dental pattern [the bite mark] is one I'd expect to find in 20 percent of the population of male Caucasians." The defense expert didn't say that the bite mark couldn't have been Bundy's, he just wasn't willing to identify the wound as coming from Bundy to the exclusion of all others. The jury found the defendant guilty, the judge sentenced him to death, and ten years later, Bundy was executed. Before his execution, he confessed to the sorority house killings and dozens of other murders. The Bundy case established the credibility and usefulness of forensic bite-mark analysis and for a while placed it on the same level, in terms of reliability, as latent fingerprint identification.

Dr. Michael West, the "Expert for Truth"

About the time of Ted Bundy's execution in 1989, Dr. Michael West, a dentist from Hattiesburg, Mississippi, began testifying as a bite-mark identification expert for prosecutors throughout the South. He would become the Dr. Louise Robbins of forensic dentistry. Like the infamous Cinderella analyst of shoe-print identification, Dr. West could see things that his colleagues could not, and the things he saw tended to send people to prison. Between 1988 and 2002, the self-proclaimed "expert for the truth" testified in seventy-two trials. His good-old-boy affability and cocksure attitude, along with his professional credentials—a dental degree, membership in various forensic organizations, published papers, and lecture appearances at the FBI Academy and New Scotland Yard—made him a highly credible witness. As Mark Hanson reported in the *ABA Journal,* defense attorneys who asked Dr. West to estimate his bite-mark identification error rate would get this answer: "Something less than my Savior, Jesus Christ."

Dr. West had invented a technique to detect bite-mark wounds not visible to the naked eye. Since no one else employed this technology, Dr. West would often be called into a case after other specialists had failed to find a bite mark on the body of a murder victim. He used a long-wave, ultraviolet light to bring bite-mark wounds into view, calling it, as Hanson reported, his "blue light technique." As Dr. West's confidence and reputation grew, he branched out from bite-mark analysis to other forms of impression identification. For example, in one case, he matched a bruise on a murdered boy's stomach to the defendant's boot. He also matched various types of murder weapons to marks they had supposedly left on victims' bodies. From impression identification, Dr. West expanded his forensic repertoire into blood-spatter interpretation, forensic photography, video enhancement, and gunshot-powder stain analysis.

Larry Maxwell and the Incriminating Blue Light

Dr. West's testimony in a 1990 triple murder case that had nothing to do with forensic odontology marked the beginning of the slow unraveling of his credibility as an expert witness. Ten days after three elderly women were stabbed to death near Meridian, Mississippi, the police arrested a suspect named Larry Maxwell. With no physical evidence linking Maxwell to the killings, the police called on Dr. West to determine, through his blue light technique, if Maxwell had held the murder weapon. According to the dentist, his ultraviolet light examination of Maxwell's right hand revealed pressure point impressions left by the rivets in the handle of the butcher knife. Maxwell sat in jail two years awaiting trial before a judge ended his nightmare by declaring the knife-handle impression identification unscientific and inadmissible.

Outraged by what he considered pure courtroom hokum, Maxwell's attorney filed a complaint against Dr. West with the International Association of Identification, the American Academy of Forensic Sciences, and the American Board of Forensic Odontology. In 1993, facing an ethics investigation, Dr. West resigned from the International Association of Identification. A year later he left the American Academy of Forensic Sciences. Suspended for a year by the American Board of Forensic Odontology, Dr. West accused his colleagues of professional jealousy. According to Michael West, the other forensic odontologists resented him for hogging all the glory.

Convictions Set Aside: Anthony Keko, Johnny Bouin, and Henry Lee Harrison

Dr. West had made a blue light misidentification in 1992 that helped hurry him off the forensic stage. In August 1991, Anthony Keko's ex-wife Louise was found stabbed and shot to death in her Plaquemines Parish home. The police suspected Keko, a Louisiana oysterman, from the beginning but had no physical evidence linking him to the murder. The investigation lay dormant until 1992, when a new sheriff took office and ordered the exhumation of the victim's body. At the request of the sheriff's chief investigator, Dr. West examined the corpse with his blue light and found, on the victim's left shoulder, a bite-mark wound. Because the forensic pathologist hadn't noticed this bite mark, and it hadn't shown up on any of the crime-scene or autopsy photographs, Dr. West's discovery amounted to a major investigative breakthrough. Sheriff's deputies arrested Keko after Dr. West matched the blue light–enhanced bite mark to a dental cast of Keko's teeth.

The case went to trial even though Dr. West's testimony could not be supported by a photograph of the crime-scene mark. (Blue light images cannot be photographed.) The jury had to take the doctor's word that (1) there was in fact a bite mark, and (2) it matched the defendant's teeth. The defense claimed this wasn't science but courtroom voodoo, but the trial judge denied Keko's motion to exclude Dr. West's testimony and the jury, relying on West's expertise, found the defendant guilty. In 1994, a Louisiana appeals court set aside the conviction on the grounds that the trial judge had erred in admitting Dr. West's testimony. Since the so-called bite mark was the only evidence against Keko, he was not retried.

The year the Louisiana appeals court set aside the Keko conviction, prosecutors in Jackson County, Mississippi, dropped charges against Johnny Bouin after fingerprint evidence proved he wasn't the man who had raped and robbed an elderly woman. The prisoner was released after he had spent eighteen months in jail awaiting trial. Bouin wouldn't have been in jail in the first place had Dr. West not linked him to a bite mark on the rape victim's body. When asked how he felt about Bouin's exoneration, according to Mark Hanson's ABA article, Dr. West said, "I know he bit that woman."

Also in 1994, the Mississippi Supreme Court set aside the conviction of Henry Lee Harrison, who had spent four years on death row for the rape and murder of a seven-year-old girl whose body had been dumped into a Jackson County swamp. At Harrison's trial, Dr. West had testified that the victim had been bitten forty-one times and that the defendant's teeth had made every wound. But having been suspended from the American Board of Forensic Odontology and no longer a member of the American Academy of Forensic Sciences or the International Association of Identification, West had lost a lot of his credibility. Moreover, he was challenged by two defense odontologists. One identified the forty-one wounds as ant bites and the other, while agreeing with Dr. West that the wounds were in fact human bite marks, declared they were too vague for forensic comparison. Notwithstanding the defense's challenge to Dr. West's testimony and his loss of professional prestige, the southern jury chose his interpretation of the evidence and found Harrison guilty. The Mississippi Supreme Court, in reversing Harrison's conviction, did so on procedural grounds unrelated to Dr. West's blue light technique or lack of professional credibility. Harrison was retried in September 1995 and found guilty again, but not on the bite-mark evidence.

The Kennedy Brewer Case: Misidentifying Insect Bites as Teeth Marks

In 1995, jurors in Lowndes County, Mississippi, after deliberating ninety-five minutes, found Kennedy Brewer guilty of raping and strangling to death his girlfriend's three-year-old baby principally because Dr. West had identified nineteen bite marks on the victim's body, five of which, according to his analysis, had been made by Brewer's upper teeth, detention that featured a gap that corresponded to the wounds. A small amount of semen had been recovered from the victim's body, but in 1992, it wasn't enough for a DNA profile. The child had been kidnapped from her home near Brookville on the night Brewer was in the house watching her and two of his own children. Because he was the only adult present, Brewer was arrested and charged with the murder. The suspect had told sheriff's investigators that the child must have been taken from the house by an intruder. Jurors evidently hadn't believed him, and they had also apparently rejected the testimony of Dr. Richard Souviron, the Miami forensic odontologist who had identified Ted Bundy's bite mark and had cofounded the American Board of Forensic Odontology, who testified that the nineteen marks on the child's body were insect bites.

The judge sentenced Kennedy Brewer to death. Dr. West, notwithstanding previous reversals of his cases and attacks on his professional credibility, was hard to beat in Mississippi. He looked and sounded believable, and for many jurors, that is enough.

Six years later, in June 2001, more sophisticated DNA analysis revealed that the victim in the Mississippi case had been raped by two men and that Kennedy Brewer was not one of them. His attorney filed a motion for a new trial based upon this exonerating evidence. The following month, a *Newsweek* article about the case featured this quote from Dr. Souviron: "His [Dr. West's] results are beyond outrageous. He has hurt a lot of people." Dr. West, also interviewed for the piece, held fast. "Just because the DNA isn't his [Brewer's] doesn't mean the bite marks aren't his. It's not an exoneration." A Lowndes County judge disagreed, ordering a new trial for Brewer in a different county. As of 2007, Kennedy Brewer has not been retried.

Dr. Raymond Rawson: Bite-Mark Identification Maverick

In 1991 Doctor West wasn't the only freelance bite-mark expert commanding big witness fees, and he wasn't the only one who got caught up in controversial cases and questionable identifications. Dr. Raymond D. Rawson, a Las Vegas dentist, odontologist, and Nevada state senator who had testified in a hundred trials, would in the 1990s become the central figure in a case that combined shoddy criminal investigation, prosecutorial misconduct, and bad forensic science. Resulting in the ten-year imprisonment of an innocent man, the Ray Krone "snaggletooth killer" case illustrates how science can be corrupted by detectives and prosecutors who remain convinced, in spite of evidence to the contrary, that they have the right man.

Ray Krone and the Snaggletooth Killer Case

The snaggletooth killer case began when an employee arriving at work at the CBS Bar and Restaurant in Phoenix, Arizona, on the morning of December 28, 1991, discovered Kimberly Ancona's body in the men's room. As part of her assignment to close the bar, the thirty-six-year-old Ancona had been cleaning the restroom. She had been raped and stabbed eleven times with a butcher knife from the kitchen. The killer had left behind follicles of his hair, bloody fingerprints, traces of blood (type O), and a bite mark on the left breast, where he had bitten the victim through her tank top. The police also found, in the kitchen, the print of the killer's shoe, a Converse sneaker size 9½. Although the victim had been sexually assaulted, the police did not recover traces of semen.

Detective Charles Gregory, a nineteen-year veteran of the Phoenix Police Department, took charge of the investigation and quickly identified a suspect, thirty-four-year-old Ray Krone, who lived nearby and was a patron of the bar. According to one of the waitresses, Kim Ancona had said that someone named Ray was going to help her close up the bar. From this, Gregory made two assumptions, both false. The detective assumed, first, that Kim was referring to Ray Krone and, second, that Krone was her boyfriend. Regarding the first assumption, Kim could have been referring to another Ray with whom she was acquainted, who had nothing to do with the murder either, it would turn out. And regarding the second, while Ray's number was in Kim's personal phone book, they had not dated. Detective Gregory, however, had another reason to suspect Ray Krone: he had crooked teeth, just like the man who had bitten Kim Ancona. Detective Gregory and Ray Krone became victims of the investigator's worst enemy, coincidence.

Questioned by Detective Gregory the day after the murder, Krone denied having anything to do with Ancona's death. A former mail carrier with no criminal record and an honorable discharge from the U.S. Air Force, he said he had been home asleep the night she was killed, a statement confirmed by his roommate. When asked by the detective to provide a dental impression, Krone voluntarily pressed his teeth into a Styrofoam picnic plate. The next day, Krone was arrested on charges of murder, kidnapping, and sexual assault.

Maricopa County prosecutor Noel Levy asked Dr. John Parkis, a Phoenix dentist who occasionally gave advice to the local medical examiner's office, to determine if the crime-scene bite mark was Krone's. Dr. Parkis made a cast of Krone's dentition, compared it to photographs of the bite mark, and concluded that Krone's teeth had made the murder-scene wound. Dr. Parkis, however, was not a certified forensic odontologist, had never testified in court, and had made only one other bite-mark identification. To obtain a second, more professional opinion, Dr. Parkis consulted with his odontology mentor, Dr. Steven Sperber, the renowned San Diego forensic dentist who had testified for the prosecution at the Bundy trial. Dr. Sperber looked at the dental evidence and determined that Ray Krone was not the source of the crime-scene bite mark. He told Dr. Parkis, in no uncertain terms, that Krone's teeth and the bite mark didn't come close to matching. Someone else had bitten the victim.

Levy, rather that step back and reevaluate the case against Krone in light of Dr. Sperber's findings, shopped around for another expert and found Dr. Raymond Rawson. For a whopping fee of $50,000, the Las Vegas dentist/state legislator identified the crime-scene bite mark as Krone's. Since none of the crime-scene fingerprints were the suspect's, the hair follicles weren't his, and he not only didn't own a pair of Converse sneakers but had size eleven feet, Dr. Rawson's bite-mark identification saved the day for the prosecutor. Krone had type O blood, but so does 43 percent of the U.S. population. If Levy could convince a jury that the defendant had bitten Kim Ancona, he wouldn't need more evidence than that.

After deliberating three and a half hours, a Maricopa County jury on November 20, 1992, found Krone guilty of the abduction and murder of Kim Ancona. They did not, however, convict him of rape. Having no idea that one of the most prominent odontologists in the country didn't believe that Ray Krone had bitten the victim, his attorney had offered no rebuttal to the testimony of Drs. Rawson and Parkis. Dr. Rawson had been impressive, with his homemade video illustrating how the crime-scene bite mark lined up with the impression on the Styrofoam plate and the dental cast of Krone's teeth. Levy, in his closing statement to the jury, had noted that "bite marks are as unique as fingerprints." Without hearing evidence to the contrary, the jurors had to assume that while the defendant had not left his fingerprints at the scene, he had left his bite mark. The judge sentenced Ray Krone to death.

The Arizona Supreme Court in June 1995 set aside the conviction on the grounds that, ignoring the rules of discovery, the prosecutor had not given the defendant enough time to prepare a response to Dr. Rawson's video presentation. The court did not base its decision on the prosecutor's failure to disclose Dr. Sperber's exculpatory opinion. As Ray Krone's new defense attorneys prepared for his retrial, they began looking for expert witnesses of their own. This led them to the shocking discovery that Dr. Sperber, who was not aware that Krone had been convicted, had already analyzed the bite-mark evidence.

Krone's defense attorneys showed Dr. Sperber the computer model and the videotape that Dr. Rawson had exhibited at the 1992 trial. Dr. Sperber said he had never seen anything like it and was more than a little disturbed. Dr. Gerald Vale, the chief forensic dentist for Los Angeles County, also viewed Dr. Rawson's material and was equally upset. The two experts later met Dr. Rawson at a forensic science conference and told him that the bite mark on Kim Ancona's body didn't look anything like Ray Krone's teeth. Dr. Rawson assured his colleagues that if they used the computer software he had invented to analyze the evidence, they would see what he had seen and change their minds. Like Dr. Michael West, Dr. Rawson had a personal technique that allowed him to see things other experts couldn't.

Before Ray Krone's second trial, Levy spoke to Dr. Sperber in San Diego. In a meeting that lasted an hour, Dr. Sperber showed the prosecutor how Krone's upper and lower teeth did not match the murder-scene bite mark.

Months after this meeting, when the defense asked Dr. Sperber to testify on Krone's behalf at his upcoming trial, the odontologist was shocked that Levy was not only going ahead with the second trial but seeking the death penalty!

In December 1996, a second jury found Ray Krone guilty of first-degree murder. Dr. Rawson, with the aid of his video presentation, had again testified for the prosecution and Dr. Sperber had taken the stand for the defense. As Jana Bommersback would later report in *Phoenix Magazine,* when it came time for sentencing, Judge James McDougall rejected the death penalty because there were "too many unanswered questions in the case" and sentenced Krone to life. Dr. Sperber's testimony had not won Krone an acquittal but may have saved his life.

On April 8, 2002, after ten years behind bars, Ray Krone walked out of prison a free man. A DNA test of the saliva on Kim Ancona's tank top revealed that she had been bitten by a convicted sex offender who was, at the time of Krone's exoneration, serving time in an Arizona prison. As Bommersback noted, in an unusual statement for a public official, the new county prosecutor, Rick Romley, said: "What do you say to him? An injustice was done and we will try to do better. And we're sorry."

For Ray Krone and his family it had been an uphill struggle. To afford the lawyers and the experts, Krone's mother had sold her home and spent all her retirement money. A cousin had contributed $100,000 to the cause. In his $100 million lawsuit against the city of Phoenix and Maricopa County, Krone accused the officials responsible for his prosecution of more than just incompetence. He alleged perjury, evidence suppression, and evidence tampering. In Bommersback's article about the case published in July 2004, Dr. Sperber recalled how shocked he had been in 1996 when he had learned that the prosecutor, Levy, was going to retry Krone. He had thought, he said: "This is so ridiculous. All the other evidence is exculpatory. It shows he couldn't have done it, and for the prosecutor to still go ahead based on a bite mark, and they know nobody agrees that the bite mark matches. It is just unbelievable that knowing what they did know, they still went ahead with the trial. Bite marks are not nearly as accurate, or scientific as fingerprints. They are probably the least accurate method of identifying people we have."

Dr. Rawson and the Tankersley Case

In 1993, a year after testifying for the prosecution in Ray Krone's first trial, Dr. Rawson was back in court identifying another Arizona murder defendant's bite marks. In *State of Arizona v. Bobby Lee Tankersley,* Tankersley was being tried for the 1991 murder of sixty-five-year-old Thelma Younkin, who had been strangled in her Yuma, Arizona, motel room. The thirty-nine-year-old defendant had lived in the same rundown motel and, according to the police, killed the woman because her grandson owed him money for drugs. Dr. Edward Blake, a pioneer in the DNA field, testified that a hair follicle in the victim's sink matched the genetic profile of 4 percent of the Caucasian population, a group that included the defendant. However, the fact that the defendant and the victim lived in the same motel tended to weaken the incriminating value of this evidence. The prosecution's strongest evidence consisted of Dr. Rawson's identification of numerous bite marks as Tankersley's—on the victim's neck, left breast, chin, and jaw.

To illustrate how he had arrived at this conclusion, Dr. Rawson showed the jury a video highlighting the similarities between the crime-scene marks and the defendant's teeth. The Tankersley trial was somewhat unusual in that Dr. Sperber, the prosecution's second odontologist, testified that he had been unable to make a positive identification. He also characterized Dr. Rawson's identification methodology and his video presentation as dubious and misleading. Tankersley's defense attorney did not put his own odontologist on the stand. After the prosecutor, Mary White, told the jury that "this defendant, in essence, signed his name to the body of his victim," the jury found Tankersley guilty. The judge, citing the bite marks as evidence of the defendant's savagery, sentenced him to death.

Tankersley appealed his conviction, alleging that Dr. Rawson had misidentified him as the source of the crime-scene bite marks. Backing up his position, Tankersley had the support of two of the country's most qualified and respected forensic odontologists. The first, Dr. Richard Souviron, the chief forensic dentist in the Miami–Dade County medical examiner's office, wrote in his affidavit for the defense that Thelma Younkin "had a human bite mark on her left breast, but Dr. Rawson came up with a bite mark on the chin and on the right breast, and other places. They were bruises. They definitely, in my opinion, were not bite marks." The second, Dr. C. Michael Bowers, an attorney and odontologist with the Ventura County, California, coroner's office and a member of the examination and credentials committee of the American Board of Forensic Odontology, also submitted an opinion affidavit on Tankersley's behalf. As the *Chicago Tribune*'s Flynn McRoberts reported on November 29, 2004, he wrote: "The serious deficiencies of the 'science' of bite marks make its use as an identifier to a single perpetrator a very dangerous proposition. The divergence of opinion between the state's own dental experts [Rawson and Sperber] in the Tankersley [trial] is another anecdotal indication of the weak reliability of bite-mark identification."

Modern DNA analysis of the crime-scene evidence failed to exonerate Tankersley. Instead of granting him a new trial, the appellate court judge granted him a new sentencing hearing. Since the trial judge had relied on the bite-mark evidence as proof of the crime's extreme brutality, which had justified the death penalty, in the face of questions regarding the reliability of bite-mark analysis, the new judge reduced Tankersley's sentence to life. Tankersley wasn't going home, but he wasn't going to be executed. In March 2004, nine months before Tankersley's sentencing hearing, Dr. Rawson retired from the business of bite-mark identification.

Dr. John Kenney and the Pushy Prosecutor

In 1994, Dan Young, Harold Hill, and Peter Williams confessed to first killing a Chicago woman, then setting her house on fire to cover up the crime. The Cook County prosecutor dropped the charges against Williams after learning he was in jail at the time of the murder. The other two took back their confessions, claiming they had been coerced by a Chicago detective with a reputation for third-degree tactics. This left the prosecutor with nothing and opened the door for Dr. John Kenney, a forensic odontologist from Park Ridge, Illinois. Dr. Kenney connected Young and Hill to the scene of the crime by identifying them as the source of bite marks on the dead woman's body. The defendants, on the strength of Dr. Kenney's testimony, were found guilty of murder and sentenced to death.

Young and Hill had been in prison for more than ten years when DNA analysis of scrapings from the victim's fingernails excluded them as participants in the murder and they were set free. Given the DNA findings, Dr. David Sweet, director of the Bureau of Legal Dentistry at the University of British Columbia, and Dr. C. Michael Bowers, the odontologist with the Ventura County coroner's office, took issue with Dr. Kenney's bite-mark identification and conclusions in the case. Dr. Sweet found that fire damage to the victim's body had made the bite marks unsuitable for comparison. Dr. Bowers had also been unable to work with the fire-altered teeth wounds. Dr. Kenney reacted to the contrary views of his colleagues by admitting that he had been pressured into the identification by the Cook County prosecutor. "You get pushed a little bit by prosecutors, and sometimes you say okay to get them to shut up. I allowed myself to be pushed," he told journalists Flynn McRoberts and Steve Mills in an extensive article on the errors of bite-mark identification published in the October 19, 2004, *Chicago Tribune.* In contrast to the self-justifications of a Dr. Michael West, Dr. Kenney's refreshingly candid remarks reveal how prosecutorial pressure can influence the testimony of an expert testifying for the state.

Dr. Lowell Levine and the Edmund Burke Case

DNA analysis has exposed the weakness of bite-mark identification in dozens of homicide cases. In Walpole, Massachusetts, a seventy-year-old woman walking through Bird Park was sexually assaulted and stabbed to death. Nine days later the police arrested Edmund Burke, a handyman who lived with his mother in a ramshackle house a quarter mile from the murder scene. An eyewitness had placed Burke in the park about the time of the attack, a tracking dog had followed his scent from the scene to his house, and a comment made by his mother led the Walpole Police to believe he had knowledge of the crime. When a bloody palm print at the scene of the crime turned out not to be Burke's, Dr. Lowell J. Levine, a pioneer odontologist, and one of the bite-mark witnesses in the Ted Bundy trial, reported that he could not exclude Burke as the source of a bite mark on the murder victim. The next day the police arrested Edmund Burke.

Dr. Levine, a prominent forensic dentist from Albany, New York, made a more detailed analysis of the dental evidence a few days later in Boston, as McRoberts and Mills reported in the 2004 *Chicago Tribune* article. The police were encouraged by his second report, in which he stated that he now believed, "to a reasonable scientific certainty," that Edmund Burke had bitten the woman murdered in Bird Park. Hoping for a confession, the police confronted Burke with Dr. Levine's findings. Burke said he didn't care what the dentist believed, he had not killed the elderly woman. He had been in jail for forty-one days when DNA analysis of the bite-mark saliva excluded him, and the authorities released him from custody. Later, when Burke filed a civil suit against the odontologist, the judge dismissed the action on the grounds that Dr. Levine had not acted with malicious intent.

Dr. Levine, whose bite-mark identification in the Burke case was supported by New York City odontologist Dr. Ira Titunik, stood by his analysis.

In explaining the DNA exoneration of Burke, Dr. Levine suggested that the crime-scene saliva might have been contaminated in some way. However, in June 2003, the Walpole police arrested a man whose DNA matched the bite-mark saliva in the Bird Park murder and who was later convicted of the crime.

Do Teeth Leave Prints? The Nonscience of Bite-Mark Identification

By 2003, the idea that bite marks were as unique as fingerprints was no longer in fashion among odontologists. Comparisons to latent fingerprint identification were by then inappropriate and susceptible to challenge in court. In recalling the Burke case, Dr. Levine backpedaled: he had never said he was 100 percent certain of his identification, he claimed, although he still defined it as involving a high probability. As an evangelist for the emerging field in 1977, Dr. Levine had perhaps oversold bite-mark identification: "Since every person's teeth are unique in respect to spacing, twisting, turning, shapes, tipping toward the tongue or lips, wear patterns, breakage, fillings, caps, loss and the like, all of which occur in limitless combinations, it is possible for them to leave a pattern which for identification purposes is as good as a fingerprint."

Dr. C. Michael Bowers, the Southern California odontologist who had questioned Dr. Rawson's analysis in the Tankersley case, had been warning against the overselling of bite-mark identification since 1996, when he wrote:

> Physical matching of bite marks is a non-science which was developed with little testing and no published error rate. . . . An opinion is worth nothing unless the supportive data is clearly describable and can be demonstrated in court. How does one weigh the importance of a single rotated tooth in a bite mark when the suspect has a similar tooth? The value judgments range widely on the value of this feature. This is not science. Instead, statistical levels of confidence must be included in the process.

In a bite-mark identification exercise Dr. Bowers conducted in a workshop at the 1999 American Academy of Forensic Science conference, 63 percent of the odontologists who participated made an incorrect identification, findings that displeased many in the field when Dr. Bowers published them. In an article published in 2003 in the *British Dental Journal,* Dr. D. K. Whittaker, a forensic dentistry professor at the University of Wales, explained why bite-mark evidence is so difficult to identify, particularly bite marks on skin:

> Human bites on skin are difficult to interpret because skin is not a good "impression" material. Moreover, victims may struggle and movement will distort the image of the bite. Skin surfaces are not flat and visual distortion may be present, often heightened by photographic distortion caused by inadequate imaging techniques. Human dentitions, whilst possibly being unique in the sense of small nuances of tooth size, shape, angulation and texture may not inflict unique bite marks which can only record gross and not fine detail. If the victim survives, the injury may change due to infection or subsequent healing and if the victim is deceased, putrefaction may introduce distortion.

Before odontologists in Great Britain can testify in court as bite-mark experts, they must have made a minimum of twenty such identifications in other cases. In the United States, an odontologist can be certified by the American Board of Forensic Odontology after two bite-mark identifications. Being certified in the field therefore doesn't carry much weight in this country. Dr. West and Dr. Rawson were both certified.

In 2004, as part of a series on forensic science, the *Chicago Tribune* examined 154 state and federal trials involving bite-mark identification testimony. In more that a quarter of these cases, the prosecution and the defense produced forensic odontologists whose expert opinions were diametrically opposed. If bite-mark identification is an exact science practiced by highly qualified experts, this many odontologists should not have been testifying against each other.

Bite-mark identification saved the day in the trial of Ted Bundy, when DNA profiling wasn't even on the horizon, showing great promise as one way to identify the worst kind of criminal, rapists who kill their victims. Today, DNA analysis can identify criminals through their blood, sweat, saliva, semen, skin, and hair follicles, a method of identification that is truly scientific. Not only has DNA profiling provided investigators and prosecutors with a better forensic tool, it has helped expose the weaknesses and unreliability of bite-mark identification.

10

Ear-Mark Identification

Emerging Science or Bad Evidence?

In the early morning hours of December 16, 1994, near Vancouver, Washington, an intruder entered James McCann's bedroom and bludgeoned him to death. In another bedroom the burglar fractured the skull of McCann's son, who managed to crawl outside, where he was discovered by a passerby at dawn. Questioned at the hospital, the boy told the police he hadn't gotten a good look at his attacker, whom he described as twenty-five to thirty years old, dark complexioned, about six feet tall, and of medium build. George Millar, a fingerprint examiner with the Washington State crime lab, lifted a latent ear print off the surface of James McCann's bedroom door. The burglar had apparently pressed his head against the door listening for signs of activity before entering the room. Millar processed the house for fingerprints as well, but they all turned out to belong to occupants of the dwelling.

Ear-Print Analysis and the David Kunze Case

Although he had red hair and didn't otherwise fit the general description of the killer, the police suspected David Wayne Kunze, the forty-five-year-old ex-husband of the woman McCann had been about to marry. When he had learned of the upcoming marriage four days before the murder, Kunze had become upset, information that led detectives to believe that he had attacked the victims out of jealousy and rage. The intruder had stolen McCann's television set, VCR, stereo speakers, and wallet, an aspect of the case detectives explained away by theorizing that Kunze had taken these things to throw them off his trail. Convinced that the scene had been staged, the police made no effort to identify a homicidal burglar through the missing property. Kunze consented to a police search of his truck, boat, storage locker, and safety deposit box. Detectives found nothing that connected him to the home invasion and murder.

Three months passed without further developments in the investigation. Then Michael Grubb, a criminalist with the Washington State crime lab, compared the partial ear-print latent with photographs of Kunze's left ear and concluded that it "could have been made by David Kunze." Six months later, on September 21, 1995, Kunze voluntarily agreed to have George Millar and Michael Grubb take seven exemplar prints of his left ear. They applied hand lotion to his ear, then placed panes of glass against it, using various degrees of pressure. They dusted the glass with fingerprint powder and lifted the prints with transparent tape.

Michael Grubb compared the seven exemplars with the crime-scene ear latent and concluded that "David Kunze is the likely source for the earprint and cheekprint which were lifted from the outside of the bedroom door at the homicide scene." George Millar, the crime-lab fingerprint analyst, declined to offer an opinion regarding the identification of the crime-scene ear latent. He said he identified fingerprints, not ear marks. In June 1996, a year and a half after the murder and eight months after Michael Grubb identified the crime-scene ear print, the Clark County prosecutor charged Kunze with aggravated murder, assault, robbery, burglary, and kidnapping.

In a pretrial motion to exclude the ear-print identification, Kunze's attorney petitioned the judge for a *Frye* hearing. Courts in many states rely not on the 1993 U.S. Supreme Court decision in *Daubert v. Merrill Dow Pharmaceutical* regarding the scientific reliability of evidence, but on the somewhat stricter standard set down in *United States v. Frye,* a U.S. district court case decided in 1923. As we have seen, in *Frye,* the district court in Washington, D.C., held that lie detection technology had not been accepted in the general scientific community as a legitimate science. This ruling became known as the "general acceptance" test. To determine if latent ear-print identification was an accepted function within the forensic science community, the prosecutor and defense attorney in Kunze's case offered expert witnesses on both sides of the issue in a *Frye* hearing in December 1996—the most thorough, in-depth judicial/scientific review of ear-print identification in legal and criminalistic history.

On the issue of latent ear-print identification as a legitimate forensic science, the prosecution presented three advocates against the defense's twelve witnesses, who in varying degrees were not enthusiastic about this form of pattern analysis. Michael Grubb, the manager of the Washington State crime lab in Seattle who had identified the crime-scene ear print as probably Kunze's, testified that comparing an ear mark to a known ear print was not unlike other forms of impression identification. A criminalist who specialized in bullet-striation and tool-mark identification, Grubb said that if you can analyze patterns made by tires, shoes, fingers, gun barrels, and tools, you can render an expert opinion on the source of an ear mark.

The next prosecution witness, Alfred V. Iannarelli, said he had studied the evidence in the McCann murder case and was certain that the crime-scene ear mark was an "exact" match to Kunze's left ear. Iannarelli had never worked in a crime lab, had not been to college, and had testified only once as an expert witness. He had been a deputy sheriff with the Alameda County sheriff's office and the chief of campus security at California State University at Hayward. From 1948 to 1962, Iannarelli had photographed seven thousand ears; from this database he had come to the conclusion that no two ears are the same. He had also devised an ear classification system based upon twelve "anthropometric measurements," a system featured in his 1964 book, *The Iannarelli System of Ear Identification*. In 1989, Iannarelli self-published a second edition of his text, titled *Ear Identification,* which included a section on latent ear-mark analysis. He was unable, however, to cite any ear-print studies other than his own, which explained why his books did not contain bibliographies.

In ear-print identification, it became clear, there were no texts other than Iannarelli's, no community of experts, no section within any crime lab that specialized in this type of work, and no professional organizations or certifying bodies. Besides Iannarelli, there was one other analyst devoted solely to this form of identification. If anyone could claim to be an internationally known ear-print identification expert, it was this police officer from Amsterdam, Cornelius Van der Lugt. It was therefore not surprising that Van der Lugt had examined the McCann murder scene evidence and was the third prosecution witness at the *Frye* hearing. Van der Lugt had become interested in the ear-print identification field after reading Iannarelli's books in the early 1990s, and since then had analyzed ear-print evidence in between 200 and 250 cases in the Netherlands, United Kingdom, and Western Europe. He had testified as an expert in six trials, all in Holland, where judges, not juries, determine a defendant's guilt or innocence.

According to Van der Lugt, many suspects when presented with his ear-print analysis had confessed and pleaded guilty. In one case, a suspect admitted putting his ear to the door but denied breaking in. Like Iannarelli, Van der Lugt had never worked in a crime laboratory, attended college, or received any kind of formal training in science. He was certain, however, that David Kunze was the source of the McCann murder ear print. As part of his *Frye* testimony, Van der Lugt praised the work done by Michael Grubb and George Millar in obtaining the seven ear-print exemplars, noting how they had varied the amount of pressure against the ear until the known and crime-scene prints looked alike. When asked if ear-print identification, as a forensic science, was accepted around the world, Van der Lugt answered that it was.

While the Kunze prosecution could not have put on a stronger case for ear-print identification, it was arguably not enough to meet the *Frye* (or *Daubert*) standards. In other words, at least in theory, Kunze's defense attorney could have won the *Frye* debate without mounting an anti-ear-print case. But leaving nothing to chance, the defense hit back with a dozen impressive witnesses, leading off with Dr. Ellis Kerley, a physical anthropologist and former president of the American Academy of Forensic Sciences and the American Board of Forensic Anthropology. Dr. Kerley said it was reasonable to assume that no two ears were the same, but he wasn't sure this uniqueness would always reveal itself in a crime-scene ear mark. He didn't consider Iannarelli's book a work of science and didn't approve of Van der Lugt's technique of getting an exemplar to match a crime-scene latent by varying the pressure against the suspect's ear. "We don't do that in science . . . because we're not trying to make them look alike," he said. In Dr. Kerley's opinion, ear-print identification had not achieved general acceptance in the forensic science community.

Andre Moenssens, a law professor at the University of Missouri at Kansas City, the author of articles and law school texts on forensic science and a former fingerprint expert in Belgium, testified that "the forensic sciences . . . do not recognize as a separate discipline the identification of ear impressions. There are people in the forensic science community, the broader forensic science community, who feel that it can be done. But if we are talking about a general acceptance by scientists, there is no such general acceptance. . . . To my knowledge, there has been no investigation in the possible rate of error that comparisons between known and unknown ear samples might produce."

Two current employees of the FBI crime lab and the former head of the FBI's latent fingerprint section testified that the Bureau has never been involved in the identification of crime-scene ear prints. Five experienced state criminalists from Arkansas, Ohio, Illinois, and Florida said they had analyzed ear-print evidence but had never identified a latent ear print in court. A Jacksonville, Florida, fingerprint analyst had once identified an ear print in court, but the defendant in that case "had a very peculiar mark in the lower lobe area of the ear." The final witness for the defense, Ron Gourley, a detective with the Sonoma County sheriff's office in California, testified that in 1984 he had called on Alfred Iannarelli to identify a crime-scene ear print in a murder case. The defense attorney had not objected when Iannarelli took the stand and identified the print as the defendant's. The jury had found this defendant guilty, Gourley said, but there had been a lot of other incriminating evidence in the case.

At the conclusion of the *Frye* testimony, the judge decided that ear-print identification had in fact gained general acceptance in the scientific community. The decision was stunning in that it was so out of sync with the weight of the expert testimony. It was certainly bad news for David Kunze, because the prosecution would have had no case if the ear-print evidence had been excluded on grounds it didn't meet the *Frye* test. Now the state could go forward against him.

The case went to trial on June 25, 1997. The prosecutor chose not to put Alfred Iannarelli on the stand, but the jury heard the testimony of state criminalist Michael Grubb and the ear-print guru Cornelius Van der Lugt. Michael Grubb identified the latent ear print to a "reasonable degree of scientific certainty." Van der Lugt said he had found a few parts of Kunze's ear that "compared completely" with the crime-scene print and found parts that revealed "differences." Van der Lugt said he wasn't bothered by the parts that didn't match because "you will never find . . . a 100 percent fit." The witness noted that these dissimilarities had been caused by "pressure distortion" in the making of the known ear exemplars. He said he was "100 percent confident" of his identification.

The prosecution ear-print analysts were followed to the stand by an equally weak witness, a jailhouse informant who claimed that Kunze had confessed to him while in custody. The prosecution rested its case without identifying the murder weapon, connecting the defendant to the crime scene through DNA or fingerprints, or linking him to any of the items taken from the house. At this point it would not have been inappropriate for the judge to have directed a verdict in favor of the defendant on the grounds that the government had not established a prima facie case. The judge, however, denied a motion to this effect, and it fell upon Kunze's attorney to put on a defense. The obvious tactic would have been to put Dr. Ellis Kerley, Professor Andre Moenssens, and the FBI lab people on the stand to question the scientific reliability of ear-print identification generally. The defense, however, did not call any of these witnesses. Perhaps the defense had run out of money, or maybe Kunze's attorney thought he had destroyed the prosecution's ear-print testimony on cross-examination. Van der Lugt's statement that "you will never find . . . a 100 percent fit" should have been enough, by itself, to discredit him.

While ear-print identification is not a recognized forensic science, it makes a lot of sense to a jury of laypeople. If fingerprints, shoe impressions, bite marks, tool marks, and tire tracks can be identified, why not ear marks? And what could be more impressive than an expert who had come all the way from Holland to make the identification? The jury found Kunze guilty of aggravated murder, burglary, and robbery. The judge sentenced him to life without the possibility of parole.

Kunze appealed his conviction, and in 1994, a three-judge panel ruled that "the trial court erred by allowing Michael Grubb and Cornelius Van der Lugt to testify that Kunze was the likely or probable source of the ear latent, and that a new trial is therefore required." The appellate court instructed the prosecutor in the second trial not to prejudice the defense by referring to the first trial and the resulting conviction. The appellate judges didn't want the second jury to know that Kunze had been found guilty once on the strength of ear-print identification.

In March 2001, ten days into the second trial, the prosecutor made reference to the earlier conviction, and the presiding judge had no choice but to rule a mistrial. The prosecutor announced that a third trial would not be scheduled. He was unwilling, however, to let Kunze completely off the hook and said he wanted to keep the option for prosecution open. As of June 2007, however, Kunze had not been retried.

The Mark Dallagher Case: Cornelius Van der Lugt's Change of Mind

The United States isn't the only country struggling with ear-print identification. In England, an intruder entered a house in Huddersfield, West

Yorkshire, on the night of May 7, 1996, and murdered ninety-four-year-old Dorothy Wood by smothering her with a pillow. The burglar had gained entry by forcing open a transom window. West Yorkshire detectives found latent ear prints on the glass below this window. The police questioned twenty-four-year-old Mark Dallagher, who lived in the neighborhood. Dallagher said that on the night of the crime he was in bed with his girlfriend. The girlfriend, however, couldn't confirm his story because she was sleeping under the influence of medication. The police eventually arrested Dallagher and made impressions of his ears. These exemplars and photographs of the murder-scene latents were sent to Cornelius Van der Lugt in Amsterdam. Van der Lugt made a comparison analysis and reported, quite emphatically, that Dallagher was *not* the source of the crime-scene ear marks. Then he changed his mind.

At Dallagher's two-week trial in the Leeds Crown Court, the prosecutor, Norman Sarsfield, put two ear-print experts on the stand, Van der Lugt and Peter Vanezis, Regis Professor of Forensic Medicine and Science at the University of Glasgow. Vanezis testified that "it was very likely that it was the defendant who had made the [crime-scene] prints," as Bob Woffinden would report in the United Kingdom's *Guardian*. He was careful to point out, however, that he was not 100 percent certain of this identification. Cornelius Van der Lugt, on the other hand, was "absolutely convinced that the prints of the defendant's left ear were identical with the prints of the left ear on the window." The police had not informed the court or the defense of Van der Lugt's initial opinion. If the prosecutor was aware of it, he didn't let on. Had the defendant's attorney known of Van der Lugt's flip-flop, the lawyer would have destroyed this witness on cross-examination.

The Dallagher and Kunze trials are similar in that the prosecution in both cases relied on Cornelius Van der Lugt, a supporting ear-print witness, and a jailhouse snitch. In both cases the defense failed to mount aggressive challenges to the validity of ear-print identification. Dallagher, like Kunze before him, was found guilty. The judge sentenced Dallagher to life in prison. According to Sarsfield, the conviction, the first of its kind in England, was "a great step forward in forensic science," Woffinden reported.

In July 2002, the Royal Court of Justice in London quashed the Dallagher conviction on grounds it was "unsafe." In setting aside the conviction, the appellate judges relied on reports submitted by Dr. Christopher Champod, formerly a professor at the University of Lausanne in Switzerland, and

Professor P. J. Von Koppen with the University of Antwerp in Belgium. Champod, an analyst with the forensic science service in Britain, had studied ear-print identification and had concluded that while ears contain features that are unique, these features are not always expressed in crime-scene latents. He had discovered that in Switzerland, where ear-print identification was fairly common, there had been many cases of misidentification. Professor Von Koppen's study of ear-print identification methodology led him to believe that the process was not scientific and therefore unreliable.

Although the appeals court had set aside the Dallagher conviction, it did not completely exclude ear-print identification testimony from the second trial. The experts, however, would not be allowed to voice their opinions regarding the likelihood that the crime-scene marks had been made by the defendant's ear but could only show the jury the similarities between the known and questioned latents. The jury would decide if the defendant had left part of himself at the scene of the crime.

Dallagher went to trial for the second time in June 2003. Ten days into the testimony, the judge called a halt to the proceeding after learning of Van der Lugt's initial and contradictory report. Dallagher was placed on bond and released from jail. Six months later the judge declared Dallagher not guilty. A DNA profile of the crime-scene ear prints had excluded Dallagher and incriminated someone else. "This most unfortunate saga at long last comes to an end," declared the judge, Sir Stephen Mitchell. In England, the Dallagher case also meant the end of latent ear-print identification.

The Outlook for Ear-Print Analysis

Dr. P. J. Von Koppen, the University of Antwerp professor whose ear-print identification report helped the Dallagher appeal, had challenged a Van der Lugt analysis in an Amsterdam case three years earlier. As Toby Egan explains in a November 1, 2000, article on Forensic-Evidence.com, the police had asked Van der Lugt to compare a suspect's ears with surveillance camera images of a man robbing several service stations. The robber's face was not exposed, but his ears were in plain view. The Dutch expert concluded that the suspect and the man in the surveillance photographs were one and the same due to ear similarities. Following the robber's conviction in 1999, the defendant appealed. Dr. Von Koppen and Professor H.F.M. Crombag of the University of Maastricht in the Netherlands submitted reports describing

Van der Lugt's ear identification as "problematical." In May 2000, an appellate court overturned the conviction on grounds that Van der Lugt's ear identification was unreliable.

Notwithstanding the unimpressive history of ear-print identification, the serious opposition to it in the scientific community, and the fact that no crime lab in the world has a recognized ear-print analyst, its proponents continue to push the idea. For example, at an American Academy of Forensic Sciences conference held in March 2004 in Dallas, Professor Guy Rutty, with the forensic pathology unit at the University of Leicester in central England, gave a talk in which he announced the development of the world's first computerized ear-impression databank. Working with K9 Forensic Services, Limited, a private firm in Northampton, England, the ear-print retrieval and comparison program would operate much like the FBI's Integrated Automatic Fingerprint Identification System. This new system of criminal identification was thus based upon the assumption that ears were capable of being forensically individualized and reliably identified, an assumption that had proven scientifically shaky in real-life applications.

But crime-scene investigators and forensic pathologists should not ignore crime-scene ear and teeth marks entirely, although they are not as reliable as fingerprint and DNA evidence. Legitimate suspects can still be developed through this form of evidence. However, only forensic scientists should identify ear and bite marks in court, and they should be required by law to appropriately qualify their identifications.

In terms of emphasis and the future of forensic science, the profession should be moving in the direction of DNA technology. Still, criminal investigators and forensic scientists have to deal with the real world of crime, which doesn't always provide evidence susceptible to cutting-edge scientific analysis. Because crimes are not committed under laboratory conditions and criminal investigators are not scientists, striking a balance between pure science and the realities of law enforcement is not easy. Nevertheless, if the criminal justice system is going to work as it should, the balance should be in favor of science and prosecutorial restraint.

PART THREE

Hired Guns, Smoke Blowers, and Phonies

The Expert Witness Problem

> The employment of private experts is open to serious objection, for they are only human, and it is natural for a man to sympathize with the side of the case from which his fee comes.
>
> –Harry Soderman, *Policeman's Lot,* 1956

Unlike ordinary witnesses such as detectives, eyewitnesses, and victims, who in criminal trials and civil litigation are required to stick to facts, experts are allowed to render their informed opinions on important questions of fact. And the opinion of a witness with an impressive educational background and professional experience greatly influences jurors. When such an opinion incriminates someone being tried for a crime, the incriminated person is in trouble. To have any chance of acquittal, the defendant in a situation like this has to hire an expert to counter the prosecution's expert. In this era of expertise, many trials are battlegrounds for competing experts. Then, instead of sorting out the truth and leading jurors to proper conclusions, forensic science is used to muddy the water and create doubt. If two experts disagree, say, on whether the defendant's handwriting is on a ransom note, one of them has to be wrong. If in trials where handwriting is an issue there is usually an expert on each side of the question, what does this say about the science of handwriting identification? What does this say about handwriting experts?

The competing-expert problem is not limited to the forensic document field, of course. It reflects a larger concern: the quality of forensic expertise and how so-called experts are certified and professionally supervised and monitored. In thinking about this problem, one that threatens the credibility of our adversarial trial process, we need to ask why a person wants to become a forensic scientist. The worst possible candidates for this field are people who are in it for the money and the glory rather than the science. Also worrisome are the ones who think of themselves as crime fighters instead of independent and objective seekers of the truth.

Replacing eyewitnesses, jailhouse informants, and biased police officers with forensic experts was supposed to make the criminal justice system more efficient and less prone to injustice. To some degree it has, but the modern expert-based system has produced problems of its own.

Chapters 11 and 12 feature the JonBenét Ramsey murder case and the role experts played in determining whose handwriting was on the crime-scene ransom note. In the end, the case degenerated into a battle of experts, exposing some of the troubling problems in the field of forensic document examination.

Chapter 13 deals with problems related to the forensic identification of hairs and fibers. For years hair and fiber experts, like their counterparts in the bite-mark field, oversold the uniqueness and identification value of this evidence. No one will ever know how many innocent people went to prison on the strength of hair and fiber misidentifications.

Chapter 14 addresses problems in the field of DNA analysis. The critical shortage of DNA personnel has led to a series of quality-control problems, the result of sloppy work, poor supervision, and misidentifications, which has led to the temporary closing of crime labs all across the country. These personnel shortages have also created a national DNA backlog, which has had a devastating effect on the quality of crime suppression involving the offenses of rape and murder. Delays in the analysis of DNA evidence allow serial killers and sex offenders more time on the street to victimize others.

Chapter 15 involves the FBI laboratory and the identification of bullet lead that was found to involve questionable science. Due to these revelations, the lab no longer employs this forensic technique. This is progress, but the question remains, How many innocent people went to prison on the strength of this flawed forensic science?

Chapter 16 uses Dr. Henry Lee's career as an example to address some of the problems associated with the so-called hired-gun celebrity expert witness. The centerpiece of the chapter is a murder case featuring crime-scene blood-spatter analysis. One of the questions raised in this chapter involves the effect of a witness's celebrity on the ability of jurors to make well-reasoned decisions.

11

Expert versus Expert

The Handwriting Wars in the Ramsey Case

The Crime

A 5:52 A.M. emergency call that a child had been kidnapped brought a pair of Boulder, Colorado, police officers to John and Patsy Ramsey's three-story house an hour before daybreak on December 26, 1996. Patsy Ramsey said she had found a handwritten ransom note inside on the stairs. Fearing that her six-year-old daughter, JonBenét, had been kidnapped for ransom, she had called 911. After a cursory sweep of the fifteen-room dwelling, the patrolmen called for assistance.

During the next two hours, amid friends and relatives who had come to console the family, police set up wiretap and recording equipment to monitor negotiations with the kidnappers. At one in the afternoon, Boulder detective Linda Arndt asked John Ramsey to look around the house for "anything unusual." Thirty minutes later, he and one of his friends discovered JonBenét's body in a small basement room. Her mouth had been sealed with duct tape, and she had lengths of white rope coiled around her neck and right wrist. The rope around her neck was tied to what looked like the handle of a paintbrush. Breaking all the rules of crime-scene preservation, John Ramsey removed the tape, carried his daughter up the basement steps, and laid her body on the living room floor. Detective Arndt picked up the child, placed her body next to the Christmas tree, and covered it with a sweatshirt. Because the police had not conducted a thorough and timely search of the house, there would be no crime-scene photographs. Moreover, the corpse, as physical evidence in the case, had been contaminated. Since John Ramsey was a parent and not a homicide detective, the responsibility for this investigative mistake was not his. This was a serious law enforcement blunder. The police had lost control of the crime scene.

The autopsy revealed that JonBenét had been strangled. The killer had also fractured her skull. The absence of semen in or on her body suggested she had not been raped. Traces of blood under her fingernails and on her underwear were determined to be unsuitable for DNA analysis.

The print and television media from the very beginning painted the Ramseys as rich and influential people who thought they were above the law. Rather than victims, they were presented as prime suspects in their daughter's murder. It became obvious that the police, and many in the Boulder County prosecutor's office, were also convinced that one or both of the Ramseys were guilty. On January 13, 1997, a supermarket tabloid, the *Globe*, published a stolen morgue photograph of JonBenét along with comments on the case from Dr. Cyril Wecht, the coroner of Allegheny County, Pennsylvania. The prominent, high-profile forensic pathologist from Pittsburgh said that his analysis of the Boulder County coroner's report suggested to him that the little girl, because of the presence of "vaginal abrasions," had been sexually abused two days before her death. From this, the inference was quite clear: Doctor Wecht did not believe JonBenét had been killed by an intruder.

The police, prosecutors, media, and most Americans agreed with Dr. Wecht that someone in the family had murdered JonBenét. But if this were the case, then who had written the two-and-a half-page ransom note? There was no question that, as ransom notes go, this one was odd. For one thing, the amount demanded by the kidnapper—$118,000—constituted a strange figure. (John Ramsey had received a $118,000 Christmas bonus from his computer company, Access Graphics.) The 350-word note was also long, rambling, and personal in tone. Quoted in the *Rocky Mountain News* on September 3, 1997, in an article by Charlie Brennan, the note began: "Mr. Ramsey, Listen Carefully! We are a group of individuals that represent a small foreign nation." The ransom note writer then warned the Ramseys not to notify the police: "If you talk to a stray dog, you die." The extortionist also taunted Ramsey: "Don't try to grow a brain, John." It was oddly and mysteriously signed: "victory, S.B.T.C." That the ransom note was strange was not evidence that it could not have been written by an intruder, however. Anyone who would enter a home and commit such a crime would not be a normal person.

The Nongraphologist Document Examiners

Determining if one, both, or neither of JonBenét's parents had written the ransom note became a key issue in the case. To resolve that question, forensic document examiners—handwriting identification experts—would be brought into the investigation. But first, detectives would have to gather two kinds of known handwriting evidence against which the crime scene writing—the "questioned document," in forensic parlance—could be compared. Handwriting samples made up of writings unrelated to the case—old letters, job applications, journals, school notes, grocery lists, and the like—are called course-of-business or conceded writing. The second type of sample, known writing, occurs in the form of requested exemplars given by suspects to the police. The benefit of course-of-business writing is that it has not been altered or disguised to make it appear different from the questioned writing. The benefit of request samples is that a suspect can be asked to replicate the contents of the crime-scene writing. In the Ramsey case, detectives acquired conceded writings when they searched the couple's vacation home in Charlevoix, Michigan. The Ramseys also consented, on at least three occasions, to provide the police with samples of their handwriting.

In February 1997, investigators asked four forensic document examiners to compare John and Patsy Ramsey's known handwriting with the crime-scene ransom note. The examiners agreed that the note had *not* been written by John Ramsey but were unable to say with absolute certainty the same thing about Patsy. On the other hand, none of the examiners were willing to state with absolute certainty that Patsy Ramsey *had* written the note. This, of course, was not what police and prosecutors wanted to hear. But, to the credit of these examiners, this is what they got.

Chet Ubowski, a document examiner with the Colorado Bureau of Investigation, having been brought into the case by the Boulder Police Department, was on the hot seat. His law enforcement colleagues were counting on him to identify Patsy Ramsey as the ransom note writer. Of the four experts, Ubowski came the closest to doing just that. On a scale of one to nine—one being that Patsy had not written the note, and nine that she had—Ubowski classified Patsy as a five. In other words, there was a fifty-fifty chance she had written the note, a forensic toss of the coin.

The other three experts—Edwin Alford, a retired Secret Service examiner in private practice; Richard Dusick, a document analyst with the Secret Service; and Leonard Speckin, a retired Michigan State Police document examiner, forensic scientists brought into the case by the Boulder County prosecutor's office—were even less certain that Patsy Ramsey had written the note. While these examiners found similarities between the known and the questioned writing, they concluded that Patsy Ramsey had *probably not* written the ransom note.

Almost buried alive by news reporting that was shoddy and aggressive even by crime journalism standards, the Ramseys hired a pair of private document examiners from California to analyze the handwriting evidence. Howard Rile, a former document examiner with the Colorado Bureau of Investigation who had been in private practice thirteen years with offices in Long Beach, was considered one of the best in the field. An active member of the American Society of Questioned Document Examiners, he had testified as an expert in more than four hundred trials. Low-keyed and not particularly interested in publicity or fame, Rile did not fully appreciate what he was getting himself into. The second examiner brought into the case by the Ramseys, Lloyd W. Cunningham, was equally qualified. A member of the Southwest Association of Forensic Document Examiners, Cunningham was a private practitioner from Alamo, California. As private, independent analysts, neither man had a stake in the outcome of the case.

In his report Howard Rile expressed his opinion this way: "The evidence argues very strongly for a writer other that Mrs. Ramsey. . . . The questioned note and the handwriting attributed to Patsy Ramsey were very probably not written by the same individual." Lloyd Cunningham, in his report, came to the same conclusion. At a meeting held at the Boulder Police Department on March 18, 1997, the two examiners informed detectives and prosecutors of their findings. To illustrate the complexity of handwriting identification, Lloyd Cunningham played a practical joke on those gathered to hear his views on the case. He showed the detectives and prosecutors a sheet of known handwriting that looked quite similar to that in the ransom document, leading his audience to believe he had identified Patsy as the writer of the ransom note. As it turned out, the handwriting sample was from a person unrelated to the case. Police detectives and representatives from the prosecutor's officer were not amused. Merely by expressing their professional opinions, Howard Rile and Lloyd Cunningham had made themselves the enemies of those who were devoted to the belief that the Ramseys had murdered their daughter. They had become unwanted interlopers helping the other side and would be treated as such. In terms of adversarial intensity, the Ramsey case would be like no other. Rile and Cunningham had walked into a media and law enforcement firestorm.

About the time Cunningham and Rile were throwing cold water on the probability that Patsy Ramsey had written the ransom note, Craig Lewis, news editor of the *Globe*, the Boca Raton tabloid that had published the morgue photograph of JonBenét, asked Thomas C. Miller, a forty-four-year-old Denver attorney, if he could get his hands on a copy of the ransom note. Lewis would pay Miller an hourly fee of $150 for his analysis of the questioned handwriting. Miller belonged to a growing cadre of handwriting analysts that functioned on the fringes of the traditional forensic document community. Instead of backgrounds in science and apprenticeships in government and private crime labs, members of this group had been graphologists or trained by graphologists, practitioners of a social science based upon the questionable assumption that handwriting reveals personality and character. This experience did not qualify them for membership in the American Academy of Forensic Sciences or document examiner organizations sanctioned by the AAFS such as the American Society of Document Examiners.

AAFS examiners like Cunningham and Rile and the four experts who had analyzed the ransom document for the Boulder authorities considered handwriting identification practitioners like Thomas C. Miller threats to the integrity of the field. Because trial judges allowed these people to testify as forensic document examiners in court, there was rarely a disputed document case without AAFS examiners on one side and graphologist examiners on the other. Denied membership in the AAFS and other professional and certifying bodies, these examiners formed their own professional organizations to produce resumes that sounded impressive in court, functioning in their own parallel forensic universe. As a result, Thomas C. Miller and his colleagues were all members of the World Association of Document Examiners (WADE) and the National Association of Document Examiners (NADE). In the traditional handwriting identification community, WADE and NADE had become code words for forensic inferiority.

On April 1, 1997, Thomas C. Miller and Craig Lewis, the forty-year-old news editor of the *Globe*, met with forensic document examiner L. Donald Vacca at his home in Evergreen, Colorado. A former document examiner with the Denver Police Department, Vacca, along with Rile and Cunningham, had analyzed the handwriting evidence for the Ramseys and had come to the conclusion that Patsy had probably not written the ransom note. The meeting had been arranged by Miller, who introduced Craig Lewis as a "representative of a large corporation." Miller pulled an envelope out of his brief case and offered it to Vacca. The envelope contained $30,000 in cash. If Vacca turned over a copy of the ransom note, that money would be his. Vacca rejected the offer and ushered Miller and the other man out of his house. Lewis returned to Evergreen on his own a week later and was thrown out of Vacca's dwelling. This time Vacca wrote down the license number of Lewis's airport rental car and called the Colorado Bureau of Investigation.

Four months after the *Globe* editor tried to buy a copy of the ransom note, *Vanity Fair* published photographs of the document. A week after that, Gerald Richards, a former document examiner with the FBI, told *Newsweek* that the ransom note writer had tried to disguise his or her writing but had not maintained consistency. Clinton VanZandt, a member of the FBI's Behavioral Sciences Unit (a psychological profiler), told the authors of the *Newsweek* piece, Sherry Keene-Osborn and Daniel Glick, that the tone of the ransom note suggested it had been written by a woman or a "genteel man."

The publication of the ransom note, along with photographs of bits and pieces of Patsy Ramsey's course-of-business handwriting, opened the door for people who fancied themselves handwriting identification experts. If one thought Patsy Ramsey guilty, one could discover her handwriting in the ransom note. Amateurs could render opinions and pass them off as professional.

Among these was Thomas C. Miller, who compared the ransom note photograph with fragments of Patsy's known handwriting and concluded she had written the crime-scene document. In an November 1997 "Questioned Document Examiner Report" sent to a New York City lawyer named Darnay Hoffman, Miller wrote: "Based upon the exemplars available, the handwriting of the 'ransom' note and that of Patsy Ramsey have numerous and significant areas of comparison. Shape of letters is one of the more telling areas of comparison, but this category would not substantiate an opinion on its own. The additional categories of size, slant, baseline, continuity and arrangement add significantly to the opinion that Patsy Ramsey wrote the 'ransom' note."

Darnay Hoffman

Miller had sent his handwriting report to Darnay Hoffman because Hoffman was planning a civil suit against the Boulder County district attorney to compel him to bring criminal charges against the Ramseys. Before filing this suit,

Hoffman had written district attorney Alex Hunter a letter in which he compared the Ramsey case to the 1932 kidnapping and murder of the Lindbergh baby. According to Hoffman, the Ramseys and the Lindberghs had orchestrated kidnap hoaxes to cover up their involvement in the deaths of their children. In the Lindbergh case, Bruno Richard Hauptmann, a German carpenter from the Bronx in the country illegally, had been found guilty of the crime and in 1936 executed. Hoffman's thesis, which had been thoroughly debunked by serious Lindbergh biographers and crime historians, existed only in the minds of a small group of Lindbergh case buffs who enjoyed weaving intricate and absurd conspiracy stories. Because the Lindbergh case had no bearing on the Ramsey investigation, Alex Hunter ignored Hoffman's letter. That's when Hoffman, a self-professed publicity hound, decided to bring his legal action against the district attorney's office. If he were to prevail in his suit, Hoffman would need experts willing to go into court and identify Patsy Ramsey as the ransom note writer.

In a letter to Thomas C. Miller dated October 31, 1997, Darnay Hoffman wrote:

> Could you please fax me a copy of your CV [curriculum vitae]? I need to begin preparing your court affidavit, and a recital of your qualifications as a handwriting expert is essential.
>
> You might be interested to know that I spoke with handwriting expert Paul A. Osborn who is, as you probably already know, the grandson of Albert S. and son of Albert D. Osborn. [The Osborns were two of the eight document examiners who had testified for the prosecution at the Hauptmann trial in January 1935. Albert S. Osborn, the author of several texts on the subject, is considered the father of modern document examination.] He [Paul Osborn] refuses to touch the Ramsey case with a ten foot pole. His reasons: he knows the handwriting experts who gave their reports to the defense team and to the C.B.I—four in all. [It's not clear if Hoffman was referring to the experts brought into the case by the police and prosecution—there were four—or to the three hired by the Ramseys.] According to Osborn these experts are supposedly top of their field with impeccable ethical credentials. Their verdict: the similarities between Patsy and the ransom note writer's handwriting is at the very lowest end of the spectrum, i.e., there is little or no basis for a match.
>
> I don't have to tell you what is going to happen when I present your

> report and affidavit to a district court judge. When Alex Hunter and Hal Haddon [an assistant district attorney] are finished with you, you will either look like Henry Lee [the famed forensic scientist who testified for the O. J. Simpson defense] or Dennis Fung [the prosecution criminalist who fell apart on cross-examination in the Simpson case]. Obviously, this is going to be a "defining moment" for both of us. . . .
>
> Trust me (as they say in Hollywood) when I tell you that if you're doing this solely for the money, then you're nuts. This is "a career move." You better be in this because you "like the action." Because you're going to see plenty of it when this report [Miller's] hits the courts.

In November 1997, Hoffman added a pair of handwriting analysts from Norfolk, Virginia, to his team. The more experienced, David Liebman, the president of NADE, had become interested in graphology in the 1960s while employed at a mental hospital. He had collected handwriting samples from patients with various mental illnesses and noticed that patients with the same mental problems had similar handwriting. His study of mental patient writing led him to believe that he could diagnose conditions such as attention deficit disorder by analyzing a person's handwriting—he once taught a course on how to detect cancer through handwriting. Liebman also claimed he could detect certain criminal traits through handwriting analysis.

Based on an examination of the same handwriting samples that had been analyzed by Thomas C. Miller, Liebman concluded that Patsy Ramsey had written the ransom note. In his November 26, 1997, "Ramsey Handwriting Report" to Darnay Hoffman, Liebman identified fifty-one points of similarity between the known and questioned writing, concluding that "there are far too many similarities and consistencies revealed in the handwriting of Patsy Ramsey and the ransom note for it to be coincidence."

The less experienced handwriting expert from Norfolk, Cina L. Wong, also a NADE member, wrote in her November 14, 1997, handwriting report to Darnay Hoffman: "The hand that authored the known standards for comparison as Patsy Ramsey appears to be very likely the same hand that authored the questioned document 'ransom note.'" On a one to ten scale—one indicating that Patsy had not written the note—Wong put the probability at 8.5. She had found thirty-two similarity points between the known and questioned writing.

In 1990, Wong had attended classes once a week for six months and received a certificate of completion from Ted Widmer's International School of Handwriting Services in San Francisco. Widmer, a graphologist, had once analyzed President Bill Clinton's handwriting and reported, in a *National Enquirer* article following the Monica Lewinsky scandal, that the president was consumed by sexual fantasies. Cina Wong had also attended a three-day handwriting course taught by Marcel Matley. (In September 2004, Matley would authenticate for CBS News the signatures on four memos damaging to President George W. Bush. The documents turned out to be phony.) Following a three-year internship with David Liebman, Wong had become a NADE-certified handwriting analyst.

Armed with the affidavits of three handwriting experts who said that Patsy Ramsey had written the ransom note, Darnay Hoffman filed his suit against Alex Hunter on November 19, 1997, in the Boulder County district court. On January 21, 1998, a judge dismissed his case on the grounds it was "premature" and in part based upon "innuendo, rumor, opinion and speculation." Hoffman and his experts had been thrown out of court, but the attorney was not about to be pushed off the Ramsey stage. There would be more lawsuits, more experts, and a lot more publicity for Darnay Hoffman.

Louis Smit: The Professional Detective

The Boulder police detectives working the JonBenét Ramsey murder were amateurs at homicide, as only sixteen homicides had occurred in Boulder during the preceding ten years. None of these officers had the specialized training and experience required for the job, and the intense media involvement in the case made sure it showed. Most Americans were frustrated because John and Patsy Ramsey were not in jail awaiting trial. The police, absolutely blind to the possibility that JonBenét had been killed by an intruder, had teamed up with the media to put pressure on the Ramseys, hoping that one of them would break down, confess, and implicate the other. The line between journalism and police work had disappeared. Cops became television celebrities, and talking heads became cops and prosecutors. No wonder the public wanted blood. The case had spun out of control into an around-the-clock television reality show.

But amid the media hysteria there was one detective who had not abandoned his role as a professional criminal investigator. This was Louis Smit, a retired detective hired in March 1997 by the Boulder County District Attorney's Office to work full time on the Ramsey case. With thirty-two years' experience in criminal investigation with the Colorado Springs Police Department, and later as captain of detectives of the El Paso County Sheriff's Office, Smit was the only professional investigator on the scene. On September 20, 1998, after eighteen months on the case, Smit resigned. He explained why in a letter to district attorney Alex Hunter, published in the *Boulder Daily Camera* under the headline "Smit Letter: Ramseys Innocent":

> It is with great reluctance and regret that I submit this letter of resignation. Even though I want to continue to participate in the official investigation and assist in finding the killer of JonBenét, I find that I cannot in good conscience be a part of the persecution of innocent people. It would be highly improper and unethical for me to stay when I so strongly believe this. . . .
>
> At this point in the investigation "the case" tells me that John and Patsy Ramsey did not kill their daughter, that a very dangerous killer is out there and no one is actively looking for him. There are still many areas of investigation which must be explored before life and death decisions are made. . . .
>
> The Boulder Police Department has many fine and dedicated men and women who also want justice for JonBenét. They are just going in the wrong direction and have been since day one of the investigation. Instead of letting the case tell them where to go, they have elected to follow a theory and let their theory direct them rather than allowing the evidence to direct them. The case tells me there is substantial, credible, evidence of an intruder and lack of evidence that the parents are involved. If this is true, they too are tragic victims whose misery has been compounded by a misdirected and flawed investigation, unsubstantiated leaks, rumors and accusations.

Lou Smit resigned because no one was listening to him. No one wanted to hear about evidence of a criminal intrusion because they had made up their minds that the Ramseys were guilty. For the past five months, deputy district attorney Michael J. Kane had been presenting a case against the Ramseys to a grand jury. The prosecution ball was rolling and there was no way to stop it. The media had already convicted the Ramseys and prosecutors were desperate to make it official. With Lou Smit off the case, investigative reason, objectivity, and professionalism got trampled in the stampede to incarcerate the Ramseys.

The Grand Jury

Michael Kane had a grand jury of eight women and four men, but no evidence. He couldn't prove that Patsy had written the ransom note. He didn't have a confession, a motive, an eyewitness, a jailhouse snitch, or one piece of physical evidence linking the Ramseys to the murder. He had nothing but his passion and his theories. He didn't have enough to indict a ham sandwich. But Kane kept going. And he had the media in his corner. As a result, the Ramseys had become the most hated couple in the United States. Justice demanded that these people be indicted, and it was Kane's job to make sure that happened.

David Liebman and Cina L. Wong, believing as they did that Patsy Ramsey had written the ransom note, wrote to district attorney Alex Hunter asking to appear before the grand jury on behalf of the prosecution. Hunter turned the matter over to Michael Kane, who wrote the Norfolk handwriting analysts asking for information about NADE certification, details regarding the International School of Handwriting Science, and other data pertaining to their qualifications as experts. Liebman and Wong complied with his request, and on January 20, 1999, received letters denying them the opportunity to appear before the grand jury as expert handwriting witnesses. In the letter each received, Kane set out the reasons for his decision, which must have felt like terrible slaps in the face:

> The primary reason that we have reached this decision is that we believe that the methodology which you have used in your reaching your conclusions does not meet the standards employed by the vast majority of forensic document examiners in this country. Most significant is your complete failure to account for or even reference any unexplained dissimilarities between the questioned and known samples. You are willing to conclude with 100% certainty that a writing was authored by a particular person based on some threshold level of similarities without any mention that there may be 10, 100, or 10,000 unexplainable dissimilarities between the known and unknown writings. I know of no reputable forensic document examiner who will not agree that the unexplainable dissimilarities between a person's natural writing and questioned handwriting will preclude a positive identification. . . . Because of this, it is not clear that your analytic methods would pass the test for admissibility in the courts of Colorado. We recognize that the rules of evidence do

> not apply to a grand jury investigation, but it would be rather pointless to allow the grand jury to indict based in part on opinion evidence that a petit jury might never be permitted to hear in trial.

In his letter to Cina L. Wong, the deputy district attorney added:

> I would note that you have engaged in a campaign of promoting your opinion in a manner that would surely open your credibility to doubt on cross-examination in a judicial proceeding. As an experienced trial attorney, I believe that an expert witness who has attempted to insinuate herself into a particular criminal investigation through a public media campaign would appear less objective and professional to a jury. It would be pointless to utilize the services of an expert who is vulnerable in this regard, given that there are hundreds of other qualified document examiners who are not tainted in this way.

David Liebman and Cina L. Wong were not the only interested parties denied access to the Boulder grand jury by Michael Kane. He kept Liebman and Wong away because he didn't consider them credible. He denied access to Lou Smit and the three Ramsey document examiners—Howard Rile, Lloyd Cunningham, and L. Donald Vacca—because he *did* consider them credible. The Ramsey legal team challenged Kane's exclusion of these four witnesses and won. In March 1999 the criminal investigator and the three document examiners appeared before the grand jury. Kane cross-examined these well-intentioned and highly qualified men as though they were enemy defense witnesses at a criminal trial. The last thing Kane needed were four credible witnesses tearing down a case that was already weak, and he took out his frustration on them. Howard Rile, the former Colorado Bureau of Investigation document examiner, was stunned and shaken by the viciousness of Kane's attack. He would later describe his grand jury appearance as a nightmare. After he had recovered from Kane's unprecedented assault, Rile asked for a second chance before the panel. Kane denied his request.

Thomas C. Miller was having a grand jury nightmare of his own. He had been indicted in Jefferson County for attempting to buy a copy of the ransom note from L. Donald Vacca. The news editor of the *Globe*, Craig Lewis, who had put up the $30,000, had not been indicted.

Despite Michael Kane's efforts, district attorney Alex Hunter, following thirteen months of twice a week grand jury sessions, 590 witnesses, and

1,058 pieces of physical evidence, decided there was not enough evidence to indict the Ramseys. In a politically courageous move, he disbanded the grand jury without allowing the jurors a chance to vote. Lou Smit had made a credible case for an intrusion, which included an unidentified crime-scene pubic hair, a mysterious boot impression, an unidentified palm print, marks on the corpse that could have been made by a stun gun, and a broken basement window. If the grand jury exercise proved anything, it was that the murder investigation had been bungled from the beginning and never got on course. At some point in the investigation, police and prosecutors should have come to their senses, regrouped behind Lou Smit, and started over. But they didn't. They couldn't. Over the past eighteen months they had fed the media so much phony, incriminating information against the Ramseys, they had painted themselves into a corner. Colorado governor Bill Owens, amid a chorus of hysterical talking heads, held a press conference to assure the public that the Ramseys were not off the hook. Although he didn't mention them by name, it was obvious whom he was referring to: "Finally, to the killers of JonBenét Ramsey, let me say this: you only think you have gotten away with murder. There is strong evidence to suggest who you are. I believe that the investigators are moving closer to proving their case. They will keep pursuing you. And I am confident that each day brings us closer to the day when you will reap what you have sown."

Wolf v. Ramsey: The Candid Mr. Epstein

In 1999, after absorbing an enormous amount of abuse from the media, the Ramseys went on the offensive. They had hired a more aggressive attorney, L. Lin Wood, and published a book, *The Death of Innocence: JonBenét's Parents Tell Their Story.* In the book, the Ramseys do more than profess their innocence. They articulate Lou Smit's intruder theory and name five suspects they believe had been ignored by the police.

One of the suspects named in the Ramsey book was a freelance writer, Robert Christian "Chris" Wolf. According to the Ramseys, Wolf once owned a sweatshirt with the acronym S.B.T.C.—like the ransom note signature—for the Santa Barbara Tennis Club. The Ramseys also alleged that Wolf's girlfriend had told Patsy Ramsey's sister that she thought Chris might have been involved in the murder. When she saw a photograph of the ransom note, the girlfriend thought she recognized Wolf's handwriting. The Ramseys were not saying that Chris Wolf, or any of the other suspects, had killed JonBenét; they were simply pointing out that, by targeting them, the police had ignored other possibilities and suspects.

By naming Chris Wolf in their book the Ramseys pulled New York attorney Darnay Hoffman back into the case and into their lives. On April 29, 2000, Hoffman filed a $50 million defamation suit against the Ramseys on Wolf's behalf. Hoffman alleged more than the mere defamation of his client. He accused Patsy Ramsey of implicating Chris Wolf to throw suspicion off herself, incorporating his pet theory that she had killed her daughter after a bed-wetting incident. In support of this accusation Hoffman had acquired the handwriting opinion affidavits of Thomas C. Miller, Cina L. Wong, and David Liebman, not a particularly impressive battery of experts. Miller had been indicted for attempted bribery and Wong and Liebman had been rejected as grand jury witnesses by Michael Kane. If Hoffman were to make a credible handwriting case against Patsy Ramsey, he would need at least one credible document examiner. About a year after he filed his defamation suit against the Ramseys, he found this expert in the person of sixty-two-year-old Gideon Epstein, a document examiner from Rockville, Maryland.

A private practitioner, Epstein had been a document examiner with the Immigration and Naturalization Service and was a member of the American Society of Questioned Document Examiners. In March 2000, six months after the Boulder grand jury adjourned without indicting the Ramseys, he had offered his services to district attorney Alex Hunter. Epstein had concluded that Patsy Ramsey, to the exclusion of anyone else, had written the crime scene note. This put him at professional odds with his colleagues Chet Ubowski, Richard Dusick, Edwin Alford, Leonard Speckin, Lloyd Cunningham, L. Donald Vacca, and Howard Rile. Until now, opinion had divided along a line that ran between the traditional experts and the NADE analysts, but at about this time, Larry Zieglar, a former FBI document examiner, showed Epstein a report he had prepared for Hoffman. Zieglar, also a member of the American Academy of Forensic Sciences and the American Society of Questioned Document Examiners, also believed that Patsy Ramsey have written the ransom note. A more ominous split had thus now opened up within the ranks of the American Society of Questioned Document Examiners.

Alex Hunter had not responded to Epstein's letter, so in January 2001, the document examiner offered his services to Mary Keenan, the new Boulder County district attorney. Keenan, who had rehired Lou Smit to continue investigating the intruder angle, wasn't interested either. This left one door into the Ramsey case and Epstein entered it.

In February 2001 Epstein submitted his Ramsey report to Darnay Hoffman, claiming that "Patsy Ramsey has been positively identified as the ransom note writer. There is no doubt in the view of this examiner that she is absolutely its author. I am 100 percent certain of my findings." Epstein also reported to Hoffman that he was not the only member of the American Society of Questioned Document Examiners who had come to this conclusion: "After making my examination of the ransom note and exemplars of Patsy Ramsey, I consulted with Larry F. Zieglar and Richard Williams, both former FBI document examiners, each with more than twenty-five years of experience identifying handwriting. They were both given the same exemplars as myself and they both independently reached the exact same conclusion: Patsy Ramsey was, without doubt, the ransom note writer."

Epstein believed that the note had been too long for Patsy to disguise her natural handwriting throughout. In other words, parts of the note revealed her fake writing style and other parts her true hand. Epstein also believed that in the samples of her handwriting she had provided, Patsy had tried to alter her natural writing.

> I disagree with Ramsey expert Howard Rile's conclusion that the handwriting exemplars of Patsy Ramsey are freely and naturally executed with no attempt at disguise. Patsy Ramsey is an accomplished writer. She is normally a rapid writer who has writing skills that are far better than the average person. She displays that in her exemplars where she demonstrates some of her other styles. If the handwriting exemplars were truly naturally and freely executed, as Mr. Rile suggests, they would be far more rapidly written and contain that quality that we call "careless abandon" which means the writer is not devoting conscious thought to the writing process. Instead, her exemplars are often very consciously executed. . . . When asked to produce her exemplars Patsy Ramsey had to write with her accustomed hand but did not write as rapidly and unconsciously as she normally would because she was attempting to avoid certain natural letter formations and habits that she was aware she made. This we call disguise and that is what she was attempting with her exemplars.

In June 2001, following a three-day trial, Thomas C. Miller was acquitted of perjury associated with the attempt by the *Globe* to purchase a copy of the

Ramsey case ransom note from L. Donald Vacca. Craig Lewis, the tabloid's news editor, had been charged in connection with the case, but the charges had been dropped after the *Globe* agreed to donate $100,000 to the University of Colorado School of Journalism.

In the spring of 2002, Gideon Epstein, a potential witness for plaintiff Chris Wolf in Darnay Hoffman's defamation suit, was in Atlanta being deposed by Ramsey attorneys James Rawls and Eric P. Schroeder. If the defamation case went to trial, Epstein, a sincere, qualified, and dedicated document examiner, could be a strong witness in support of Hoffman's allegation that Patsy Ramsey had killed her daughter. In an effort to attenuate the value of Epstein's role as a plaintiff expert, the Ramsey attorneys would try to paint handwriting identification as a scientifically unreliable endeavor.

When asked why he had offered his expertise to district attorneys Alex Hunter and Mary Keenan, then signed on with Darnay Hoffman, Epstein gave an answer that would have shocked and upset many of his colleagues:

> I feel that the questioned document profession let the criminal justice system down in this particular case, and I feel very strongly that I would, if possible, like to set that straight.
>
> I don't believe that the forensic reports that have been rendered in this case thus far by those document examiners who earlier examined these documents were correct, and I don't believe that justice has been served, and that's my only reason for becoming involved in this case.

Epstein believed that the Ramseys had escaped indictment because the grand jury had heard from Howard Rile and Lloyd Cunningham, instead of from himself and the two other ASQDE members who believed that Patsy Ramsey had written the note—former FBI examiners Richard Williams and Larry Zieglar (who chose to stay out of the case). Epstein also felt that the opinions of the four examiners brought into the case by the Boulder County District Attorney's Office and the Boulder Police Department had been influenced by the Ramseys' experts. When asked why he thought Chet Ubowski, Leonard Speckin, Edwin Alford, and Richard Dusick had gotten it wrong, Epstein replied:

> First of all, I'd like to say that the field of forensic document examination in the United States is a very small profession, as you may have

> found out, especially within the ranks of those people who are board-certified and who are the mainstream examiners in this country. Everyone knows everyone else. There are certain document examiners who, because of their exposure in the profession, because of the workshops that they may present, are looked upon by other examiners as leaders in their field.
>
> A lot of these examiners are in private practice, and they're retained oftentimes by one side or the other. In this particular case I think the fact that Howard Rile and Lloyd Cunningham, who became involved in this case very early on, and who were retained by the Ramsey family, coupled with the fact that Howard Rile came out of the Colorado Bureau [of Investigation], I believe that the connection was very instrumental in the Colorado Bureau coming to the conclusion that they did, because Howard Rile had come to the conclusion that he did.
>
> Lloyd Cunningham works very closely with Howard Rile and they were both on this case, and then it was a matter of chain of events, one document examiner after another refusing to go up against someone who they knew, someone who was large in the profession, for fear they would be criticized for saying something that another examiner—it's sort of like an ethic within the medical community, where one doctor protects the other doctor. . . . I feel personally that the other examiners were simply afraid to state what they believed to be the truth. . . . And I just don't believe some of these people devoted the necessary amount of time to the case to come up with the correct conclusions, and I think they simply went along with what had been previously said because it was the most expedient thing to do.

When asked by Rawls if his opinion regarding why he was at odds with the other examiners was based on speculation, Epstein said, "That's my opinion. My opinion is based on my knowledge of the profession and the people involved."

Epstein did not speak highly of his field in response to a question by Darnay Hoffman regarding the meaning and context of the term "blow smoke":

> It's often used in the handwriting profession when a document examiner comes into a case to cast doubt or dispersions on a particular finding, primarily to inject into the minds of the jury or to attempt to

> add some sort of confusion to the case, without specifically saying that the previous examiner was wrong, in other words, they simply say there wasn't enough evidence or the evidence was not comparable, that kind of thing. So, the term "blow smoke" comes from that particular type of testimony, if you will.

Objections from the opposing attorneys kept Epstein from answering another question from Hoffman: "Did you find any examples of 'blowing smoke' in either the report of Howard Rile or Lloyd Cunningham."

Rawls asked Epstein to read part of a letter New York City document examiner John Paul Osborn had written to someone (the name had been obliterated) in the Boulder Police Department about whether or not Christian Wolf had written the ransom note. Osborn had written:

> There are some similarities between the handprinting in the ransom note and the writing you submit as that of Christian Wolf. However, there are also similarities between my own handprinting and that of the questioned note, if you disregard the poor line quality. Similarities, while playing a role in the process of examination and comparison of writing, are not as significant as fundamental or significant differences. Many people share common handwriting characteristics and even some distinctive handwriting characteristics. The proper weight must be given to differences which cannot be accounted for by natural variation of a single writer.

Since John Paul Osborn said the same thing about Wolf's writing that Howard Rile and Lloyd Cunningham had said about Patsy Ramsey's, Rawls wanted to know if Epstein agreed with Osborn's comments about the importance of dissimilarities over common similarities. Epstein's answer: "I agree with the paragraph [in Osborn's letter], but not with its application in this case. . . . There are no significant differences [between Patsy Ramsey's writing and the ransom note]. There are variations to the same basic handwriting patterns, but there are no significant differences."

Having established that Darnay Hoffman had sent Epstein samples of Christian Wolf's handwriting, Rawls asked Epstein if he had eliminated Wolf as the author of the ransom note. Epstein replied that he hadn't studied Wolf's handwriting because he was 100 percent sure that Patsy Ramsey was the one who had written the note.

While Gideon Epstein was hard on members of his own profession, he wasn't any kinder to Cina L. Wong, one of Hoffman's NADE analysts from Norfolk. When Rawls referred to Wong as Epstein's coexpert, Epstein set him straight: "I don't consider her as my co-expert in this case. . . . Someone else may consider her as a co-expert, but I don't. . . . I don't believe that she meets what I and the profession consider to be the necessary qualifications for forensic document examination."

Cina L. Wong, in her deposition a week earlier, had mentioned a textbook, *Attorney's Guide to Document Examination* by Katherine Koppenhaver, and Rawls asked Epstein if he was familiar with it. Epstein said he was not but knew of its author. Asked to elaborate, Epstein said: "I consider her [Koppenhaver] another one of the people practicing on the fringe of forensic document examination who probably have a background in graphology, who have opposed a number of qualified examiners in the Washington [D.C.] area. . . . I have absolutely no respect for her or her work, so there's no reason for me to read her book."

Pointing out that Koppenhaver was the immediate past president of the National Association of Forensic Document Examiners, Rawls asked if that was a bona fide certifying body within the field. Epstein:

> The only recognized body for the forensic profession in document examination in North America is the American Board of Forensic Document Examiners. They're the only recognized board to certify document examiners. [There are] a lot of these fringe organizations, and there are others besides that certify their own members . . . so the individual can go into court and state that they're board-certified, and the court doesn't know the difference between one board and another.

Taking advantage of Epstein's candor about the various problems with the document examination field, including the infiltration of practitioners with graphology backgrounds, Rawls asked if he considered David Liebman, the other Norfolk handwriting analyst on Hoffman's team, qualified. As was his custom, Epstein didn't mince his words:

> He is not, no. . . . I don't believe that he ever went through any kind of recognized or accepted training program. The profession requires that for a person to be qualified to do this work, they must complete, at a minimum, a two-year resident training program in a recognized laboratory or by a recognized forensic document examiner. Many of

> these people [the graphologists] are self-taught. This is not a profession that you can learn by yourself. I mean, this is an apprenticeship type of profession. You have to learn from others, people who have been doing this for years.

The Carnes Decision

Fortunately for both factions of the handwriting identification profession—examiners with and without forensic science backgrounds—Gideon Epstein would not get the opportunity to testify in open court. His low opinion of the graphology-oriented analysts and his groupthink portrayal of his forensic science colleagues would not become public knowledge, because on March 31, 2003, federal district court judge Julie Carnes of Atlanta dismissed the Christian Wolf suit. Because it was not based upon procedural grounds, the ruling was more than just a defeat for Hoffman and his client—it comprised the closest thing the Ramseys would get to being judicially exonerated. As revealed in the ninety-three-page order dismissing the action, Judge Carnes was clearly sympathetic to Lou Smit's view of the case and to the idea that Christian Wolf could have been considered a viable suspect by the police.

In making her decision in *Wolf v. Ramsey*, Judge Carnes weighed the handwriting evidence and came down on the side of the six document examiners who could not identify Patsy Ramsey as the ransom note writer. Moreover, in assessing whether the Ramseys had been reasonable in characterizing Christian Wolf as a potential murder suspect, the judge noted that Patsy Ramsey wasn't the only person some of these experts hadn't been able to eliminate as the writer of the ransom note:

> For example, forensic document examiner Lloyd Cunningham cannot eliminate plaintiff [Wolf] as the author of the ransom note. Plaintiff's ex-girlfriend has also testified that she was "struck" by how the handwriting in the note resembled [Wolf's] own handwriting. Further, to the extent that the use of a single [copy] editing mark [symbol] might suggest to plaintiff's experts that Mrs. Ramsey was the author, given her bachelor's degree in journalism, one should also note that plaintiff himself has a master's degree in journalism.

In defending the Ramseys against the defamation suit, Rawls and Schroeder had challenged the admissibility of questioned document examination generally as not sufficiently scientific and reliable under the guidelines of the *Daubert* decision. Judge Carnes, however, was not about to throw the baby out with the bathwater:

> The field of forensic document examination is premised on the assumption that no two persons' handwriting is exactly alike; instead, each person has a unique handwriting pattern that allows the person to be identified through a comparison of proper handwriting specimens. Forensic document examination involves the subjective analysis and assessment of writing characteristics found in a person's handwriting or handprinting style, by examination of subtle and minute qualities of movement such as pen lifts, shading, pressure and letter forms. Two or more handwriting experts can reach different conclusions of authorship, even when examining the same questioned document and handwriting exemplars. . . . The only recognized organization for accrediting forensic document examiners is the American Board of Forensic Document Examiners. . . . The court concludes, as a general proposition, that forensic document examiners, who are equipped with the proper background qualifications and who employ the accepted methodology in their analysis, can serve to assist the trier of fact, in some regards, through providing reliable testimony about similarities or differences, or both, between a questioned writing and comparative exemplars.

While the judge did not rule in favor of the Ramsey *Daubert* challenge, she did agree that handwriting analyst Cina L. Wong was not qualified as an expert witness:

> Although the court has concluded that a proper expert may assist a jury in a comparison of handwriting between a known and an unknown piece of writing, that conclusion does not mean that a person can be deemed as an expert in forensic document examination merely by announcing himself as such. Indeed, defendants assert that plaintiff's experts, in particular Ms. Wong, lack the necessary credentials to qualify as experts. . . . The court agrees with defendants that Wong is not qualified to provide expert testimony. . . . In stark contrast to Epstein, Wong has never taken a certification exam, completed an accreditation course in document examination, been an apprentice to an ABFDE certified document examiner, or worked in a crime lab.

Judge Carnes, while recognizing Gideon Epstein as qualified document examiner, cites two reasons for giving his handwriting opinion less weight than the conclusions of his colleagues: (1) he didn't have an opportunity to examine the original documents, and (2) he failed to explain the methodology behind his identification of Patsy Ramsey as the ransom note writer. Regarding the methodology problem, Judge Carnes wrote:

> Epstein bases his conclusion on perceived similarities between [the ransom note and Patsy Ramsey's known writings]. Yet, as noted by defendants, Epstein never indicates how many similarities or what kind of similarities are required before he can reach absolute certainty, 50% certainty, or no certainty at all. . . . Whenever encountering any differences between the known writing of Mrs. Ramsey and the ransom note, Epstein finds refuge in the explanation that Mrs. Ramsey must have been trying to disguise her handwriting. While it is, of course, possible that differences between known writing and questioned documents are the result of a known writer's efforts to disguise her handwriting, it is just as plausible that the differences can occur because the known writer is not the author of the questioned matter. On that issue, Epstein offers no hint of the methodology that he employs to distinguish between disguised writing and writing that is simply being provided by two different people.
>
> The underlying notion behind *Daubert*, and all good science, is that a given premise or principle should be capable of being tested to determine whether the principle is, in fact, sound. Thus, if Epstein indicated, for example, that whenever a writer of known material has X number of similarities, there is a given probability that the writer wrote the note—and if this methodology had been tested by reliable means in the past—then Epstein would have shown reliability in the methodology that he used to reach a determination of the likelihood of his conclusion. As it is, however, Epstein's explanation for his conclusion seems to be little more than "trust me; I'm an expert." Accordingly, the Court concludes that while Epstein can properly assist the trier of fact by pointing out marked differences and unusual similarities between Mrs. Ramsey's writing and the ransom note, he has not demonstrated a methodology whereby he can draw a conclusion, to an absolute certainty, that a given writer wrote the note.

Judge Carnes had obviously given a lot of thought to forensic document examination. Her detailed rationale makes her ruling important on the general question of the admissibility of handwriting testimony. It appears that Judge Carnes applied a higher standard to Epstein than she did to Howard Rile and his colleagues. But if she did so, it was because Epstein had identified Patsy Ramsey positively as the ransom note writer while Rile and the others simply declared they could not make such an identification. If Epstein says he's 100 percent certain of his identification, Judge Carnes wrote, he's got to tell the court *how* he arrived at that conclusion. That's a heavy burden, one that Judge Carnes didn't think he had met.

In summing up her reasons for dismissing Darnay Hoffman's suit, Judge Carnes wrote:

> Plaintiff has failed to prove that Mrs. Ramsey wrote the ransom note and has thereby necessarily failed to prove that she murdered her daughter. Moreover, the weight of the evidence is more consistent with a theory that an intruder murdered JonBenét than it is with a theory that Mrs. Ramsey did so. For that reason, plaintiff has failed to establish that when defendants wrote the book, they "in fact entertained serious doubts as to the truth of the publication."

The Aftermath

The national media, once obsessed with the Ramsey case, showed little interest in the dismissal of Darnay Hoffman's suit and what that dismissal implied. The media circus had packed up and left town two years earlier, leaving behind an unsolved murder, ruined careers, tarnished reputations, and heartbreak.

Within the forensic document community, the debate over who wrote the ransom note was not over. In August 2004, at the American Society of Questioned Document Examiners annual convention held that year in Memphis, Howard Rile and Gideon Epstein presented their contradictory findings to the hundred or so document examiners in attendance. Using comparison word charts and other exhibits, Rile carefully pointed out the dissimilarities in the known and questioned writing that had led him to the conclusion that Patsy Ramsey had probably not written the ransom note. Gideon Epstein, without word charts or exhibits to support his conclusion, spoke about the need for independent thinking and the value of an open mind. He thought that perhaps his colleagues were a bit too critical of their graphology-oriented counterparts, going so far as to say that in some cases the graphological point of view might be helpful in the analysis of a document. This was not what most of those in the audience wanted to hear. With forensic handwriting identification being regularly challenged in court as unscientific and unreliable under *Daubert* criteria, this was not the time to associate it with graphology or graphologist-trained document examiners.

There are Gideon Epsteins in all the forensic sciences, practitioners who are mavericks but qualified, courageous, and sincere. Free thinking, open-mindedness, and independence are not bad traits for a forensic scientist. The questions that have to be answered, however, are these: Is Epstein (and the two former FBI examiners) correct and the majority wrong? Is there in fact too much groupthink in forensic science? Could Epstein be wrong about the Ramsey document but right about his profession, generally? Or is he wrong on all counts?

The JonBenét Ramsey case, like the Lindbergh kidnapping trial earlier in the century, publicized forensic document examination. The Lindbergh case, considered a landmark in the history of this criminalistic field, did not involve competing experts. That case put forensic handwriting examination on the map, while the Ramsey case, featuring credible experts on both sides of the issue, put its future in danger. And there was more to come in the JonBenét Ramsey case. With respect to the field of forensic document examination, the second act would not be helpful.

12

John Mark Karr

DNA Trumps the Graphologists in the Ramsey Case

A Murder Suspect Emerges

After a thirteen-year-battle with ovarian cancer, Patsy Ramsey died on June 24, 2006. She was forty-nine. The media that had helped police and prosecutors portray the Ramseys as child murderers treated the death as a one-day news event, giving it less attention than the passing of a supporting actor on an old television sitcom. But in April 2006, two months before her death, the Ramseys flew from their home in Michigan back to Boulder where they met with district attorney Mary Keenan (now Lacy), who asked them if they had ever heard of a man named John Mark Karr. The Ramseys said they had not—neither the name nor the description of this man rang a bell. What did he have to do with the case?

Karr, a forty-one-year-old itinerate elementary school teacher, an American living in Bangkok, Thailand, since 2002 had been corresponding with Michael Tracey, a journalism professor at the University of Colorado. Karr's interest in the JonBenét murder had drawn him to the Boulder professor, who had produced three television documentaries favorable to the theory that the crime had been committed by an intruder. The e-mails from Karr, sent under the pseudonym Daxis, had recently become quite bizarre, reflecting more than just a morbid interest in the case. After receiving a series of disturbing phone calls from this man, Professor Tracey alerted the district attorney's office. The calls were traced to John Mark Karr in Bangkok. After Daxis had confessed to Tracey that he had accidentally killed JonBenét while inducing asphyxia for his sexual gratification, he became a suspect in the murder. Karr had revealed over the phone that when he couldn't revive

JonBenét, he had struck her in the head with a blunt object. He told the professor that he had engaged in oral sex with the victim but had not performed sexual penetration. Aware that Tracey was writing a book on the Ramsey case, Karr offered the author the inside story from the killer's point of view. In the event the book became a movie, Karr wanted to be played by Johnny Depp.

Having taken over the Ramsey case investigation from the Boulder Police Department, the district attorney's office began investigating John Karr. DA investigators spoke to the authorities in Bangkok and read hundreds of the e-mails Karr had sent to Tracey. A sampling of this correspondence, quoted in "John Mark Karr: In His Own Words" in *People* magazine on August 24, 2006, reveals the common theme behind Karr's obsession with the case:

> [The Ramseys] feel their daughter was brutally murdered, and she wasn't. It looks like that but she wasn't. I want them to hear the truth. . . . I need closure and [JonBenét's] family need closure . . . all of us have gone through enough pain.
>
> We all need to put this chapter in place, which is a conclusion, and then start the healing. The healing is what I want to start happening. To heal and feel better about what happened.
>
> People will say I am a monster and a horrible person. They don't know me. . . . They won't approve of what happened. I don't approve of what happened.

One of the messages suggested that Karr had a general knowledge of forensic science and had made all of this up: "The DNA might not match, but you can't trust the test."

As investigators gathered details of Karr's life and background, it became clear that he was not an ordinary man and that his strangeness was not inconsistent with the profile of a person who might commit a Ramsey-type crime. After Karr's parents divorced when he was nine, he went to live with his grandparents in Hamilton, Alabama. In 1983, one year after graduating from Hamilton High School, Karr, then twenty, married a thirteen-year-old girl. The marriage ended nine months later in an annulment. In 1989, Karr married sixteen-year-old Lara Marie Knutson. In four years, he and his wife had three sons. While pursuing a teaching degree through an online teacher's college, Karr opened a licensed day-care center in his home in 1997. Although he didn't have a teaching degree, he also worked as a substitute teacher at Hamilton High School. He acquired a college degree in 1999 and that year closed his day-care business. A year later he and his family were residing in Petaluma, California, where he taught as a substitute in six schools in the Sonoma Valley Unified School District.

One year after arriving in Petaluma, while teaching at the Pueblo Vista Elementary School, Karr was arrested by investigators from the Sonoma County Sheriff's Office. They had found child pornography on Karr's computer and arrested him on five misdemeanor counts of possessing such material. The charges carried a maximum sentence of one year in prison for each count. Karr's bail was reduced after he had spent six months in the county lockup awaiting trial, and he was released in October 2001. While in custody, Karr had written letters to Richard Allen Davis, who had been convicted of kidnapping and murdering Polly Klaas in Petaluma. When Karr failed to show for a court appearance in the pornography case, the judge issued a bench warrant for his arrest, making him a California fugitive from justice.

During the child pornography investigation, detectives in Sonoma County came across writings and notes Karr had made pertaining to the murder of JonBenét Ramsey. In these musings, Karr had speculated on the killer's thoughts as he committed the crime. Although these were not confessions, the Sonoma detectives took the writings seriously enough to notify the authorities in Boulder. There were follow-up discussions between investigators in California and Colorado, but nothing came of this discovery.

Karr was now divorced. His children and former wife had moved back to Hamilton, Alabama, and following his release from the Sonoma County Jail, Karr fled the country. He taught in Honduras and Costa Rica and worked as a children's nanny in Germany, the Netherlands, and South Korea. In December 2005, he arrived in Bangkok, where he had landed a grade-school teaching position.

The Arrest and Confession

On August 11, 2006, four months after district attorney Mary Lacy learned that the Ramsey e-mail writer and telephone confessor was John Mark Karr, police and immigration authorities in Thailand informed her that Karr was living in a downtown Bangkok apartment. In less than a week Karr would be starting a new teaching job at the New Sathorn International School in the city. Because they didn't want this man interacting with young girls at that school, the Thai authorities planned to arrest and deport him within the next five days. This development presented Lacy with a dilemma. If she did nothing, a man who had confessed to killing JonBenét Ramsey could slip away upon his return to the United States. If she filed charges against Karr and had him extradited back to Colorado, the probable cause supporting the arrest warrant would be based entirely on his e-mails and his telephone confessions. Lacy's investigators had not linked Karr to the ransom note through his handwriting, could not place him in Colorado on or about December 26, 1996, and had not matched his DNA to a pair of bloodstains on JonBenét underwear.

In 1996, the underwear evidence had been placed into storage, deemed insufficient for DNA analysis. In 2003, taking advantage of advances in DNA science, Lacy had sent the evidence to the Denver police crime laboratory for analysis. The major component of the blood traces belonged to JonBenét; a minor component—ten of thirteen markers necessary for a full profile—belonged to an unidentified white male. Lacy sent the incomplete profile to the FBI's DNA computer bank, CODIS, where it was scanned against the 1.5 million profiles on file in dozens of statewide DNA repositories without finding a match.

Operating on the theory that Karr was not a false confessor and that his DNA would eventually connect him to the victim, Lacy presented her case to a Boulder judge, who issued a warrant for Karr's arrest on charges of first-degree murder, kidnapping, and sexual assault. Lacy also dispatched one of her investigators to Bangkok. At this point, the public was not aware of John Karr or the new developments in the case. This was the calm before the media storm.

After surveilling Karr's apartment building for five days, police and immigration officials took Karr into custody on Wednesday, August 16, 2006. In response to a Thai police officer who had informed Karr that he had been charged with first-degree murder in Boulder, Karr declared that his killing of JonBenét had been accidental, and therefore the charge should more appropriately be second-degree murder. Without being interrogated, John Karr had just confessed to the ten-year-old homicide. Before being taken to his holding cell, Karr asked if he could make a brief statement to members of the news media gathered at the immigration detention center. Standing before a group of print and television reporters, he said, "I was with JonBenét when she died." MSNBC carried the exchange that followed:

REPORTER: Were there other people with you?

KARR: No.

REPORTER: What happened to her?

KARR: No comment.

REPORTER: What happened in the last moments? Were you playing with her?

KARR: Her death was an accident.

REPORTER: In the basement?

KARR: Yes.

Over the next twelve days, one couldn't turn on CNN, Fox-News, MSNBC, or Court TV without seeing this narrowed-shouldered, pencil-necked man with his oversized head and expressionless, feminine face. And the more the media learned of John Karr, the weirder he became—including the news that he had recently been to a clinic where technicians had permanently removed his facial hair and given him estrogen pills in anticipation of a sex-change operation in Bangkok. The tabloid press and cable news networks lost their journalistic minds over this man. Without waiting for the evidence, it seemed that every criminal justice talking head in the country had an opinion regarding Karr's guilt or innocence. Syndicated columnist Marie Cocco quipped that Karr's arrest had "rejuvenated the Roman circus of cable coverage" of the JonBenét Ramsey case.

The day following his arrest, in an exclusive interview with Associated Press reporter Marcus Wohlsen, Karr said, "I am so sorry for what happened to JonBenét. It is very important for me that everyone knows that I love her very much and that her death was unintentional, and that it was an accident." The next day, Mary Lacy's office released a few of Karr's e-mails to Professor Tracey. The media was not being told, however, of Karr's phone confessions that included details of the killing. In an e-mail sent on May 22, 2006, Karr had written: "Her and I were engaging in a romantic and very sexual interaction. It went bad and it was my fault." In another e-mail Karr wrote: "I am trapped in a world that does not understand. I can understand people like Michael Jackson and feel sympathy when he suffers as he does. . . . I do think that he is sexually attracted to certain children, but could never divulge this." And finally: "JonBenét, my love, my life. I love you and shall forever love you. If there is to be a life for me after this one, I pray that it will be with you—together forever with you and other little girls who are gone from my life forever. That would be Heaven." For believers in an afterlife, this must have sounded like hell.

The consensus in the media and among the talking heads was that Karr was an emotionally dysfunctional false confessor who was obsessed with JonBenét and wanted to insert himself into the history and fabric of the case.

Karr's ex-wife, by insisting that he had never been to Colorado, lent credence to the false confession theory. She said she was certain that when JonBenét was murdered, Karr was celebrating Christmas with his family in Atlanta, Georgia. Moreover, no one could figure out how JonBenét would have crossed Karr's path in the first place.

Shortly after Karr's arrest the *New York Times* published a letter attorney Darnay Hoffman had just written to Mary Lacy. In 1997, as we have seen, Hoffman had written Boulder district attorney Alex Hunter that the Ramseys, like Charles and Anne Lindbergh, had murdered their child and tried to cover it up with a fake kidnapping. Hoffman now warned Mary Lacy that she might be the victim of a false confession, a hoax.

The Graphologists Surface

Karr had been in custody two days when David Liebman, appearing on local television in Norfolk, Virginia, told reporter Stacy Wiggins that in his opinion Patsy Ramsey had written the ransom note. "It doesn't mean she was the murderer," he said, "but I believe she knew who did it." The NADE analyst based this opinion on his previous work in the case, not an analysis of John Karr's handwriting. When one of Karr's high school classmates came forward with a 1982 yearbook bearing his inscription, the public got a look at Karr's known writing. The yearbook inscription generated quite a buzz due to the way in which Karr had signed-off: "Though, deep in the future, maybe I *Shall be the conqueror* and live in multiple peace" (italics mine). Had the meaning of the ransom note writer's S.B.T.C. finally been revealed, or was this one of those coincidences that regularly plague criminal investigators? "I shall be the conqueror" was now on the list of possibilities, along with Saved by the Cross and the Santa Barbara Tennis Club.

The public display of the twenty-four-year-old sample of Karr's handwriting, like the 1997 publication of the ransom note and snippets of Patsy Ramsey's known writings, became a career opportunity for a handful of fourth-tier handwriting analysts willing to risk their professional reputations. Two days after declaring that Patsy Ramsey had written the ransom note, David Liebman was on *The Today Show* pointing out similarities between Karr's yearbook inscription and the same ransom note. If Patsy Ramsey had written the ransom note, of what relevance were the similarities in the note to Karr's known writing?

Don Lehew, the second handwriting expert to break out of the gate into the Ramsey second act limelight, told local Jacksonville, Texas, newspaper reporter Kelly Young that he had compared Karr's yearbook writing to the ransom note and had found "13 similarities that are, in many cases, very unusual." Without requiring more known handwriting before drawing a conclusion, Lehew said, "It certainly indicates to me that John Karr did write the ransom note." Besides functioning as a forensic document examiner, Lehew was also a practicing graphologist, which prompted him to note that Karr's handwriting also revealed several "hallmarks of a child molester." Lehew told the reporter: "I can't look at someone's handwriting and say, 'That's a pedophile.' What I can do is look at it and say, of all the handwriting samples that I have looked at, about 10,000 in the past 25 years, this one closely approximates and relates to the handwriting of known pedophiles."

Lehew, a graphologist and licensed private investigator, had entered the questioned document field in 1986 at the age of forty-six. Because he had not worked in a public or private crime laboratory, he did not qualify for membership in the American Society of Questioned Document Examiners and belonged only to an organization called the American Board of Forensic Examiners. According to Young's newspaper article, he had given handwriting identification opinions in two hundred questioned document cases and had "served as a handwriting expert in county courts, state district courts and U.S. District Courts." He also claimed to have been certified by the Texas Department of Public Safety to "teach questioned document examination to law enforcement throughout Texas." Lehew said he was an instructor at Handwriting University International, an online school founded by another self-made handwriting expert, Curtis Baggett.

Having identified John Karr as the Ramsey case ransom note writer, Lehew found himself on August 21, 2006, on the CNN true crime show *Nancy Grace*. Although serious doubts were being raised regarding the validity of Karr's confessions, Lehew, finding himself on national television, stuck his professional neck out as far as it would go. A partial transcript of the show:

GRACE: Joining us tonight is a veteran handwriting comparison expert. His name is Don Lehew and he believes Karr did, in fact, write the ransom note. . . . What do you think?

LEHEW: I found significant comparison documents that were adequate to convince me that he is the one who wrote the ransom note.

GRACE: At first I thought . . . there's no way this guy was even remotely involved except in his own dreams. But if you take a look at his handwriting, especially the "a," with the umbrella over it . . . you don't often find that. And the “d” is an upside-down candy cane. It’s a very unusual marking, Don.

LEHEW: That’s true.

GRACE: And what do you make of it?

LEHEW: Well, that’s part of what we look at to make the comparison to make the match on the handwriting. The lower case ‘l’s” are a little bit strange, to say the least . . . and certainly the “a” and the “d” are out of character from what we learned in school, typically. The “w” is a good match. And the lower case “t” has a tail that comes off to the right in both the known and questioned document.

GRACE: Don, when you make a handwriting comparison, how many similarities do you look for?

LEHEW: Well, in this case as many as I can find.

GRACE: When you say it’s a match, how many matches do you have to have? How many points of match do you have?

LEHEW: In this case, I had 13 letters that matched significantly. Typically, in a forgery, you’re looking for differences, whereas in this case, I was looking for similarities.

GRACE: How many similarities does it take before you will deem it a match?

LEHEW: I don’t have a specific number. It’s kind of like fingerprints. They don’t have a specific number, and neither do we. But I like to get at least ten or better. And once I’ve got ten or better, I’m pretty convinced.

In an August 22, 2006, *Rocky Mountain News* article in which he was identified as a “well-known national handwriting expert,” Curtis Baggett, who started Handwriting University International, not only weighed in with a premature Ramsey case handwriting identification, he boasted about it. Quoted by Lou Kilzer in “Handwriting Expert Points Finger at Karr,” Baggett claimed that “most” document examiners “are riding the fence. But there are at least a dozen traits that match up perfectly, when comparing a yearbook signed by Karr and the ransom note.” Based on that comparison, Baggett, revealing that he didn’t even know how to sound like a forensic scientist, declared that he was “99.9 percent certain” that John Karr had written the ransom note.

Meant to entice trial lawyers looking for a dependable hired-gun expert, Baggett’s Web site actually reveals an unprofessional attitude toward questioned document work that would by itself disqualify him as an expert witness if brought to the attention of a presiding judge. For example:

> [Curtis Baggett is] court qualified and certified by the American Bureau of Document Examiners and the American Institute of Applied Science.

> Curt has examined documents and/or testified in every state in the USA and over a dozen countries—over 2,000 cases since 1990. His television appearances include programs on CNBC and "Texas Justice." He is the consulting forensic document examiner for the #1 television show, "CSI: Crime Scene Investigation." . . . As Texas forensic document examiners, we know how tricky the opposition can be. We can *guide you to the truth* about the handwriting and hopefully to victory. Attorneys love us! Unlike many "document examiners" who routinely charge $1,200 or more to just look at your document, I am available right now to provide a "Quick Verbal Opinion" of your case via fax for just $295. For most people, this is all they need. Please call to discuss your case now. [italics mine]

Baggett didn't tell the *Rocky Mountain News* that in recent cases judges had refused to accept him as a qualified document examiner. In a *Daubert* motion to disqualify him in a federal case in Maine, *United States v. Bourgois,* the opposing attorney had written:

> In an attached sworn statement, Baggett asserts that he was certified in 1983 as a "master handwriting analyst" and in 1987 as a "questioned document examiner" by Dr. Ray Walker, "a leading authority in the field of handwriting analysis and document examination." Baggett also asserts that he has "taught handwriting analysis classes for over 20 years in approximately twenty states." . . . Ray Walker, the individual identified by Baggett as the source of his certification, was as of 1992, well after the certification, not a member of the American Board of Forensic Document Examiners. And a federal district court's decision not to qualify him [Walker] as a handwriting expert was upheld on appeal.

The U.S. magistrate in the Maine case granted the *Daubert* motion to exclude Baggett's handwriting testimony, calling it "unfounded and unreliable" and not based on forensic science expertise. This would not be one of the cases featured on Baggett's Web site.

In an Arkansas disputed last will and testament case, *Wael Abdin v. Delores Abdin,* in which Baggett had testified that a "lost" will bore the legitimate signature of the deceased, he faced, as an opposing handwriting witness, Linda Taylor, who opined that the questioned signature was not genuine. Taylor had been employed as a forensic document examiner in the FBI lab and in the Arkansas state crime laboratory. She was certified by the American Board of Forensic Document Examiners and had published articles in peer-reviewed journals. The trial court's verdict supported Taylor's testimony, which caused Baggett's client to appeal the decision. The appellate judge, David M. Glover, in affirming the lower court's decision, compared Baggett's professional credentials with Taylor's:

> By contrast . . . Baggett admitted that he was not board-certified. He testified that he had studied under Dr. Ray Walker, who was a doctor of divinity. He also said that he taught document examination through Handwriting University, a mail-order school, and that his son [Bart Baggett] owned HandwritingUniversity.com, "the largest handwriting analysis school in the world." He admitted that he had taken no continuing education classes, had never published in any trade journals, and that he had once been convicted of felony theft.

The day after the *Rocky Mountain News* published reporter Lou Kilzer's article featuring Baggett's 99.9 percent identification of Karr as the Ramsey case ransom note writer, Kilzer reported that Baggett had been disqualified by a federal judge in Georgia. In that case, Judge Clay Land ruled that Baggett was not professionally certified, had not undergone proficiency testing, and had not published anything in the document examination field. Judge Land labeled Baggett's qualifications as "clearly paltry."

In response to these revelations, Baggett reminded the *Rocky Mountain News* reporter that he had testified as an expert in more than two thousand trials in all fifty states and twelve countries. During this period he had only been disqualified four times. In defending Baggett, a student enrolled at Handwriting University was quoted by Kilzer on August 29 as follows: "Every time you're on the top, someone is after you. Curtis does more document exams per month than most people do all year." It may have been true that Baggett was a handwriting-analyzing machine, but in forensic science it's quality, not quantity, that counts.

DNA Trumps Karr's Confession

After being flown to Los Angeles from Bangkok, Karr arrived in Colorado on August 24 and was incarcerated in the Boulder County Jail. Four days later, the John Karr phase of the Ramsey case came to an abrupt end when Mary Lacy announced that because Karr's DNA didn't match the crime scene evidence, the charges against him had been dropped. For a while it looked like Karr would walk free, but the authorities in Sonoma County, California, asked to have him held in connection with their child pornography case. As the army of television and print reporters glumly marched out of town, Colorado governor Bill Owens took a shot at district attorney Lacy: "I find it incredible that Boulder authorities wasted thousands of taxpayer dollars [$9,300] to bring Karr to Colorado given such a lack of evidence. Mary Lacy should be held responsible for the most extravagant and expensive DNA test in Colorado history." This same tough-talking governor hadn't seem bothered in 1999 by an equal lack of evidence against the Ramseys, when he had implicitly accused them after the grand jury refused to indict them. In defense of her actions, Lacy replied: "We felt we could not ignore this, we had to follow it, [and] . . . there was a real public safety concern here directed at a particular child." Lacy added that a forensic psychologist had advised her that Karr was a dangerous person and growing more dangerous with the passage of time.

The day after Karr's DNA exoneration, Curtis Baggett gave his take on the development to *Rocky Mountain News* reporter Lou Kilzer. As the headline "Handwriting Expert Holds to Opinion" suggests, he was unswayed: "It's my belief that there were so many matches that he [Karr] wrote the [ransom] letter. He may not have murdered the girl. We don't know that. I believe he wrote that letter." In the same article was a quote from Don Lehew: "If the DNA evidence does not put [Karr] at the scene of the crime, then he had a hand in writing the note and didn't do the crime." In other words, the two men agreed, Karr wasn't a child-molesting killer, just a man who broke into homes and left bizarre ransom notes.

On October 6, 2006, the authorities in Sonoma County, California, had to drop the child pornography charges against Karr because sheriff's office investigators had lost vital computer evidence needed to prove the case. Karr walked out of jail a free man. He would later resurface as a guest on numerous national television shows, but the public had little interest in what he had to say.

The Ramsey case reappeared in the news a few months later on the tenth anniversary of the murder. There seemed to be general agreement among those who had played a role in the case that JonBenét's murderer would never be brought to justice. The issue of whether or not that murderer had been an intruder had obviously not been resolved. For example, on December 26, 2006, former Allegheny County coroner Dr. Cyril Wecht told Fox News: "You are not going to find the killer because there is no killer in the sense of a third-party intruder. Such a person does not exist." Dr. Wecht, the coauthor of *Who Killed JonBenét Ramsey,* a mass-market paperback that came out in July 1998, had apparently not changed his view that John Ramsey had been involved in the death of his daughter. At one time, a majority of Americans would have agreed, but ten years later, with no evidence, physical or otherwise, connecting John Ramsey to the murder, this had become a minority opinion. In ten years not one handwriting expert had identified John Ramsey as the writer of the ransom note. The dwindling few who thought the case might still be solved cited DNA analysis as the only hope.

IN THE RAMSEY CASE, DNA analysis played what has become its frequent role in the administration of U.S. justice. Rather than corroborating the identity of the killer, it revealed that the authorities had arrested the wrong person in John Mark Karr. Moreover, the John Mark Karr chapter in the Ramsey case has provided further evidence that in the United States there are so many unqualified handwriting experts rendering unscientific opinions, the entire field of forensic document examination may lose its credibility.

In the expert witness business, hired guns who don't shoot straight are a problem that can be fixed by judges who have learned how to identify the phonies and keep them off the stand. Because the pretenders have created their own pseudoprofessional organizations and are not above embellishing their résumés, this is not an easy task. Even among legitimate forensic science organizations, weeding out the bad ones has been difficult. The problem with experts, it seems, will not be solved anytime soon.

13

Hair and Fiber Identification

An Inexact Science

Forensic analysts who microscopically compare crime-scene hair follicles with samples from a suspect's head or other part of the body note similarities or differences in hair length, thickness, texture, curl, color, and appearance of the medulla, the strip of cells that runs up the center of the shaft. A hair follicle, however, cannot be individualized like a fingerprint. A hair identification expert can declare, for example, that the defendant's hair looks like a crime-scene follicle, or it is consistent in appearance with the questioned evidence, but they are not supposed to testify that a follicle at the scene of a crime could have come only from the defendant and no one else. In essence, the expert should merely say that the crime-scene follicle and the defendant's hair look the same. What nobody knows about forensic hair identification is this: if two follicles look alike in all respects, what are the chances they have come from the same person? Just how strong an identification is this, and how incriminating?

Hair identification experts also analyze crime-scene strands of fiber and compare them with samples of clothing, carpets, blankets, and other fabrics associated with a defendant. Fibers can be distinguished by material, shape, and color—there are seven thousand dyes used in the United States. A fiber expert can testify, for example, that a fiber on a murder victim's body is consistent in appearance with carpet fibers from the trunk of the defendant's car. To go further than that is crossing the line, scientifically.

Up until the mid-1990s, hair and fiber experts were pushing the envelope by identifying crime-scene follicles and fibers the way one would identify a latent fingerprint. In hundreds if not thousands of cases, defendants went to prison on the strength of this form of expert testimony. When DNA came on the scene, abuses in hair and fiber identification were exposed, and the scientific unreliability of these matches was dramatically revealed. In Texas alone, between 1995 and 2002, DNA analysis exonerated thirty men who had been convicted solely on crime-scene hair identification. Dr. Edward Blake, the Berkeley, California, DNA pioneer, put hair identification in perspective: "You can never say that a hair came from a particular person, but people have been basically promoting that kind of misconception for the past 25 years. They did it because they could get away with it. A defendant in Idaho and another in Florida were sent to death row in cases where the only evidence against them were jailhouse informants and crime-scene hair identifications."

The Michael Rivera Case: DNA Exposes Erroneous Hair Identification

On July 10, 1985, a man crept up behind an eleven-year-old girl at the Green Glades apartment complex in Coral Springs, Florida, choked her into unconsciousness, then fled the scene. The victim was not sexually assaulted but had come within seconds of death. She recovered but was unable to provide a description of her assailant. Without an eyewitness or physical evidence, the police had few leads and no suspects. Within a week, the investigation was dead in the water.

Four months after the Green Glades attack, Coral Springs police arrested twenty-two-year-old Michael Rivera for exposing himself at the apartment complex. Rivera had a history of sexual exhibitionism, obscene phone calls, and window peeping. He had also been arrested for indecently touching young girls and in 1982 had been convicted of indecent assault. Rivera had spent two years in prison and, when released in July 1984, started using crack cocaine. When the police questioned Rivera about the attack at the Green Gables apartments, he denied any connection to the assault and, after promising not to expose himself again, was released.

Six months after the Green Glades assault, eleven-year-old Staci Jazuac was found dead in a vacant lot in Coral Springs. She had been abducted a few days earlier while riding her bicycle in Lauderdale Lakes, a twenty-minute drive from the lot. According to the Broward County medical examiner, she had been strangled but not sexually molested. Crime-scene investigators found six hair follicles on her body that were not hers. Unfortunately, the police failed to submit this evidence to the crime lab.

Eight days after the discovery of Staci Jazuac's body, the Broward County Sheriff's Office received a telephone call from a woman named Star Peck. According to Peck, a man who said his name was Tony had called her to say he was the one who had killed the girl from Lauderdale Lakes. He said he had abducted the girl in his van, rendered her unconscious with an ether-soaked rag, raped her, then tossed her body into Lake Okeechobee. Since Tony's description of the site where the body was found didn't fit the Jazuac case, police didn't think he was the killer. Tony told Star Peck that he had once worked for her, information that led the sheriff's office back to Michael Rivera, who at one time sold pots and pans for Peck.

While Rivera, with his history of petty sexual crimes in and around Coral Springs, was an excellent suspect, there was no solid evidence that he was the man who called himself Tony. Moreover, he didn't own a van. Nevertheless, six days after receiving Star Peck's tip, detectives arrested Rivera. Over the next six days, as interrogators from Broward County subjected Rivera to an intensive grilling, no one bothered to warn him of his *Miranda* rights. Detectives also failed to videotape or audiotape the interrogation-room sessions. Despite their efforts to get him to confess to the killing of Staci Jazuac, the most incriminating thing that came out of Rivera's mouth was that *if* he had killed the girl, he must have blacked out.

Rivera agreed to take a polygraph test, and although he continued to deny he had killed Staci Jazuac, he confessed to creeping up behind and choking into unconsciousness the eleven-year-old girl at the Green Glades apartments. A few months after this confession, Rivera was convicted of attempted murder and sentenced to life.

In 1987, even though Rivera was already in prison for life and there was no credible evidence against him in the death of Staci Jazuac, he was brought to trial for her murder. If convicted, he could be sentenced to death. Not only did the prosecution have a weak case, but also there was evidence that someone else had committed this homicide. In February 1986, when Rivera was already behind bars, another eleven-year-old girl had been found strangled to death and dumped not far from where Staci Jazuac's body had been discovered. Broward sheriff's office detectives made no effort to connect the 1986 murder to the Jazuac case. That case remained unsolved.

At Rivera's trial, his interrogators patched together what at best could be characterized as a series of mildly incriminating remarks by stringing together a series of loose statements the defendant had supposedly made during the six days they had questioned him. Equally weak testimony followed from Rivera's one-time cellmate, Frank Zuccarello. After refusing to confess to the Jazuac murder, Rivera had shared a cell with Zuccarello, one of South Florida's most notorious jailhouse informants. When Rivera went to trial, Zuccarello was facing twenty-three felony charges of burglary, theft, and kidnapping that could put him away for life. He therefore was eager to help several South Florida prosecutors with some of their more difficult murder cases. At Rivera's trial, Zuccarello told the jury that the defendant had confessed to killing the victim in a blue 1971 Ford van he had borrowed from a friend. (The owner of this vehicle had sworn to the police that Rivera didn't have access to it.) Perhaps realizing that witnesses like Zuccarello were not particularly credible, the prosecutor put two more jailhouse snitches in the witness box—a professional burglar and a convicted child molester. Both swore that Rivera had confessed to them as well.

To connect Rivera to the 1971 Ford van, the prosecutor needed more than the testimony of a notorious jailhouse informant. To strengthen this link, he called upon a Broward County hair analyst who identified a hair follicle in the van as possibly coming from Staci Jazuac's head. With that evidence—a phantom confession, three jailhouse informants, and a possible hair identification—the prosecution rested its case. The judge rejected the defense motion for a directed verdict, and the case went to the jury, which found Rivera guilty of murder. Without the forensic hair identification, it's hard to imagine this jury returning this verdict. Rivera was later sentenced to death.

In return for his testimony against Rivera, Frank Zuccarello received a sentence of less than three years for his own crimes. Sometime later, after being released on parole, Zuccarello admitted to falsely accusing an innocent man in a 1984 murder case in Hollywood, Florida.

In 1990 and again in 1997, bodies of two more young girls turned up at dump sites in Coral Springs. Both of these victims had been strangled to death. While the murder modus operandi and location of the bodies fit nicely with the 1985 Jazuac killing and the unsolved case in February 1986, the Broward sheriff's office did not treat the four homicides as the work of a single killer. Having already convicted Rivera of the Jazuac murder, they couldn't.

In March 2003, DNA testing of the hair follicle found in the 1971 Ford van established that it had not come from Staci Jazuac. That meant Michael Rivera had been sent to death row on the weak testimony of county interrogators and three jailhouse informants. The sheriff's office testimony looked weaker than ever when DNA exonerated two other men who had supposedly confessed to these detectives in other cases. But not all the detectives on the Jazuac case were happy with the way Rivera had been treated. Detective Robert Rios had objected to the prolonged interrogation in violation of Rivera's *Miranda* protection and said he was not surprised when DNA analysis invalidated the hair identification. Several jurors who had found Rivera guilty said that without the hair testimony they would have acquitted him. But other jurors disagreed, finding the jailhouse informants credible.

Just months after DNA testing revealed the hair follicle in the van wasn't Jazuac's, a DNA analyst in the Broward County crime lab mixed biological samples in a murder and a rape investigation, contaminating the evidence in both cases.

Following the DNA revelations in the Jazuac case, Michael Rivera declared that he was also innocent of the Green Glades attempted murder and petitioned an appellate court for new trials in both cases. The Broward County prosecutor, claiming that the evidence against Rivera was overwhelming, is fighting both appeals. As of June 2007, these cases are pending.

The Charles Fain Case: DNA Invalidates FBI Hair Testimony

In 1983, a man abducted a nine-year-old girl as she walked to school in Nampa, Idaho. Three days later her body was found along the Snake River a few miles from where she had been abducted. She had been raped and murdered. From the scene, police recovered three hair follicles believed to have come from her killer. During the next seven months, the police obtained hair samples from dozens of men with sex crime backgrounds. The samples were sent to the FBI lab for comparison with the crime-scene evidence. When none of the hair matched, the police kept looking—until they found Charles Fain. A local man who had served with the 101st Airborne in Vietnam, Fain lived down the street from the victim when she was abducted. He had no history of crime but since the war had not been able to hold down a job. When the police got around to him, he willingly gave up samples of his hair. He said he had been visiting his father in Redmond, Oregon, a town about 360 miles from Nampa, when the crime took place.

In Washington, D.C., FBI hair and fiber analyst Michael P. Malone compared the murder-scene hair with the Fain samples and reported similarities between the questioned and known evidence. Malone's report made Fain an instant suspect, and he was subjected to an intense interrogation. After refusing to confess, Fain was given a polygraph test by police. According to the polygraph examiner, Fain was telling the truth when he denied involvement in the crime. The police arrested him anyway and placed him in the Canyon County Jail.

At Fain's trial the following year, Michael Malone testified that the crime-scene hair follicles could have come from the defendant's head. Taking no chances, the prosecutor put a pair of jailhouse informants on the stand who swore that Fain had confessed to them in the Canyon County lockup. The jury took less than an hour to find Fain guilty, and the judge sentenced him to death.

After eighteen years on death row, Fain was released in August 2001 after DNA analysis excluded him as the source of the murder-scene hair follicles. The man who had prosecuted Fain and the judge who had sentenced him were not apologetic. They both insisted that the jailhouse snitches had told the truth. Michael P. Malone, the FBI hair and fiber expert whose testimony helped send Fain to death row, had retired two years earlier.

The Controversial Career of Michael P. Malone

Until the mid-1990s, all the forensic scientists in the FBI crime lab had at least three years' experience in the field as ordinary special agents. Staffing the lab with former criminal investigators was supposed to make them better forensic scientists. However, critics of the policy believed it made them part of a law enforcement team instead of independent forensic scientists. Moreover, by basing the hiring criteria on special agent qualifications, the FBI lab was not attracting or being staffed by first-rate scientists.

Michael P. Malone had earned his bachelor's and master's degrees in biology and taught high school science for two years before he joined the FBI as a special agent in 1970. After four years in the field as a criminal investigator, he was assigned to the bureau's crime lab. During his twenty-five years there as a hair and fiber analyst, Malone testified in five thousand criminal trials. He became extremely popular as a prosecution expert, testifying in dozens of high-profile cases where the fate of the case depended upon his identification of a crime-scene hair or fiber. As a witness he was confident and hard to rattle, and he knew how to impress a jury.

Some years after Malone testified in Charles Fain's trial, his career took an abrupt turn for the worse when one of his FBI colleagues started blowing the whistle on questionable practices and procedures within the nation's largest and most prestigious crime lab. In the early 1990s, Frederic Whitehurst, a bomb-residue analyst who identified chemical components of explosive substances, alerted supervisors to problems in the trace-evidence section of the lab. Whitehurst complained that the laboratory was so dirty that the physical evidence was always in danger of being contaminated. He was especially critical of hair and fiber analyst Michael Malone, whom he accused of allowing his loyalty to police and prosecutors to attenuate his independence and objectivity as a forensic scientist. In memos to the director of the lab, Whitehurst pointed out that hair and fiber identification was an inexact and subjective process, making this form of crime-scene identification highly unreliable. The whistleblower noted that Malone's testimony had sent many defendants to prison, some of whom might have been innocent.

When Whitehurst's internal complaints fell on deaf ears, he began writing long, detailed letters to Michael Bromwich, the Justice Department's inspector general. Between 1991 and 1994, Whitehurst wrote Bromwich 237 letters. In September 1995, the inspector general launched an investigation after ABC's *Prime Time Live,* having gotten hold of some of these letters, aired a story about Whitehurst's campaign to improve the FBI lab. In April 1997, almost six years after Whitehurst began documenting problems in the facility, Bromwich issued a 517-page report critical of the laboratory. As Sudmeu Freedberg would later report under the headline "Good Cop, Bad Cop" in the *St. Petersburg Times,* the inspector general singled out eleven lab employees, including Michael Malone, whom he accused of having provided "false testimony" and recommended for disciplinary action.

Two years later, a second Department of Justice investigation revealed that Malone had made hair and fiber identification errors in four homicide cases in the Tampa Bay area. In the same report detailing these findings, Department of Justice investigators also criticized Whitehurst for overstating the forensic implications of his scientific analysis in some of his own cases. Whitehurst, who had been transferred to the paint identification unit of the lab, was suspended. After the Bureau denied his petition for reinstatement, Whitehurst retired and entered private practice. Later, as Freedberg reported, a U.S. senator called Whitehurst a "true national hero" during a congressional inquiry into FBI activities. To the FBI, however, Whitehurst was a traitor, a whistleblower, the lowest form of bureaucratic life.

Michael Malone denied lying under oath or playing fast and loose with hair and fiber evidence. He blamed the FBI lab scandal on jealous colleagues whom he described as incompetent. Regarding those cases in which DNA analysis had exonerated defendants whose hair he had identified as being at crime scenes, Malone blamed overzealous prosecutors who overstated the implications of his findings. Following the inspector general's investigations and recommendations, Malone was reassigned back to the field. He retired in December 1999. To the FBI, he was a hero.

The Bromgard Case: Delayed Justice

Before DNA, defendants were routinely sent to prison on the strength of hair follicle identification. From the defendant's perspective, it really didn't matter who—the prosecutor or the expert—had hyped the value of this evidence. For example, in 1987, a jury took less than an hour to find Jimmy Ray Bromgard guilty of raping an eight-year-old girl in Billings, Montana. Arnold Melnikoff, a hair and fiber expert with seventeen years' experience, had identified Bromgard as the source of crime-scene head and pubic hairs. As director of the Montana State Crime Lab, Melnikoff was an impressive witness. At the Bromgard trial he had testified that a double hair match like this could happen in only one out of ten thousand instances. This statistic, coming from a respected expert witness, virtually guaranteed the conviction. The judge sentenced Bromgard to three forty-year sentences.

In August 2002, thanks to DNA analysis of the crime-scene hairs and the efforts of Peter Neufeld of the Innocence Project at Yeshiva University's Benjamin N. Cardozo Law School in Manhattan, Bromgard was set free after serving fifteen years in prison. When asked by *New York Times* reporter Adam Liptak how he felt about the Bromgard case, Melnikoff said, "I did the best I could with the technology that was available at the time. I'm really quite disturbed that this person is really innocent."

Following the Bromgard exoneration, the attorney general of Montana ordered an audit of Melnikoff's work in 250 cases. No longer employed in the Montana state lab, Melnikoff was head of the Washington State Patrol Crime Lab in Spokane, where he did drug analysis. He was no longer identifying hairs and fibers because, in 1989, he had failed the Washington State hair and fiber proficiency test. His employers in Seattle placed him on administrative leave pending the results of the Montana audit and started an inquiry into the quality of his work in that state.

In March 2004, Melnikoff's employer fired him from the Washington State crime lab. Auditors had found he had failed to follow proper procedures in thirty of a hundred felony drug-testing cases. Characterizing his work as "sloppy," the auditor determined that in fourteen instances, Melnikoff had identified substances as methamphetamine on insufficient data, *Seattle Post-Intelligencer* reporter Ruth Teichroeb reported on March 24. There is no way to know how many innocent defendants were sent to prison on the basis of this expert's work, particularly in the field of hair and fiber identification.

The Rocky Career of Charles A. Linch

The emergence of forensic DNA analysis in the late 1990s created problems for hair identification experts like Michael Malone and Charles Linch. With a bachelor's degree in science from the University of Houston, Charles A. Linch went to work at the Southwestern Institute of Forensic Sciences in Dallas in 1990, where he served a one-year apprenticeship and attended FBI classes on hair and fiber identification. He would later receive FBI training in forensic serology and DNA analysis and, in the course of his career, would appear in court as an expert in the fields of blood-spatter interpretation, DNA analysis, and hair and fiber identification. One of these appearances came in 1997, when he testified for the prosecution in the Darlie Lynn Routier case.

The Routier Case: The Depressed Expert

Darlie Lynn Routier was charged in 1996 with murdering her two sons, ages five and six. She was accused of stabbing the boys as they slept in her three-story mansion in Rowlett, near Gatesville, Texas. Darlie Routier had been stabbed as well, wounds the police believed had been self-inflicted. According to her account, an intruder had broken into the house through a screen window and attacked the family. After the stabbings, the intruder fled the house through the garage. There were signs of a struggle, but according to the police, the scene had been staged.

At Routier's 1997 trial, Charles Linch, called as a blood-spatter interpretation expert, testified that the crime-scene blood evidence did not support the intruder theory. Had the killer fled through the garage as the defendant claimed, he would have left a trail of blood. Linch didn't find such evidence. He did find the blood spatter on the defendant's nightshirt consistent with a "cast-off" pattern he believed made when the defendant stabbed her children. In the role of trace-evidence analyst, Linch identified residue on a knife from the defendant's kitchen as similar to tiny metal fibers from the broken window screen that was the alleged point of entry. Although Linch identified a hair follicle attached to this screen as the defendant's, it was later determined to have come from a police officer. The Tarrant County jury, relying heavily on Linch's expert analysis of this evidence, found Darlie Routier guilty of double murder.

In 2003, Routier filed from prison a habeas corpus petition for a new trial, along with a motion for DNA testing of crime-scene evidence she claimed supported her account of a home invasion. In connection with these motions, her legal team attached affidavits from forensic experts who took issue with Linch's crime-scene analysis. Regarding the defendant's bloodstained nightshirt, an expert for the defense challenged the cast-off interpretation, noting that Linch had not conducted reenactment tests to determine the bloodstains' source. He had also failed to eliminate the defendant as the donor of this blood. If the nightshirt blood was the defendant's, Linch's cast-off theory didn't work. A second defense expert believed that the residue on the knife, identified by Linch as coming from the broken window screen, was in fact traces of fingerprint powder.

In support of a motion for a new trial, the Routier defense accused the prosecution of withholding relevant information about Charles Linch that would have cast doubt on the validity of his findings and testimony. In 1994, two years before the Routier case, Linch had checked himself into a psychiatric hospital in Dallas, suffering from clinical depression exacerbated by alcoholism. Not only was he was under medical care when he testified at the Routier trial, but he had twice given testimony in other cases while still a patient in the psychiatric unit. The defense also claimed that when Linch had filed a job grievance against Southwestern Institute of Forensic Sciences, he had questioned his own qualifications as a blood-spatter analyst and hair identification expert.

The defense had learned of Linch's mental and emotional problems, as well as his doubts regarding his own competence, from two *Dallas Morning News* articles published on May 7 and 10, 2000. Linch had responded to the pieces by admitting his clinical depression. He insisted, however, that his psychiatric problems had not affected his trustworthiness as an expert witness, describing his hospitalization as a mere career interruption.

Routier's petition for a new trial and DNA testing was denied in 2004. On January 25, 2005, at a hearing held in Dallas, her attorneys asked for additional DNA testing of the crime-scene evidence, which they believe would exonerate her. As of June 2007, no such analysis has been done.

The Michael Blair Case: Multiple Hair Misidentifications

Meanwhile, there had been developments in a Plano, Texas, case Linch had testified in back in 1994, the year he had spent time in a Dallas psychiatric hospital. Police had suspected Michael Blair, a convicted child molester, of murdering seven-year-old Ashley Estell, found dead along a highway six miles from where she had been abducted from a soccer field. A witness placed Blair near the soccer field that day in 1993, and the police had found the victim's clothing near a carpet-cleaning establishment where Blair was employed. Blair denied committing the crime and said he had been nowhere near the soccer field on the day in question. At the place two and a half miles from the soccer field where investigators believed the victim had been raped and murdered, the police found a clump of hair believed to have come from the victim and her killer. This evidence, along with the victim's underwear, a fiber found on her body, hair follicles recovered from Blair's car, a stuffed rabbit taken from the suspect's vehicle, and hair samples from the suspect, was sent to Charles Linch in Dallas.

At Blair's trial, Linch testified that the hair clump consisted of follicles consistent with hairs from the victim's head and the defendant's, placing them together at the possible site of the crime. Linch identified a follicle recovered from the waistband of the victim's underwear as Blair's, and three follicles from Blair's car as the victim's. And finally, to remove any doubt that the victim and Blair had been together, Linch testified that a fiber on the girl's body was similar to fibers from the stuffed rabbit found in Blair's car. In his testimony, Linch had been careful to qualify his identifications by using expressions such as "consistent with" and "similar to," but the prosecutor, in his closing remarks to the jury, used the terms "link" and "match" instead, as Mark Wrolstad reported in the *Dallas Morning News*. The jury, finding the hair and fiber evidence compelling, particularly in the light of defendant's background, found Blair guilty. He was sentenced to death.

Seven years later, in 2001, DNA analysis of the hair evidence in the Blair case revealed that the follicle on the victim's underwear had not come from Blair and the hairs in Blair's car hadn't been the victim's. A year later, another DNA test showed that the hair clump found two and a half miles from the soccer field did not originate from Blair or the murder victim. When asked to comment on three misidentifications in one case, Wrolstad reported, Linch said, "Hair should never be the only thing" and blamed the prosecutor for overselling the results of his analysis to the jury.

In 2006, DNA tests on traces of blood and tissue from under the victim's fingernails proved not to have come from Blair. His attorneys, still attempting to get him a new trial, believed this evidence would match the DNA of a convicted child molester who had been one of the soccer referees the day Ashley Estell was murdered. While the police had questioned this man, they had overlooked him as a suspect. Further DNA testing could lead to the identity of the actual murderer. And a new trial and acquittal in this case could get Blair off death row. John Roach, the Collin County district attorney, fought Blair's quest for a new trial, insisting that he was guilty notwithstanding the DNA exonerations. The prosecutor wasn't fighting to keep Blair behind bars—Blair was never getting out of prison, new trial or not. He was serving three consecutive life sentences for abusing children.

In his effort to win a new trial, Roy Greenwood, Blair's head attorney, had petitioned a judge for the right to depose Charles Linch, the expert who had tied Blair to the murder through the physical evidence. When contacted in October 2006 by a reporter with the *Dallas Morning News,* Linch said he wasn't aware that Blair's attorneys wanted to talk to him about his work on the case. In an article headlined "Michael Blair Sits on Death Row," Linch denied that his personal problems had affected his scientific analysis of the evidence, claiming that his hospitalization, an overreaction to a drinking problem and mild depression, "turned into an expensive place to have a hangover."

Although the district attorney promised to retry Blair in the event he won his appeal, Blair's attorneys didn't believe this would happen. Without the hair identification, there was no case against their client.

WITH ONE EXCEPTION, U.S. courts have held that hair and fiber identification meets the standards set out in the *Daubert* case. In 1995, however, an expert's hair and fiber identification was challenged under *Daubert* in a federal habeas corpus case filed in the Eastern District of Oklahoma. In the state criminal trial, the prosecution hair expert had found the defendant's hair "microscopically consistent" with the crime-scene follicle and had told the jury that "hairs are not absolute identification, but they either come from this individual or there is, or could be, another individual somewhere in the world that would have the same characteristics to their hair." This expert had testified that he had found twenty-five points of similarity between the known and questioned evidence. The U.S. district court judge held that because the expert had not explained these areas of similarity and had not said how many people were likely to have similar-looking hair, his testimony did not rise to *Daubert* standards. The decision in this case, *Williamson v. Reynolds*, was reversed by the Tenth Circuit Court of Appeals. A few years later, DNA testing of the crime-scene hair exonerated the man convicted of this crime.

In this case, like so many others, the prosecutor oversold the identification value of hair evidence by telling the jury that it had placed the defendant at the murder scene. But until hair identification experts, even careful and cautious ones, can tell a jury just how rare or common a hair match-up is—with scientific proof to back them up—it might be a good idea to exclude this kind of testimony altogether.

14

DNA Analysis

Backlogs, Sloppy Work, and Unqualified People

For police and prosecutors, DNA science has been a double-edged sword. Since the mid 1990s, thousands of rapists and killers have been identified by DNA and sent to prison. But DNA technology has also revealed flaws in other forensic sciences such as bite-mark and hair-follicle identification. It has also exposed weaknesses and corruption in the way crimes are investigated. As of 2007, 190 prisoners nationwide have been exonerated by DNA testing and released from custody. In Florida alone, DNA has freed twenty-five men who were awaiting execution. Wrongful convictions involve a combination of factors such as tunnel vision, poor interrogation tactics, bad forensic science, prosecutorial overzealousness, mistaken eyewitnesses, and the ever-present jailhouse snitch. Extrapolating from the number of prisoners set free by DNA evidence, it is not unreasonable to assume that thousands of wrongfully convicted inmates whose cases do not lend themselves to DNA exoneration remain behind bars.

In an effort to improve the crime-fighting potential of DNA profiling, the FBI in 1990 initiated a pilot project called Combined DNA Index System (CODIS). The program would link data banks across the country housing computerized collections of DNA profiles of arrested felons. Investigators would be able to submit an unknown DNA profile for identification by activating one computer instead of running the evidence through dozens of statewide systems. An evidence submission that matches a DNA profile in one of the databases is called a hit. When such a computer match is made, it's tantamount to solving the crime *and* proving who committed it. CODIS promised a crime-fighting potential equal to the FBI's Integrated Automatic

Fingerprint Identification System. Even better, the criminals caught by CODIS would be the worst of the worst—rapists, child molesters, and sexually motivated killers—serial offenders all.

In 1998, with one million DNA profiles on file, CODIS evolved from a pilot project to a permanent law enforcement program. By 2006, the data bank held the DNA profiles of 3.6 million arrestees. Since its inception, the system has produced 39,241 hits, a source of pride for the FBI and for the forensic science community in general. But CODIS might not be as effective as advertised.

On November 21, 2006, *USA Today* published the results of an extensive journalistic investigation that revealed the disturbing and surprising fact that an unknown percentage of CODIS hits had not resulted in arrests. For a variety of reasons, the police had not pursued these hot investigative leads, reported Richard Willing in "Many DNA Matches Aren't Acted On." *USA Today* researchers had uncovered, in several jurisdictions, numerous cases in which identified rapists had committed sexual assaults that would have been prevented had police responded to the DNA hit in a timely fashion. Because there is no governmental agency responsible for tracking the disposition of CODIS match-ups, no one knows how many DNA computer identifications die on the vine. Based upon the anecdotal evidence uncovered by *USA Today* investigators, one could reasonably assume that this percentage is not low.

Beyond police ineptitude, laziness, and investigative understaffing, the reasons why these important leads were ignored are a mystery. It is also difficult to understand why the results of the CODIS program are not monitored for quality control. One aspect of the *USA Today* study is easy to comprehend: the frustration of victims in cases where science holds the solution and the police, for whatever reason, have done nothing to bring the DNA-identified offenders to justice.

In Virginia, as in several states, law enforcement and corrections personnel are required by law to take DNA cheek swabs from every person arrested for rape, aggravated assault, and criminal homicide for inclusion in a statewide database incorporated into CODIS. The law also requires that DNA samples be taken from every convicted felon. By 2006, there were 253,000 profiles in Virginia's DNA repository, but a review conducted in 2005 by the Virginia crime laboratory revealed that DNA samples had not been submitted to the database from 3,149 of the state's 13,000 registered sex offenders. Because samples were not taken or processed, DNA from a quarter of the state's convicted sex felons was not on file. A year later, an audit of

12,000 general felony cases revealed that 20 percent of these offenders' DNA was not in the Virginia data bank. The results of these reviews prompted an audit of the state's 54,000 felons who were either on parole or probation. Public safety administrators wanted to know how many of these offenders had slipped through Virginia's DNA net.

Given the enormous promise of DNA science to fight crime and prevent injustice, it is also disappointing that within the field itself, there have been serious and persistent problems. DNA analysts have been caught lying about their job qualifications, cheating on proficiency tests, contaminating evidence, and not following scientific protocol. FBI analyst Jacqueline Blake failed to conduct control tests in a hundred cases, work that called into question the credibility of the DNA analysis done in the FBI crime laboratory. Over the years, dozens of DNA sections in crime labs all over the country have been closed down as a result of misidentifications, poor quality control, questionable hiring and training procedures, and weak job supervision.

Karla Carmichael: A Question of Competence

In October 2002, the Fort Worth Police Department suspended DNA testing by its crime lab after the findings by an analyst with the Tarrant County Medical Examiner's Office conflicted with the results of the crime lab's DNA expert, Karla Carmichael. In April 2003, after the DNA unit was back in operation, Carmichael failed a DNA proficiency test, which prompted DNA experts at the University of North Texas Health Sciences Center to conduct a review of her work. Following the review, the outside experts concluded that Carmichael was not qualified to perform DNA examinations.

Carmichael, with the title senior forensic scientist, had worked at the lab three years. She had been promoted to her position simply because she had worked there the longest. She was fired.

Cutting Corners: Indianapolis–Marion County

Dissention among DNA analysts that creates scandals in government crime labs erodes public confidence in forensic science. In September 2000, Kuppareddi Balamurugan, one of four DNA analysts at the Indianapolis–Marion County Forensic Service Agency, received a reprimand for mishandling DNA evidence in a rape case in which the charges eventually had to be dropped. A few months later, Mohammad Tahir, one of Balamurugan's colleagues, told the director of the lab that Balamurugan had been cutting corners in his work, for example, by failing to analyze control samples to insure the reliability of his findings. Instead of investigating Tahir's allegations, the director of the laboratory accused the whistleblower of vindictiveness motivated by his recent demotion. It was no secret, however, that there were problems in the crime lab. Prosecutors in Marion County had been frustrated by how long it took to get DNA work done and were not pleased with the quality of the DNA results. These prosecutors were constantly being challenged in court by defense attorneys who were embarrassing the crime lab's DNA witnesses on cross-examination.

Two years after being accused by the whistleblower of taking shortcuts, Balamurugan resigned—after six years with the lab—when an analyst claimed he had tampered with a DNA test. In 2003, as the Indiana State Police were auditing the Indianapolis–Marion County Forensic Services Agency, the Marion County prosecutor ordered the retesting of sixty-four of Balamurugan's cases. The prosecutor's office then acquired a December 2003 sworn statement by the whistleblower, Mohammad Tahir, that detailed problems within the crime lab, including what he believed was an ongoing cover-up.

Two weeks later, the mayor of Indianapolis placed the director of the lab on administrative leave and appointed a special prosecutor to investigate the possibility of criminal wrongdoing. While the yearlong probe did not result in criminal charges, it did expose a serious lack of quality control in the Indianapolis–Marion County crime lab. As it turned out, Balamurugan's DNA work held up under review, but the scandal triggered a flurry of defense motions for new trials.

Cutting corners has been a reoccurring problem in DNA analysis. Lab personnel take shortcuts because they are overwhelmed with evidence, and this creates backlogs and delays that frustrate victims, police, and prosecutors. According to a study conducted in 2003, at any given time, there were 550,000 DNA samples waiting to be analyzed. According to the U.S. Bureau of Statistics, in 2005, crime labs were understaffed by 1, 900 employees. The greatest shortages involved DNA analysts. DNA backlogs were so bad in Massachusetts, where it could take up to a year to get results back, district attorneys were not allowed to submit more than four samples per case. In Tennessee, detectives submitting DNA evidence have to wait six months to get results in rape cases, and eight months in homicide investigations.

In 2005, the Tennessee Bureau of Investigation spent $500,000 in outsourcing DNA and toxicology tests in a failed attempt to eliminate the backlog. A few years earlier the state had closed two of its five crime labs as part of a budget-cutting plan. In 2006, the state hired seventeen additional DNA analysts and began building a new crime lab in Knoxville.

An analyst for Cellmark, the nation's largest private DNA laboratory, revealed in testimony at a 2003 trial in Maryland that a colleague in the Germantown, Maryland, lab had falsified the results of twenty control tests in cases being handled for the FBI and the Los Angeles Police Department. At the time, the Los Angeles Police Department had a three-year, $2.7 million contract with Cellmark to perform DNA analyses until the police lab could train more analysts. The Cellmark employee in Germantown, fired for "professional misconduct," denied any wrongdoing. Although there was no indication that procedural shortcuts had resulted in DNA misidentifications, the Los Angeles County Public Defender's Office announced that it would be reviewing all its cases involving Cellmark analysts.

In Texas a 2003 audit of crime labs run by the Department of Public Safety (DPS) revealed shoddy DNA work in DPS labs in Houston, Austin, El Paso, Garland, Lubbock, Corpus Christi, McAllen, and Waco that might have sent innocent people to prison and threw thousands of criminal cases into doubt. Crime lab auditors had found DNA analysts who didn't know how to interpret test results, who didn't test blank samples to ensure against contamination, and who had failed to include important statistical probabilities in their lab reports.

A year before the audit of the DPS labs, the Houston Police Department Crime Lab had been forced to close its DNA operation and Houston auditors ordered DNA retesting in four hundred cases. As of April 2005, the reexamination of this evidence has led to the freeing of two men who had been convicted of rape.

Because it's so fragile, the improper handling of DNA samples in a crime lab can easily lead to contamination, which in turn invalidates the test results. Lab technicians in Broward County, Florida, and in Tacoma, Seattle, and Maryville, Washington, accidentally contaminated DNA evidence by commingling samples from different cases. If DNA facilities are not spotless, secure, and properly designed, contamination is likely to occur. Understaffing, poor training, and lack of supervision also contribute to the problem.

Fred Salem Zain: The Forensic Serologist from Hell

Overworked, poorly trained, and undersupervised DNA technicians who take shortcuts and occasionally make honest mistakes are a problem, but analysts who are knowingly incompetent, regularly negligent, and dishonest can create criminal justice havoc. Fred Salem Zain was one of these analysts. Between 1977 and 2002, Zain testified as an expert witness in the fields of serology, hair-follicle identification, and DNA profiling in more than a thousand trials. He did a lot of harm.

A West Virginia state trooper, Zain began working in the state police crime lab in 1977 as a forensic serologist, a position for which he was not qualified. He had a degree in biology from West Virginia College, where he had been a C student and flunked organic chemistry. He would also fail an FBI course in general forensic science. In court under oath, Zain regularly lied about his professional qualifications, telling jurors that he had majored in biology and chemistry.

Besides being unqualified and incompetent, Zain was outlandishly dishonest in his eagerness to please prosecutors with testimony tailored to their demands. As his fame grew as a prosecutor's expert, he began testifying in courts outside West Virginia, eventually rendering opinions in nine states. Zain made himself particularly useful in rape cases where the victim could not provide a good description of her attacker, and the defendant had not confessed, had a solid alibi, and was not connected to the crime through fingerprints, shoe impressions, bite marks, or tire tracks. All Zain needed was a spot of blood, a trace of semen, or a hair follicle, which he would without fail identify as coming from the defendant. His testimony was based not upon science or expertise, but on assurances from police and prosecutors that they were trying the guilty man. Zain was nothing more than a prosecution prop, as bogus as planted evidence. Over the years, as serious questions were raised regarding the authenticity of his work, prosecutors turned a blind eye. Fred Zain was simply too valuable a tool for them in getting convictions.

In 1987, Zain's testimony in a double rape case put Dale Woodall, a twenty-nine-year-old from Charleston, West Virginia, in prison for life. In that case, a man wearing a brown and yellow ski mask had grabbed two women outside a Huntington, West Virginia, shopping mall and raped them in the car of one of the victims. The women, unable to provide detailed descriptions of the rapist, were forensically hypnotized to enhance their memories, a tactic that would today render their testimony inadmissible in most courts. Dale Woodall was put on trial, and Fred Zain identified one of his pubic hairs as being in the rape car and a semen stain as matching his blood type. He testified that one in ten thousand people had Woodall's grouped and subgrouped blood type, a statistic he had pulled out of his hat. The jury found Woodall guilty.

Five years later, after Zain quit the West Virginia crime lab and moved to San Antonio to take a position with the Bexar County Medical Examiner's Office, Dr. Edward Blake, a DNA expert from Berkeley, California, reexamined the Woodall evidence and concluded that Zain had incriminated an innocent man. DNA lawyers Barry Scheck and Peter Neufeld with the Innocence Project had become involved on Woodall's behalf. The case also prompted the American Association of Crime Lab Directors to form a panel of experts to review a sampling of Zain's work. Of thirty-six cases selected at random, the panel found that in all thirty-six Zain had fabricated his data and had committed perjury. In 1992, Woodall was set free when a West Virginia court set aside his rape conviction. To head off a lawsuit that would have exposed the state's complicity in allowing this charlatan to function as an expert witness for a decade, the West Virginia Insurance Board agreed to pay Woodall $1 million, the maximum settlement allowed. This payment offended the speaker of the West Virginia House of Representatives, who didn't believe in making millionaires out of people who just happened to have been imprisoned for crimes they didn't commit.

The Woodall case caused the West Virginia Supreme Court to create its own panel of experts to review all of Zain's West Virginia cases. In 1993, this panel reported that 134 people had been sent to prison on Zain's testimony, much of which was highly questionable. According to this report, quoted by journalist Sau Chan in the *Los Angeles Times,* Zain had routinely "overstated the strength of results, . . . reported inconclusive findings as conclusive, repeatedly altered laboratory records, . . . and reported scientifically impossible or improbable results." In 1994, as Chan reported, the West Virginia Supreme Court designated Fred Zain's expert testimony "invalid, unreliable, and inadmissible."

In 1994, the Bexar County Medical Examiner's Office fired Fred Zain after a Uvalde County court overturned a 1990 rape conviction that had been based on Zain's testimony. Zain had been on the job six years. In the 1990 case, he had testified that semen stains on the victim's clothing could only have been deposited by Gilbert Alejandro, when in fact his findings had been inconclusive. Zain knew that his analysis failed to incriminate the suspect, but he went ahead with his incriminating testimony. A few months after

Alejandro's conviction, two DNA experts excluded him as the source of the semen. Released after serving four years in prison, Alejandro received $250,000 in compensation from the taxpayers of Bexar County.

In August 1994, sheriff's deputies in Bexar County arrested Zain in Hondo, Texas, on three counts of aggravated perjury, one count of document tampering, and one count of evidence fabrication. If convicted, Zain could have been sent to prison for ten years, but the jury found him not guilty on all counts. Perhaps they felt uneasy about sending a crime fighter to prison for what might have been honest mistakes that stemmed from his eagerness to put rapists in jail. Maybe they believed that the defendants his testimony had sent to prison were guilty.

The next year, in March, Zain went on trial in Fairmont, West Virginia, again charged with three counts of perjury. Again, the jury was sympathetic, acquitting him on two of the charges and deadlocked on the third. Prosecutors wanted to retry Zain on the third count, but a judge threw the case out for lack of evidence. Two years later, Zain was again charged with perjury in Texas, but a judge dismissed that case on procedural grounds.

These repeated charges of perjury and evidence tampering didn't stop Zain from presenting himself as a forensic expert and testifying in court. By 1996, he was in South Florida working in the private sector as an expert for hire. Back in West Virginia that year, a grand jury indicted him for defrauding the state by earning a salary on the pretense that he was a forensic scientist. In 1998, a judge ruled that a governmental entity cannot be the victim of fraud, rendering the indictment invalid. A higher court overruled that decision, however, in 2001. Zain was indicted again, but before the case came to trial, he died at the age of fifty-two of colon cancer.

Like Dr. Louise Robbins, the phony shoe-print expert, Fred Zain died before having to face the full consequences of his actions. His death, however, did not erase the problems he had caused. Efforts to help his victims continued, and concerns regarding his legacy were still being studied and debated in 2006. That year, Kanawha County public defender George Castelle petitioned the West Virginia Supreme Court to declare a legal presumption that the work performed by technicians at the West Virginia State Police Crime Lab who had been trained by Zain was unreliable and invalid. Castelle also recommended that the state's crime lab be removed from state police supervision and operated by an independent agency overseen by a supervisory board of scientists, judges, and lawyers.

In support of Castelle's petition, Barry Scheck and Colin Starger of the Innocence Project filed an amicus curiae brief. Included was a report by an independent forensic biologist from Maryland, Mark D. Stolotow, and Ronald L. Linhart, the former head of the Los Angeles County Sheriff's Office Crime Lab Serology Division, which showed a pattern of errors in the West Virginia lab that "represent[ed] a divergence from good science and on occasion ethical conduct," the staff of the *Charleston (W. Va.) Gazette* reported on May 10, 2006. The Stolotow-Linhart report noted further that the Zain-trained serologists "willfully violated standard protocols" and that the crime lab reported "scientifically impossible or improbable results" and had engaged in the repeated altering of laboratory notes. Stolotow and Linhart concluded there was "a significant danger that probative errors occurred that would have changed the results of the reviewed prosecutions."

In April 2006, three out of the five state supreme court justices decided not to accept the friend-of-the-court brief as part of Castelle's petition. On May 10, the West Virginia Supreme Court, sans the Innocent Project brief, began hearing arguments regarding how to respond to the lingering fallout from the Fred Zain era. On June 16, 2006, the court found insufficient evidence of intentional misconduct by assistant lab personnel to warrant an invalidation of their serology evidence and a systematic review of each case in which this evidence was offered.

Joyce Gilchrist: The Prosecutor's Best Weapon

The damage a single discredited forensic scientist can do to the criminal justice system is enormous. Such is the case of Joyce Gilchrist. Gilchrist was working in the Oklahoma City Police Department's lab analyzing bodily fluids and identifying crime-scene hair follicles even before she graduated from the University of Central Oklahoma in 1980 with a degree in forensic science. She had made a name for herself as an expert witness whom prosecutors could rely on to bolster their cases with testimony that physically linked defendants to scenes of crimes. But by 1987, questions were being raised regarding the scientific reliability of her work, and the challenge came from within the forensic science community in the form of a letter to the Southwestern Association of Forensic Scientists. Quoted by *Washington Post* journalist Lois Romano, the letter was from a Kansas City, Missouri, police chemist, who complained that Gilchrist was rendering "scientific opinions from the witness stand which in effect positively identified the defendant based on the slightest bit of circumstantial evidence." A year later, this letter writer's concerns were taken more seriously when, as Romano reported, the Oklahoma Criminal Court of Appeals overturned a murder conviction on the grounds that Gilchrist had given "personal opinions beyond the scope of scientific capabilities."

The action of the appellate court in Oklahoma did not derail Joyce Gilchrist's forensic science career. Like Fred Zain, she had made herself an extremely valuable tool for law enforcement, too important to be decommissioned on mere allegations of incompetence and dishonesty. However, by 2001, it had become difficult, even for her strongest supporters, to ignore the obvious. Persistent questions regarding the accuracy of her work and the credibility of her testimony led to a review by FBI forensic scientists of eight of her cases. According to the FBI, in five of the eight cases, Gilchrist had either erred or overstated her scientific findings in favor of the prosecution. The crime lab had no choice but to suspend her with pay pending further investigation. After she had spent more than two decades as a forensic scientist, her supervisors had finally gotten around to checking her work. At this point, however, to find her work unprofessional and unreliable would reflect badly on the people commissioning the investigation. Perhaps for this reason, the matter was swept under the rug. But in July 2001, a crime lab whistleblower pulled up the rug.

In a memo obtained by the local press, the whistleblower recalled a 1982 murder case that led to the execution of Malcolm Rent Johnson. At his trial, Gilchrist had identified the defendant's blood type in the crime-scene semen. According to the whistleblower, there had been no spermatozoa to analyze. This evidence had therefore been fabricated out of whole cloth. In responding to the memo, police and prosecutors who had been involved in the case pointed out that Johnson had been convicted on a lot more evidence than Gilchrist's testimony. These statements, however, did little to blunt the criticism of the Oklahoma police crime lab and didn't change the fact that if the allegations were true, Gilchrist had committed perjury in a capital case. Although she denied any wrongdoing, Gilchrist was fired. Since she had testified in twenty-three cases in which defendants had been sentenced to death, and eleven of these had already been executed, the accusations against her were extremely serious. Following her termination, some men sent to prison on her testimony were released after being exonerated by

DNA. Nevertheless, Gilchrist continued to stand behind her work in all her cases.

In 2004, the Associated Press acquired a confidential Oklahoma City Police memo written by then deputy chief William Citty, dated September 21, 2001; in it, Citty detailed how Gilchrist might have intentionally lost or destroyed the crime-scene hairs used to convict Curtis Edward McCarty of murdering Pamela Willis so that they could not be DNA retested. The evidence was therefore not available to possibly exonerate McCarty, who was on death row. The memo also accused Gilchrist of altering her notes in the McCarty case.

On May 10, 2007, after an independent crime lab in New Orleans found that the major DNA component on fingernail clippings taken from Pamela Willis did not match Curtis McCarty, Judge Twyla Mason released the death-row inmate from prison. "I'm very troubled with the case," the judge said, according to Scott Cooper in the *Oklahoma Gazette*. "Twice this case has been reversed because of state misconduct. . . . Joyce Gilchrist said she found a coup de grace (incriminating DNA evidence against McCarty) but what she really found was a way to undermine everything I believe in. I want to know, where is Joyce Gilchrist and why isn't she in prison?"

Dr. Pamela Fish: The Defendant's Nightmare

About the time questions were being raised in Oklahoma about Joyce Gilchrist's work, prosecutors in Cook County charged John Willis with several counts of rape in connection with a series of sexual assaults committed in the late 1980s on Chicago's South Side. Willis, a petty thief, and illiterate, denied raping the women even though several of the victims had picked him out of a police lineup. The only physical evidence in the case was a piece of toilet paper containing traces of semen. Police took this evidence to the Chicago police lab where it was examined by Dr. Pamela Fish. Dr. Fish had come to the lab in 1979 with bachelor's and master's degrees in biology from Loyola University. Ten years later, after taking courses at night, she earned a Ph.D. in biology from Illinois Institute of Technology. According to her handwritten lab notes, Dr. Fish determined that the secretor of the semen had type A blood. John Willis had type B blood. Dr. Fish had reported, in contradiction of her lab notes, that the semen donor possessed type B blood and testified to this fact at Willis's trial in 1991. The jury also believed the eleven eyewitnesses and found Willis guilty. The judge sentenced him to one hundred years, and he was packed off to Statesville Penitentiary.

Eight years later, a south Chicago rapist confessed to these sexual assaults after being linked to the crimes through DNA. A judge set aside the Willis conviction and he was freed. On the day of his release, Dr. Fish, now head of biochemistry testing at the state crime lab, spoke at a DNA seminar for judges. (The Chicago police lab had been incorporated into the Illinois crime lab system in 1996.)

The Willis reversal led to a 2001 review by DNA expert Dr. Edward Blake of nine cases in which Dr. Fish had testified that her blood-typing tests had produced inconclusive results. In all nine cases, as Margaret Fisk reported in the *National Law Journal,* Dr. Blake found that Dr. Fish's tests had actually exonerated the defendants and that she had given false testimony. Dr. Blake characterized what she had done as "scientific fraud." In the summer of 2001, a state representative in a legislative hearing on prosecutorial misconduct suggested to the head of the Illinois State Police that Dr. Fish be transferred out of the crime lab job into a position where she could do less harm. The suggestion was ignored.

In 2002, three more Illinois men, in prison for rape since 1987, were exonerated by DNA. Dr. Fish had testified for the prosecution in all three cases. Two years later, after the state paid John Willis a large settlement for his wrongful prosecution and incarceration, the state refused to renew Dr. Fish's employment contract. Rather than firing her, the state simply refused to rehire.

RECURRENT PROBLEMS in the nation's crime labs, particularly in the DNA field, have prompted the establishment of watchdog commissions. In 2004, the Indiana State Police Crime Lab at Fort Wayne began doing DNA analysis work. The operation was shut down a year later when auditors found mold on five out of twenty refrigerated evidence packages. The discovery placed four rape cases and one homicide investigation in jeopardy. Dr. Edward Blake, who wondered about the competence of the people in charge of Fort Wayne's DNA operation, noted that the first rule of collecting and storing biological evidence such as blood, semen, or saliva is to dry it. Mold can eat away and destroy this kind of evanescent evidence.

Former U.S. Department of Justice inspector general Michael Bromwich, in a January 2006 report on the Houston Police Department Crime Lab, found problems in 40 percent of its DNA cases between 1987 and 2002. The lab had been closed in 2002 after an outside auditor discovered serious problems throughout the facility.

At a three-day conference held in April 2006 in Greensboro, North Carolina, the American Judicature Society, a research group made up of judges, lawyers, and academics, recommended that DNA watchdog commissions be established in every state. Such bodies were already in operation in Virginia, Texas, and New York. A law had been passed in Oklahoma requiring state and private crime labs to be certified by a professional group of forensic scientists, and a similar law was being considered in Illinois. These are small steps in the right direction, but complete DNA upgrading and reform may be a long way off. Not much will be accomplished until more firewalls are built between the practitioners of law enforcement and the practitioners of forensic science.

15

Bullet Identification, FBI Style

Overselling the Science

In the early 1980s, forensic chemists in the FBI crime laboratory began matching crime-scene bullet fragments to unfired ammunition through a trace identification process called comparative bullet lead analysis (CBLA). The process first used a trace detection technique, neutron activation analysis, and later used optical emission spectroscopy. Normally, suspects are linked to shootings through the guns that fired the crime-scene bullets. Standard firearms identification involves the microscopic comparison of rifling (barrel) marks on the crime-scene bullet with the striations (scratches) on a bullet that has passed through the barrel of the suspect's gun. If the marks on the crime-scene bullet line up with those on a bullet fired through the suspect's gun in a test, the weapon used in the crime has been identified. But when the crime-scene bullet is badly mangled or fragmented, there are no bullet striations to compare with a test-fired counterpart. That's when CBLA comes into the picture.

Instead of identifying the suspect's *gun* as the murder weapon, the CBLA expert links the crime-scene bullet or bullets to ammunition in the suspect's possession. This is done by determining that bullets in the suspect's possession had been manufactured on the same day, and from the same batch of lead, as the crime-scene fragment. The forensic identification process is based on two scientific assumptions: (1) that every batch of lead is perfectly consistent in its chemical composition (CBLA experts use the term "homogenous" in this context), and (2) that every manufacturer's batch of lead has a unique chemical composition, different from all other batches.

Since the early 1980s, forensic chemists from the FBI lab, the only crime laboratory in the world that provided CBLA, have analyzed bullet lead in

2,500 cases. FBI CBLA experts have testified for the prosecution in more than five hundred trials. No court has ever ruled CBLA evidence inadmissible on the grounds that the process is either not supported by science or overreaching in the conclusions drawn from that science.

Bullet manufacturers use lead from recycled car batteries to which they add various amounts of antimony to harden the alloy. Lead comes to the factory in ingots (sixty-five- to eighty-pound lots), billets (100- to 300-pound lots), and sows (2,000-pound lots). The lead is remelted into seven- to ten-ton pots along with rejected bullets and other lead waste. The molten material is poured into molds, then formed into bullets. Once attached to cartridge cases, they are packed into boxes, which are stamped with a packing code. A large batch of lead can, for example, produce millions of .22-caliber bullets. In a small town, almost every bullet sold could come from the same manufacturing batch.

The Earhart Case: A Bullet Match that Led to Lethal Injection

On May 27, 1987, deputies with the Walker County, Texas, Sheriff's Office found forty-three-year-old James Otto Earhart sleeping in his 1975 Oldsmobile in the Sam Houston National Forest and arrested him for kidnapping Kandy Kirkland. The nine-year-old girl had been last seen two weeks earlier getting off a school bus in Bryon, Texas. Shortly after Earhart's arrest, a jogger came across the victim's body in the forest not far from where the police had found Earhart. The girl had been shot twice in the back of the head. Both of the .22-caliber slugs were so badly mangled, they could not be matched to a test-fired bullet.

Earhart was not what you would call a prominent citizen of the community. He was five-foot nine, weighed four hundred pounds, repaired used appliances, lived with his mother, had some kind of nervous condition, and drank a case of beer a day. He admitted giving Kandy a ride home from the bus stop on the day she went missing but denied abducting or killing her. On May 4, Earhart had been to the Kirkland home in response to an ad placed by Kandy's father, who was selling a paint gun. Journalist Michael Graczyk reported that according to Mrs. Kirkland, who described Earhart as "obese, filthy, and unkempt," he kept looking her daughter "up and down."

The police found a partially loaded .22-caliber handgun in Earhart's car, a tiny spot of blood on the gun, and traces of blood on his shirt. There was not enough blood to type, but the jailer at the Walker County Jail reported that Earhart had confessed the crime to him. Two weeks after Earhart's arrest, a grand jury indicted him of capital murder. The defendant was so unfavorably portrayed on local television that his attorney, William Vance, won a change of venue to Giddings, Texas, in Brazos County.

In many ways, the Earhart case represents a prosecutor's nightmare. The district attorney was confronted by a terrible crime with a suspect who had admitted being with the victim on the day of her disappearance. Moreover, the suspect had been arrested not far from where Kandy's body was found and was in possession of a blood-spotted gun that could have fired the fatal bullets. It added up to this: Earhart had probably kidnapped and murdered Kandy Kirkland. But probability, as a standard of proof in a criminal trial, is not enough. A prosecutor needs more, and in this case, it would not take much more evidence to reach a more appropriate level of proof.

In Giddings, Texas, Brazos County district attorney William Turner transformed probability into guilt beyond a reasonable doubt through the testimony of a forensic chemist from the FBI lab. The CBLA expert linked the defendant to the murder scene by identifying the bullet fragments in the victim's head as coming from the same batch of lead found in the bullets in Earhart's car. Since all the CBLA analysts worked for the FBI, there was no way for the defense attorney to counter this testimony with an expert of his own. The jury found Earhart guilty of first-degree murder. Eleven years later, he was executed by lethal injection. He went to his death maintaining his innocence.

The Michael Behn Case: Two Trials, Same Verdict

A defendant faced with CBLA testimony, because all of the examiners worked for the FBI, had a hard time finding a credible witness to counter this evidence. Detectives from South River, New Jersey, believed that on July 19, 1995, Michael Behn shot to death a coin dealer named Robert Rose. According to the victim's wife, on the night of the shooting, Rose went to his office to sell coins to a customer named Mike. Because Michael Behn had been talking to Rose about buying $40,000 worth of coins, he became the suspect. The victim's son, however, told the police that his father had a 7:45 PM appointment with someone else, and two people who worked in the coin dealer's office confirmed this. Rose had been shot several times, and from the marks on both his wrists, it looked as though the killer had handcuffed him. He was believed to have been murdered between 9:25 and 10:30 that night.

Brought in for questioning, Michael Behn told detectives that on the night of the killing, around 5:30 PM, he went to Rose's place and purchased the $40,000 worth of coins. When he left the office, Rose was alive. The detectives didn't believe him. They accused Behn of killing Rose because he wanted the coins but didn't have the money. Behn insisted that he had the funds and had been somewhere else when Rose was shot. In September 1995, a Middlesex County grand jury indicted Behn on charges of murder, felony murder, and armed robbery. The case was headed for trial without a confession, an eyewitness, a jailhouse informant, a murder weapon, fingerprints, or a DNA identification.

The trial commenced on April 7, 1997, and featured the prosecution's attempt to connect Behn to the murder scene through a box of .22-caliber bullets found in his possession. The state put two firearms identification experts on the stand who said that the victim had been shot four times by a gun firing .22-caliber longs. One of the experts said he believed the crime-scene bullets had been fired by a Marlin rifle or a Jennings semi-automatic pistol. The other expert agreed that the murder weapon could have been a Marlin rifle but wasn't sure about the handgun. A third prosecution expert, Dr. Peter R. DeForest, a professor of criminalistics at John Jay College in New York City with a Ph.D. in biology from the University of California at Berkeley, testified that the marks on the victim's wrists could have been made by handcuffs of a design similar to a pair owned by the defendant.

Charles Peters, one of the FBI's most active CBLA experts, comprised the centerpiece of the case against Behn. Peters had testified in dozens of trials, which included a North Carolina murder case where he had established that bullets in the defendant's possession had been made on the same day as the crime-scene bullet fragments. As reported by Charles Piller and Robin Mejia in the *Los Angeles Times*, when asked the likelihood of this occurring, he had said, "I can't put any statistics, but I could spend a lifetime looking for that." At the Behn trial Peters testified that from the code number on the box of bullets in the defendant's possession, he knew they had been manufactured on April 1, 1988. By comparing the chemical makeup of the crime-scene fragments with one of the bullets in the ammunition box, Peters had determined that the murder-site lead had been made into slugs on that date as well. When asked to explain to the jury how CBLA worked, Peters said:

> The basis of bullet lead analysis is that when bullets are manufactured they start out with a molten pot of lead and this molten pot of lead will

> have elements added to the lead to make the lead hard and then sometimes they'll even take out elements, trace—trace elements if they don't want it in there, but basically you would have this large pot of lead and everything in that large pot of lead is mixed and it has its own unique composition, so every time they make up this batch it will be unique. And how do we know that? Because over the years we've analyzed tens of thousands of bullets. These are single bullets, partial boxes of bullets, full boxes of bullets. And every box of bullets that comes from different sources of lead has its own unique composition, that is, if you can characterize enough of the trace elements that are in there, and we look at things like copper, antimony, arsenic, tin, and bismuth in the lead and if we get enough of these elements we can actually source it to the source of lead. And we know from our analysis at the FBI over—we've been doing this over thirty years, that there's millions of these compositions out there.

Michael Behn's defense attorney, Paul Casteleiro, knew that any chance of keeping his client out of prison depended on challenging the science behind CBLA. When he asked Judge Barnett Hoffman to adjourn the trial until he could find a witness with this expertise, the judge denied his motion and the trial moved forward. Forced to do the best he could with what he had, Casteleiro found a firearms identification expert from Fredericksburg, Virginia, and a tool-mark analyst from New York City. The firearm witness, William E. Conrad, testified that in his professional opinion, there was no way to determine what brand of gun had fired the slugs based on the crime-scene bullet fragments. Nicholas D. K. Petraco, a professor of chemistry at the John Jay College in New York City, said he did not believe that the bruises on the victim's wrists had been made by the defendant's handcuffs.

On May 5, 1997, obviously influenced by the testimony of the CBLA expert from the FBI lab, the jury found Michael Behn guilty on all three counts. On June 23, Judge Hoffman, calling the CBLA testimony "particularly significant," sentenced Behn to life in prison. Defense attorney Costeleiro, skeptical of CBLA, told journalists, as Joyce Gramza reported on *ScienCentral Video News*, "I believe the only reason he got convicted was because of the fact the court admitted this bullet lead composition analysis evidence."

Michael Behn went to prison, but his sister Jacqueline, a criminal justice professor at Bergen County Community College, kept hope alive for him. She believed he was innocent and questioned the reliability of CBLA. In 1999, she spoke to Frederic Whitehurst, the former FBI lab whistleblower, who referred her to William Tobin. Tobin had retired from the FBI lab a year earlier as the Bureau's chief metallurgist. He had been an FBI agent for twenty-seven years, twenty-four of which he spent in the crime lab. Like Frederic Whitehurst, Tobin had been critical of the hair identification work of Michael P. Malone, and he had always had doubts about the scientific assumptions behind CBLA. Tobin was bothered by the fact that Charles Peters and the other CBLA experts were not metallurgists, but chemists, and that the FBI was the only place this kind of analysis was done. Because the Bureau had refused to share its data and technology with outsiders, CBLA had never been subjected to peer review.

Tobin's interest in the Behn case led to his decision to assemble a team of scientists to study and test the science underlying CBLA. The team he picked included Erik Randich, a metallurgist at Lawrence Livermore National Laboratory in Livermore, California, and a pair of scientists who worked for companies that provided lead to bullet manufacturers. The self-funded research team completed their study in 2003 and published the results in *Forensic Science International.* Tobin and his colleagues found that CBLA experts often testified that bullets are smelted from batches of lead as small as seventy pounds, enough to make ten thousand .22-caliber bullets. In fact, bullets were usually created out of lots of lead that weighed fifty tons or more, enough to make up to 17 million .22-caliber bullets. The more bullets that have the same chemical composition, the less incriminating the crime-scene match to ammunition in the defendant's possession. The research team didn't find fault with the FBI's method of lead analysis but criticized the way CBLA witnesses interpreted their findings.

Tobin and his team also questioned three of the assumptions upon which CBLA was based. First, they learned that each batch of lead was not necessarily unique in composition. Second, they found that lots of lead were not always uniform in their chemical makeup. This meant that two bullets that did not match could have come from the same batch of lead. Third, and perhaps most important: the fact that a crime-scene bullet and the defendant's ammunition had been made on the same day was not absolute proof that the crime-scene evidence had originated from the defendant's bullet supply. Because bullets are made, stored, and then mounted on cartridge cases as needed over a period of years, bullets made on the same day did not necessarily come from the same box. Such bullets could have been purchased years apart, from anyplace in the world.

The results of the Tobin-Randich study caused the FBI to ask the National Research Council (NRC) of the National Academy of Science to conduct its own study of CBLA. In November 2004, the Academy published the results in a report entitled "Forensic Analysis: Weighing Bullet Lead Evidence." The NRC researchers had come to the same conclusion William Tobin and Erik Randich had arrived at, that is, a CBLA match was about as incriminating as a footwear expert's identification of a crime-scene print that had the general size and style of the defendant's shoe. In other words, a CBLA match was a mere class or group identification, hardly enough evidence by itself to sustain a conviction. The Academy did not recommend that the FBI discontinue CBLA, just that it alter the manner in which CBLA evidence was presented in court.

On September 1, 2005, following the issuance of the NRC study results, the FBI announced that it was closing its CBLA operation because the cost of the process did not justify the evidentiary value of the work. If CBLA could no longer support a conviction on its own, what was the point? The Bureau sent letters outlining its decision to three hundred law enforcement agencies that had received positive CBLA identifications from the FBI lab since 1996. Prosecutors in these cases would have to reevaluate the evidence, and charges in many of these cases would have to be dropped.

So after twenty-five years of overselling the ramifications of their findings, the FBI's CBLA experts had packed up and left town. But what about all of those people who had been sent to prison on the strength of their testimony? Speaking on behalf of the National Association of Criminal Defense Lawyers, as the Associated Press's Michael Sniffen reported, Jack King addressed this point: "Now, we need the FBI to provide a live witness, a scientist from the FBI lab, to testify at post-conviction hearings on these old cases. Some of these guys [convicted defendants] have sat in jail for decades, and it's about time they got a fair hearing." It was too late for James Otto Earhart, executed in 1999, but it wasn't too late for Michael Behn.

Paul Casteleiro, Behn's attorney, filed an appeal, and on March 7, 2005, three appellate court judges set the conviction aside on the grounds it had been based on "erroneous scientific foundations." Unable to make his $400,000 bail, Behn remained in jail awaiting his new trial, which assistant Middlesex County prosecutor Neil Casey III said would go forward without the CBLA evidence.

In June 2006, a jury sitting in New Brunswick, New Jersey, without the benefit of CBLA testimony, found Behn guilty of first-degree murder and robbery. The judge sentenced him to life in prison with eligibility for parole in thirty years. Behn still maintains his innocence and with the help of his sister Jacqueline continues to fight for vindication.

The Shane Ragland Case: Stress in the Crime Lab

The Behn conviction was only one of many affected by the end of CBLA. Shane Ragland, serving a thirty-year sentence in Kentucky, would get a new trial as well. His ordeal began on July 17, 1994, when a sniper shot football player Trent DiGiuro in the head as he stood outside a rental house in Lexington, where he lived with some of his University of Kentucky teammates. He and the others were celebrating his twenty-first birthday. The shooter got away, and the case remained unsolved for six years until Shane Ragland's former girlfriend told the police he had confessed the crime to her. Ragland was taken into custody after CBLA expert Kathleen Lundy linked lead fragments from the sniper's bullet to a box of ammunition found in Ragland's possession. An experienced CBLA expert, Lundy had testified in eighty trials.

At a pretrial *Daubert* hearing in 2002 requested by Ragland's attorney, reported *Lexington (Ky.) Herald-Leader* journalist Louise Taylor, Lundy testified that the murder-scene bullet fragments were "analytically indistinguishable" from the bullets in the box seized in the search of Ragland's residence. William Tobin, having completed his CBLA study with Erik Randich, testified for the defense that millions of bullets could be "analytically indistinguishable" from each other. The judge nevertheless ruled that the CBLA evidence met the *Daubert* criteria and could therefore be introduced at Ragland's trial. A few months later a jury found Ragland guilty.

Ragland's attorney, J. Guthrie True, appealed the conviction in July 2002, in part because at the *Daubert* hearing, Kathleen Lundy had made a statement related to her CBLA testimony that she knew was false. Lundy would later plead guilty to a charge of false swearing. In November 2004, the Supreme Court of Kentucky granted Ragland a new trial, but not on that issue. The justices found that while Lundy had lied under oath, the falsehood did not directly affect the credibility of her scientific findings. The court granted Ragland a new trial because the prosecutor made reference to the fact that the defendant had not taken the stand on his own behalf, a Fifth Amendment violation and standard ground for appeal.

On March 26, 2006, the Kentucky Supreme Court, in a five-to-two decision, reversed itself on the CBLA issue. Citing the NRC study and the FBI's discontinuation of CBLA work, the court ruled that in Ragland's retrial, the commonwealth could not present lead composition evidence. The prosecutor said he'd go ahead with the retrial. Ragland, in August 2007, pled guilty to second-degree manslaughter. Given credit for time served, he completed his sentence.

In an internal FBI memo explaining why she had failed to correct a statement she knew was false at the Ragland *Daubert* hearing, Kathleen Lundy reveals how some forensic scientists react to the pressures of the job, as Taylor reported: "I was stressed out by this case and work in general. I had been under a great deal of professional pressure for over a year and had considered resigning. This pressure has increased by new and repeated challenges to the validity of the science associated with bullet-lead composition analysis. These challenges affected me a great deal, perhaps more than they should have. I also felt that there was ineffective support from the FBI to meet the challenge."

ALTHOUGH LYING UNDER OATH is hard to justify, it's understandable how an expert would be under terrible stress, testifying in a field whose scientific basis is being questioned by reputable critics. But when even well-meaning forensic scientists knowingly or unknowingly mislead a jury or a judge, the results can be devastating for the defendant. Science is in constant flux, but a defendant convicted on evidence considered reliable at the time will remain in prison—notwithstanding new developments in the field—unless someone in a position to help takes an interest in the case. And there aren't enough people with the time and resources to help all the victims of bad forensic science.

16

The Celebrity Expert

Dr. Henry Lee

Dr. Henry Lee has come as close to becoming a household name as any forensic scientist in U.S. history. He has achieved fame in a profession whose practitioners generally operate behind the scenes. In the criminal justice field, it's usually the defense attorneys who get the headlines, and in forensic science, it's often forensic pathologists like Dr. Michael Baden and Dr. Cyril Wecht. In the 1930s, a pair of criminalists in the Seattle area, Oscar Heinrich and Luke May, achieved celebrity status by solving a number of celebrated murder cases in the Northwest, as did Clark Sellers, a handwriting expert from Los Angeles, who made headlines with his testimony at the Lindbergh kidnapping trial in Flemington, New Jersey. In the 1960s, Dr. Paul Kirk, a forensic chemist from Berkeley, California, the protégé of the well-known police reformer August Vollmer, became something of a celebrity. The peak of his notoriety came in 1965 when he analyzed the blood-spatter patterns for attorney F. Lee Bailey in the Dr. Sam Sheppard murder case near Cleveland, Ohio.

But because he rose to prominence in the era of crime television, Dr. Henry Lee has enjoyed a fame more intense and more intimate than his celebrity predecessors.[1] He has made hundreds of television appearances and hosts a show on Court TV called *Trace Evidence: The Case Files of Dr. Henry Lee.* Dr. Lee's personality, demeanor, and life story have helped make him a bigger-than-life character. Like sports stars and major film and television actors, he tends to be vain and dramatic. On the witness stand, he informs jurors and, as a charismatic courtroom showman, entertains them. When Dr. Lee testifies for the prosecution, he's the defense attorney's worst nightmare. When he's appearing on behalf of the defense, it's not good news for the prosecutor. In either case, the media loves it, and so do the jurors.

Dr. Henry Chang-Yu Lee was born in Rugao City, China, on November 22, 1938. When Henry was four, the Chinese communists murdered his father, and two years later his family fled to Taiwan to avoid the communist revolution. After graduating from the Taiwan Central Police College in 1960 with a degree in police science, Henry joined the Taipei Police Department. Six years later, after rising to the rank of captain, he came to the United States, where he graduated from New York City's John Jay College of Criminal Justice with a bachelor of science degree in forensic science in 1972 and earned a master's degree in biochemistry from New York University in 1974. A year later, he was awarded a Ph.D. in biochemistry. In 1979, Dr. Lee was named director of the Connecticut State Police Forensic Laboratory, where he also held the title of chief criminalist. Following his retirement from the lab in 2000, Dr. Lee began teaching at the University of New Haven, where he founded the Henry C. Lee Forensic Institute. According to his résumé, Dr. Lee has been awarded several honorary degrees, written twenty books, published numerous scientific articles, given hundreds of speeches, investigated four thousand homicide cases, testified in hundreds of trials, and consulted with more than three hundred law enforcement agencies.

The Wood Chipper Case

Dr. Lee vaulted onto the national stage in 1986 when an airline pilot named Richard Crafts went on trial in Connecticut for murdering his wife, Halle. Having incurred her husband's wrath by announcing her plans to divorce him, Halle Crafts had covertly audiotaped his threats to kill her. Perhaps even more incriminating, Richard Crafts was seen by a motorist, on the night of Halle's disappearance, operating a commercial-grade wood chipper in the midst of a blizzard along the bank of the Housatonic River. The audiotaped threats and the wood chipper sighting led the police to suspect Crafts of murdering his wife. But investigators had a serious problem: they didn't have a corpse. Faced with one of those maddening cases of a good suspect but no physical evidence, investigators called on Dr. Lee.

In the couple's bedroom, Dr. Lee found traces of the victim's blood. When he examined a chainsaw that had been in the suspect's possession, Dr. Lee discovered hair follicles, traces of blood, and tissue that he identified as the victim's. In the rented wood chipper, Lee recovered the same, and at the spot where Richard Crafts had been seen operating the equipment, Dr. Lee found fragments of the victim's teeth and bones, along with follicles of her hair. It wasn't much, but it was enough to establish that Halle Craft had been murdered. From this evidence, Dr. Lee was able to reconstruct the crime, theorizing that the defendant had bludgeoned his wife to death in their bedroom, frozen her body in a home freezer, cut her into pieces with the chainsaw, then shoved the body parts into the wood chipper, which sprayed her remains into the river. The jury, obviously impressed with Dr. Lee and his evidence, found Craft guilty. A few years later, while serving his life sentence, Richard Craft confessed to murdering his wife. Featuring blood and gore, an attractive victim, a suburban killer, a dramatic trial, and a scientific investigator in the mold of Sherlock Holmes, the wood chipper case turned Dr. Henry Lee into a celebrity forensic scientist.

William Kennedy Smith Case

Five years later Dr. Lee took the stand on behalf of William Kennedy Smith, on trial in 1991 for date rape in a case that dominated crime news because of the Kennedy family connection. According to Smith, following a night of drinking with his Palm Beach accuser, the two had engaged in consensual sex on the lawn of the Florida Kennedy estate. Apart from evidence of a struggle, physical evidence doesn't play a role in date-rape cases, because the defendant is not denying having sexual intercourse with the woman. To help prove Smith's accuser had consented, Dr. Lee testified that he found no grass stains on the woman's pantyhose, suggesting she had not struggled against an attacker. Had there been resistance, there would have been grass stains on the victim's undergarment. To illustrate his point, Dr. Lee produced a grass-stained handkerchief he had rubbed against his own lawn. Although Dr. Lee's testimony was also consistent with the victim's having been in a drunken stupor or too afraid to move, the jury found Smith not guilty.

Dr. Lee's testimony in this case drew criticism from John Hicks, the director of the FBI lab, who called it "outrageous," as Christopher Ruddy reported on November 24, 1995. Hicks characterized Lee's handkerchief experiment as unscientific and labeled the conclusions drawn from it speculative. He pointed out that the handkerchief was not made of the same fabric as the pantyhose and the conditions that had created the handkerchief stains did not necessarily replicate the environment at the alleged crime site. Criticism of this kind—that Dr. Lee's testimony is more theater than science—has followed him throughout his career.

The Death of Vincent Foster: Suicide or Murder?

In the spring of 1995, Dr. Lee was called on in another politically sensitive case. Independent counsel Kenneth Starr asked Dr. Lee to render an opinion as to whether President Clinton's White House counsel, Vincent Foster, had taken his own life in Marcy Park on July 20, 1993, or had been murdered and dumped there. Investigators with the park police, and later the FBI, had concluded that Foster had gone to the park for the purpose of shooting himself in the head with a handgun. Conspiracy theorists, however, believed that Foster had been murdered, and that forces within the Clinton administration were behind the cover-up. In support of their theory that Foster had been shot elsewhere—perhaps in his White House office—and dumped in Marcy Park, they pointed out that the bottoms of Foster's shoes were clean. Had he walked the seven hundred feet into the park, his shoes would have been at least slightly soiled. The conspiracy proponents also noted that the fatal bullet had never been found. If Foster had shot himself in the park, where was the bullet? The scene would also have produced traces of Foster's blood, tissue, brain matter and hair. According to police reports, nothing of that nature had been recovered from the death site.

In a report made public in November 1995, Dr. Lee concluded that Foster's death had been a suicide, basing this opinion on his examination of death-site photographs that showed a blood-spatter pattern that was consistent with a self-inflicted gunshot wound. The report wasn't made public until three months after Dr. Lee's highly publicized defense testimony in the O. J. Simpson trial, a role that has drawn criticism from inside and outside the forensic science community. By testifying on behalf of a man most people believe guilty of double murder, Dr. Lee opened himself up to the charge he is a hired gun addicted to the limelight.

The O. J. Simpson Case

In general, Dr. Lee's testimony at the Simpson trial helped the defense in five ways. It depicted Los Angeles police investigators and crime-scene technicians as incompetent; it suggested that blood evidence had been contaminated; it supported the theory that evidence had been planted; it pushed the time of the crime forward forty-five minutes, which accommodated the defendant's alibi; and it laid the groundwork for the theory that Nicole Simpson and Ronald Goldman had been murdered by more than one person.

On this last point, Dr. Lee's testimony contradicted that of the FBI's renowned footwear identification expert, William Bodziak. Dr. Lee identified a bloody stain on an envelope and scrap of paper found in Nicole Simpson's house as a shoe print that didn't match the footwear—the Bruno Magli Italian designer shoes—prosecutors believed the defendant was wearing when he committed the murders. Bodziak testified that this bloody print had not been made by a shoe at all. Douglas Deedrich, also from the FBI crime lab, testified that the bloody pattern was in fact a fabric print. Dr. Lee also raised the possibility that a bloodstain on Ronald Goldman's blue jeans had been made by a shoe that was not a Bruno Magli. On cross-examination, when pressed about his blood print identifications, Dr. Lee said that *if* these patterns were footwear marks, they were not made by the Bruno Magli brand. Gideon Epstein, the Ramsey case handwriting expert, might have accused Dr. Lee of blowing smoke—providing vague testimony to cloud the issue, to create reasonable doubt.

In the wake of the Simpson trial, Douglas Deedrich called Dr. Lee's shoe-print testimony "irresponsible." John Hicks, as the former director of the FBI lab, said this about Dr. Lee, Ruddy reported: "I think he is definitely a hired gun." Dr. Richard Saferstein, the head of the state crime lab in New Jersey, commenting on Lee's fifty-page curriculum vitae, was reported to have said that "he has an ego as big as his résumé." Dr. Lee seemed unfazed by such criticism.

Although a team of overachieving lawyers and their expert witnesses had given a defense-oriented jury a reason to acquit O. J. Simpson, the trial hooked the American public on forensic science like no other case before it. And no matter which side of the Simpson case an expert was on, or how well or poorly he or she had performed, the trial also turned a lot of obscure criminal justice practitioners into celebrities and made figures like Dr. Lee, already well-known in their fields, even better known.

The Ramsey Case

On March 10, 1997, two and a half months after JonBenét Ramsey's body had been found in the basement of her parent's home, Boulder County district attorney Alex Hunter asked Dr. Henry Lee to analyze the evidence and reconstruct the crime. In his 2004 book, *Cracking Major Cases,* Dr. Lee includes a discussion of the Ramsey case, a crime, one could argue, that still hasn't been cracked by anyone. Although he has acknowledged that since his work on the case there might be evidence pointing to an intruder, Dr. Lee initially believed that the victim could have been accidentally killed, and that members of the family had staged an elaborate crime scene that included the writing of the ransom note. This is a theory not unlike the one floated by New York attorney Darnay Hoffman.

When Patsy Ramsey died in 2006, a *Rocky Mountain News* article by Charles Brennan quoted Dr. Lee: "If she had a secret, then that secret is going to be gone forever. . . . But on the other hand, if she did share a secret with anybody else, it may be easier for that person to talk." Responding to the implications of Dr. Lee's statement, Lin Wood, the Ramsey family attorney, replied: "Patsy Ramsey was not involved in the death of her daughter. There was never going to be any sort of death-bed confession to a crime she did not commit. It's another example of some of the injustices inflicted on the family over the years."

The Scott Peterson Case

In 2003 and 2004, Dr. Lee appeared on dozens of television news and talk shows devoted to the big case of that period, the Modesto, California, murders of Laci Peterson and her unborn child. On April 21, 2003, Scott Peterson was charged in Stanislaus County Court with murdering his pregnant wife and dumping her body into San Francisco Bay on or about December 22, 2002. In August 2003, four months after the bodies washed up on shore not far from where Peterson said he had been fishing just prior to his wife's disappearance, Peterson's attorney, Mark Geragos, called in Dr. Lee and Dr. Cyril Wecht, the forensic pathologist from Pittsburgh. The fact that two of the country's most famous forensic scientists had been brought into the case by the Peterson defense set the tongues of TV's talking heads wagging. Would Dr. Lee, as in the O. J. Simpson case, be instrumental in getting Scott Peterson acquitted? This was crime entertainment at its best.

On August 8, a few days before Drs. Lee and Wecht traveled to Martinez, California, to examine the remains, a TV panel hosted by CNN anchor Anderson Cooper discussed this exciting new development in the case. What follows is an exchange between the host and Harvey Levin, the executive producer of a reality television series about celebrities in trouble with the law:

COOPER: We knew this was going to happen [the Lee/Wecht examination]. . . . Legally, how significant?

LEVIN: I think it's pretty significant. I've seen Henry Lee work on the stand before. . . . He can do several things. Number one, if he can make it so clear that it's impossible to know the cause of death and that people are speculating, juries tend to be very uncomfortable about speculating when it comes to somebody's life, namely, Scott Peterson's. So that works in favor of the defense. . . .

COOPER: Let me just interrupt. In a sense, you're saying it's not so much what they [Lee and Wecht] find; it's what they don't find?

LEVIN: Oh, I absolutely think that's the M.O. here, that what they don't find will create the uncertainty that makes juries uncomfortable. However, there's something else that I've seen Henry Lee do before. And that is, if there is something that would be consistent with an injury that might exclude Scott Peterson. I'll give you an example. Suppose he [Dr. Lee] says that there's some sign that would suggest that the fetus had been attacked. And there has been this devil-worshipping theory floated out there. [Mark Geragos had raised the possibility that the murders had been committed by members of a satanic cult in a ritualistic killing.] As long as he [Dr. Lee] can find something to suggest, well, that doesn't exclude the possibility of that theory and he can create that and reinforce it for the defense. . . . It's not that he's [Dr. Lee] going to show anything to a certainty. It's that he's going to suggest possibilities that might tend to exclude Scott Peterson.

As it turned out, Dr. Lee's examination of the remains produced no evidence whatsoever, and he was not called as a trial witness for the defense. On November 12, 2004, the jury found Scott Peterson guilty. A few months later he was in San Quentin prison, maintaining his innocence from death row.

The Other Peterson Case

On December 9, 2001, almost a year before Laci Peterson went missing in Modesto, California, Michael Peterson (no relation) made a late-night 911 call in Durham, North Carolina. The frantic-sounding caller reported that his wife had fallen down a flight of stairs. Fire department paramedics arrived at the stately house on Cedar Street at 2:40 AM, just minutes after the emergency call. A pair of officers from the Durham Police Department rolled up to the scene shortly afterward.

When the paramedics entered the house, they found Mr. Peterson beside his dead wife, who lay on her back in a pool of blood at the foot of the stairs, wearing a sweatshirt and sweatpants. The paramedics were immediately taken aback by the quantity of blood. In a shirt and a pair of shorts, and in his bare feet, Michael Peterson was covered in it. It was pooled beneath the woman's head, on the stairs, and splashed up and down the stairwell wall.

About six thousand people a year fall and die in their homes, but there isn't much blood, and most are elderly. Moreover, the stains on the wall didn't seem that fresh, certainly not just five or ten minutes old. How long had this woman been dead before the 911 call? To the paramedics, this looked more like a homicide, a beating, than an accidental fall.

The victim's head rested on a stack of blood-soaked bathroom towels. (Before the body was photographed as it lay, someone removed the towels and placed them nearby.) Next to the body lay the victim's clear plastic sandals, a pair of Converse sneakers (Michael Peterson's), some bloody paper towels, and a pair of socks.

The first responders, and those who followed, realized that the people involved in this bloody death were not ordinary citizens of Durham, North Carolina. The dead woman was forty-seven-year-old Kathleen Peterson, a Research Triangle executive who worked for Nortel and was active in the local art community. Her fifty-nine-year-old husband, Michael, had published a pair of best-selling Vietnam War novels and had run, unsuccessfully, for mayor and city council. Both previously divorced, the Petersons had been married since 1997. He had two grown sons and a college-aged daughter. He also wrote a column for the *Durham Herald,* a forum he had used to criticize the Durham Police Department, particularly the detective bureau. In one of his pieces, he likened the chance of a criminal being identified and caught in Durham with the chance of being struck by lightning. He had repeatedly pointed out that the local police had the lowest crime solution rate in the state.

The police officers in the Peterson house that night either didn't know how to manage the scene of a potential homicide or didn't think that a crime had been committed. Peterson, with blood on his hands and on the bottoms of his feet, was allowed to wander about the house touching and moving items that might have had evidentiary value. His son Todd, upon returning from a party at 3:00 AM, was allowed to enter the house. Not only that, the police let Todd visit the body and pick up a cordless telephone that had been on the steps, leaving behind its outline in blood. Like his father, Todd had free access to the scene, tracking blood about the house, and at one point drawing a glass of water from the kitchen sink.

Instead of restricting entry to authorized personnel, the police officer posted at the front door acted more like a greeter. A friend of Todd's, along with neighbors, friends, and total strangers, were allowed to enter the house and satisfy their curiosity. Even the mayor of Durham showed up for a death-scene tour.

Dr. Kenneth Snell, the medical examiner of Mecklenberg County, came to the house at 7:40 that morning, five hours after Michael Peterson's 911 call. After examining the body at the foot of the stairs, Dr. Snell told a police officer that it appeared that the lacerations on the back of Kathleen Peterson's head had been caused by a fall. He ordered the removal of the body to the state's medical examiner's office in Chapel Hill, where it would be autopsied. Dr. Snell noticed that one or two of the head wounds appeared deep and, in case the death was more than an accident, suggested that the police search the house for an object that could have made these lacerations.

By the time the body was loaded into the ambulance and driven off to Chapel Hill, the only people left in the house were uniformed police officers and a couple of detectives. At 11:00 AM, one of these detectives, convinced that Kathleen Peterson had been murdered, tossed Michael Peterson's bloody garments into a paper evidence bag. He did the same with the stained clothing left behind by Peterson's son, Todd. The detective did not make sure that these garments were completely dry, nor did he place each item of clothing in a separate bag. Moreover, he failed to seize, as potential crime-scene evidence, the cordless telephone, the victim's sandals, the bloody paper towels, the Converse sneakers, or the pair of eyeglasses that had been lying on the steps above the victim's body. As crime-scene investigators, the Durham police were worse than amateurs. Even civilians who watch crime television know about the need to protect a potential crime-scene from contamination and alteration, and how important it is to gather anything and everything that might be evidence and to make sure bloody clothing is completely dry and separately bagged.

Just prior to the autopsy, scheduled for noon that Sunday, Dr. Snell spoke on the phone to a Durham detective who told him that the police strongly suspected that Kathleen Peterson had been murdered. Dr. Snell, however, was not the one who would perform the autopsy. He looked on as his assistant, Dr. Deborah Radisch, did the job. According to the autopsy report by Dr. Radisch, Mrs. Peterson had not suffered a fractured skull, a broken neck, or bleeding in the brain. The victim did have seven scalp lacerations on the back of her head, bruises on her arms, wrists, and back, and contusions and abrasions (bruises and scratches) on her face. A blood-alcohol content of .07 percent showed she had been drinking before she died. Dr. Radisch summarized her findings:

> In my opinion, the cause of death in this case was due to severe concussive injury of the brain caused by multiple blunt force impacts of

> the head. Blood loss from the deep scalp lacerations may also have played a role in her death. The number, severity, locations, and orientation of these injuries are inconsistent with a fall down stairs; instead, they are indicative of multiple impacts received as a result of beating.

Later that day, Duane Deaver, a crime-scene technician who specialized in blood-spatter analysis for the North Carolina State Bureau of Investigation (SBI), spent several hours at the Peterson house studying the patterns on the staircase wall. The amount and location of the blood, as well as the size and shape of the drops, led Deaver to conclude that the victim had been beaten to death. In his estimation, someone had struck her in the head at least four times. Deaver's analysis supported the theory held by the Durham police that Michael Peterson had assaulted his wife with a fireplace tool called a blow poke. Several years earlier the victim's sister had given the family a fireplace poker that was now missing from its place on the hearth.

On December 20, 2001, district attorney James Hardin Jr. presented the case against Peterson to a grand jury. After four hours of deliberation, the jury returned an indictment charging Peterson with first-degree murder. That day, accompanied by his newly hired lawyer, David Rudolf, Peterson turned himself in. Three weeks later, after the prosecution announced that it wasn't going to seek the death penalty, Peterson was released on $850,000 bail.

At this point, Jim Hardin had a case riddled with holes. The prosecutor didn't have a confession, an eyewitness, or the murder weapon. The defense would argue that the victim had been drinking and had fallen down the stairs and, to back that up, would produce Dr. Henry Lee, the world-renowned blood-spatter expert. Moreover, the defendant was a prominent citizen who had angered the police with his newspaper column. By bungling the crime-scene investigation, the police had proven that his criticisms were correct. In the motive department, the prosecutor was working on a pair of theories. The first had to do with money, and the second, with sex. The state hoped to prove that Michael Peterson was facing financial ruin, a problem that his wife's life insurance and other corporate postmortem benefits would solve. To show that the Peterson marriage was not as perfect as it looked from afar, the prosecution would produce a series of e-mail messages between the defendant and a male prostitute.

The Peterson case, featuring a dead socialite, a well-known suspect, a lot of blood, a male prostitute, and Dr. Henry Lee, had more than enough to satisfy crime buffs and provide fodder for crime television's army of talking heads. The trial would be televised, gavel to gavel, on Court TV.

On October 20, 2002, ten months after Michael Peterson's indictment, district attorney Jim Hardin jacked up interest in the case with a bit of startling news. The police were investigating the 1985 death of a woman who had lived next door to Peterson and his first wife, Patricia, in Germany. The forty-three-year-old widow, Elizabeth Ratliff, with daughters aged three and four, had been a second-grade teacher at a school operated by the Department of Defense. On November 25, Ratliff was found dead at the bottom of her stairs by her nanny, who was reporting for work at 7:15 in the morning. The nanny summoned Michael Peterson, who returned with her to the apartment. At the scene, Peterson told the nanny to cover the body with a sheet. Because the body was still warm and possibly alive, the nanny hesitated. "She's dead," Peterson was reported to have said. "The warmth comes from the floor heating." The victim's head had been framed by a pool of blood, and she was dressed in what she had been wearing the previous evening when she and her daughters were dinner guests at the Peterson house. After walking her home the previous night, Michael had helped Elizabeth get her girls ready for bed. After their mother's death, the daughters, now twenty and twenty-one, had been raised by Peterson.

The local authorities in Germany had ruled Elizabeth Ratliff's death accidental following an autopsy by a U.S. Army pathologist, Dr. Larry A. Barnes. Dr. Barnes believed the death had been caused by a brain hemorrhage and had concluded that the victim had suffered a stroke, lost consciousness, and tumbled down the stairs. An inquiry by the Army's Criminal Investigation Division (CID) turned up nothing to contradict this combined natural/accidental manner-of-death ruling.

On April 16, 2003, Judge Orlando F. Hudson Jr., noting similarities between the deaths of Peterson's wife and his former neighbor, ordered the exhumation of Elizabeth Ratliff's body. Dug out of a cemetery in Bay City, Texas, the remains underwent a second autopsy two days later in Chapel Hill. Those attending the procedure performed by Dr. Deborah Radisch included Dr. John Butts, the state medical examiner, forensic pathologist Dr. Wemer Spitz, neuropathologists Dr. Aaron Gleckman and Dr. Stephen Smith, and representatives of the Durham Police Department and the district attorney's office.

Dr. Radisch must have been amazed by the similarities between the wounds on the seventeen-year-old corpse and the trauma she had seen on

Kathleen Peterson. Both women had been struck seven times on the back of the head, and one of the blows suffered by Elizabeth Ratliff had fractured the left base of her skull. The defendant's former neighbor had bruises on her left arm and hand, contusions on her face, and bruises on her chest, abdomen, and back. Doctor Radisch, in summing up her findings, wrote in her autopsy report:

> The decedent was a 43-year-old white woman who was found dead at the bottom of stairs in her apartment residence in Germany on the morning of November 25, 1985. A military autopsy was performed by a non-forensically trained pathologist who attributed her death to "intracranial hemorrhage, cerebellar brainstem secondary to von Willebrand's coagulation abnormality." He [Dr. Larry A. Barnes] noted "multiple deep lacerations over the posterior and parietal areas" of the head but neither described the size of these lacerations nor their number. He concluded that the scalp trauma was the result of a "fall onto steps" following an intracranial hemorrhage. . . . The most significant finding [in the reautopsy] was the presence of multiple blunt force injuries of the head. Injuries of the left hand, forearm, and back were also present. Seven distinct lacerations were identified on the head. These were located in different areas and a number extended completely through the scalp to the skull with accompanying bruising. Beneath one laceration of the left posterior scalp was a non-displaced skull fracture. In my opinion, the location, number, and severity of these lacerations are inconsistent with a fall down a set of stairs; rather they are indicative of multiple blunt force impacts, either from blows of the head caused by a blunt object or by the head being forcibly struck against a hard surface. It is further my opinion that these injuries were incurred while Mrs. Ratliff was alive and are of sufficient severity to have caused her death. It is further my opinion that the intracranial hemorrhages noted in the first autopsy were primarily the result of blunt trauma rather than any underlying natural disease process.

The testimony phase of the Peterson trial got under way at the end of June 2003. During the next three months, the jury of seven women and five men would hear from fifty-five prosecution witnesses. According to the roles they played in the case, and the nature of their testimony, these witnesses fell into six categories: first responders to the scene, crime-site evidence technicians, crime-lab personnel (this testimony was largely irrelevant since connecting the defendant to the death site was not an issue), forensic pathologists and their colleagues, motive witnesses, and people connected to the death of Elizabeth Ratliff.

The fire department paramedics described the death scene as too bloody to have been the result of an accidental fall down a flight of stairs. The police officers described the defendant and his son as uncooperative and inappropriate. One officer saw the defendant checking his e-mail, writing in a notebook, and sleeping with his dogs. Another watched Todd Peterson and a friend conversing furtively on the patio; when they ignored orders to stop talking to each other, they had to be separated. At one point Todd insisted upon consulting an attorney.

Dan George, the Durham detective who had been in charge of the crime scene, was on the stand four days, which included seven hours of cross-examination from defense attorney Rudolf, who did his best to make the Peterson case investigators look like fools. Duane Deaver, the SBI blood-spatter analyst, was put through the wringer as well. Rudolf attempted to disqualify Deaver as an expert witness by pointing out that he lacked an advanced degree and hadn't published any professional articles. In recognition of the fact Deaver had been working in the field fifteen years and had testified as an expert in sixty cases, the judge denied the defense attorney's motion.

On the stand for six days, Deaver began by describing how he had duplicated the blood-splash patterns at the death site through the use of a model of the staircase, wet sponges, Styrofoam heads, wigs, and a variety of weapons. He identified cast-off stains high on the wall in the foyer as coming from the motion of the murder weapon. He also stated that someone had tried to clean up some of the wall stains, leaving them smeared. On cross-examination, Rudolf mocked Deaver's crime-lab blood-spatter experiments as unscientific and biased in favor of the prosecution's theory of the case. Rudolf pointed out that if the five-foot-nine defendant had wielded the fireplace poker, it would have come into contact with the overhead molding. With Dr. Lee in the wings, though, Rudolf had to be careful not to disparage blood-spatter analysis as a science. He simply wanted to make the prosecution witness look incompetent.

The most important witnesses for the prosecution were the forensic pathologists. The case hinged on how Kathleen Peterson had died. If the prosecution could convince the jury that she had been murdered, the defendant would go to prison. The outcome of the trial therefore rested on the forensic pathology, and in this department, the prosecution had the advantage. Dr. Radisch testified that a fall down a flight of stairs would not have produced the seven deep lacerations on the back of the victim's head. The pathologist had also found that the cartilage on the left side of the victim's neck had been broken, leading her to believe there had been an attempted strangulation. Dr. Kenneth Snell, the forensic pathologist who had been to the scene that morning, and Dr. John Butts, the state's chief medical examiner, also testified for the prosecution. Neuropathologist Thomas W. Bouldin, professor of pathology at the University of North Carolina, opined that the victim had started bleeding several hours before her death, a reality that didn't square with the defendant's account of the event. A second neuropathologist, Aaron Gleckman, an associate medical examiner for the state, testified that the victim's facial contusions were more consistent with a beating that with an accidental fall. An injury biomechanics expert from Duke University, Dr. James McElhaney, testified that the victim's injuries could have been made by an instrument like a fireplace poker.

The prosecutor, through a series of witnesses, tried to establish a motive for the murder. Several witnesses testified that the defendant, plagued with money problems, gained financially from his wife's death. The male prostitute testified that a September 5, 2001, meeting arranged by him and the defendant never took place.

Having made a strong case that Kathleen Peterson had been the victim of homicide, Jim Hardin was allowed to introduce evidence that Elizabeth Ratliff had been murdered as well. Although the defendant hadn't been charged with that killing, he was, in essence, being tried for that crime, too. The similarities between the two deaths, right down to the fact both women had seven lacerations on the backs of their heads, played into the strength of the prosecution's case. The star witnesses in this phase of the trial were Dr. Radisch, who had performed the Ratliff reautopsy, and Dr. Gleckman, who had attended the procedure.

The defense put only nine witnesses on the stand, two of whom were vital to Michael Peterson's case. Jan Leestma, a forensic neuropathologist from Chicago, and Dr. Henry Lee, the famous criminalist, comprised the core of Peterson's defense. These two experts would have to convince the jury, against the weight of three forensic pathologists, two neuropathologists, and an injury biomechanics analyst, that Kathleen Peterson had *not* been murdered.

For Dr. Leestma, testifying for the defense in a murder case was a familiar role. During his thirty-five-year career, he had examined five thousand brains and testified in more than a hundred trials. He was best known as a defense expert in cases involving defendants charged with shaking babies to death. In Massachusetts, he had testified in the 1997 SBS trial of British au pair Louise Woodward. The jury had found Woodward guilty of second-degree murder, a verdict later reduced to involuntary manslaughter.

Dr. Leestma testified that after visiting the Peterson house, reviewing the crime-scene work, the police files, the brain tissue slides, and the autopsy report, he had come to the conclusion that Kathleen Peterson had fallen twice, each time hitting her head two times on the steps. The doctor did not believe that either this victim or Elizabeth Ratliff had been murdered.

Dr. Henry Lee, looking as comfortable in the courtroom as the judge, took his seat in the witness box. Although no one was challenging Dr. Lee's status as an expert, Rudolf allowed him thirty minutes to impress the jury with his professional achievements. Referring to a one-inch-thick sheaf of paper held together by an alligator clip, Dr. Lee said: "That is a copy of my résumé, but it's a couple of years old. I did not have time to update it." Dr. Lee informed the jury that during his career he had visited more than a thousand death scenes involving hundreds of sites where the victim had been beaten to death. He had been to the Peterson house and had reviewed Deaver's blood-spatter work, referring to Deaver's crime lab recreations as "so-called experiments," and "child's-play." Dr. Lee said that the ten thousand or so medium-velocity blood spots on the staircase wall did not present a blood-spatter pattern consistent with a beating. In his expert opinion, the blood on the wall could have gotten there in a variety of ways including being coughed up, flung off moving hands, and sprayed by blood-soaked hair, events consistent with a person's falling down a flight of stairs. To illustrate how the shape and size of blood spots revealed distance and directionality, Dr. Lee used an eyedropper to splash red ink onto a white surface from various heights and angles. Some of the spots were round with smooth edges, others were round and ragged, and others, coming from an angle, formed the shapes of teardrops and other formations. To show the difference between low-velocity stains and those caused by medium-velocity impact, Dr. Lee used his fist to smash an ink-soaked sponge. To replicate what he called aspirated bloodstains, Dr. Lee spewed a mouthful of watered-down ketchup onto white poster board, producing stains he said matched some of those on the Peterson staircase. This was courtroom theater at its best.

District attorney Hardin opened his four-hour cross-examination of Dr. Lee with a question prosecutors like to ask experts for hire: How much was the doctor getting paid for his testimony in this case? Dr. Lee said that while he normally charged clients $500 per hour, his bill in the Peterson case, on which he had worked 240 hours, was only $27,000, and that included expenses. Hardin asked Dr. Lee if, to support his aspirated blood theory, he had tested the wall stains for traces of the victim's saliva? "That's not my job," Lee replied. "If they [the police] did their job properly, I do not have to come here." When asked to explain how a fall down a flight of steps could have produced the seven lacerations on the back of the victim's head, Dr. Lee said that wasn't his job either. He was a death-scene reconstruction expert, not a forensic pathologist.

On October 10, 2003, after deliberating fifteen hours, the Peterson jury found the defendant guilty of first-degree murder. The testimony of Dr. Leestma and Dr. Lee had not, in the minds of jurors, created reasonable doubt. The judge sentenced Peterson to life without parole.

In the Peterson case, it was the nature and number of the victim's head wounds that mattered, not the blood on the wall. Forensic pathology had rendered the debate between Dr. Henry Lee and Duane Deaver irrelevant. As a crime-scene reconstruction expert, it seemed that Dr. Lee had blocked out or wished away the forensic pathology. He had blown a little smoke, but it wasn't enough to cloud the mind of a single juror.

The Phil Spector Case: Dr. Lee and the Missing Fingernail

In the morning of February 3, 2003, Los Angeles County Sheriff deputies responded to a call from the Alhambra mansion owned by Phil Spector, the sixty-seven-year-old music producer who became famous in the sixties for his "wall of sound." In the foyer they found forty-year-old actress Lana Clarkson slumped in a chair. She had been shot once in the mouth by the .38 caliber Cobra revolver lying on the floor under her right hand. When the fatal shot was fired, she and Spector were the only ones in the house. Spector's chauffeur told the police that at five in the morning he heard a noise that sounded like a gunshot. Shortly after that, he said, Spector came out of the mansion carrying a handgun. According to the driver, Spector had said, "I think I killed somebody."

Spector had met the victim the previous night at the House of Blues on the Sunset Strip, where the struggling actress worked as a hostess for nine dollars per hour. When the nightclub closed for the night, she had accompanied Spector back to his house for a drink. According to Spector's account of the death, Lana Clarkson had committed suicide.

The crime-scene investigation and the analysis of the physical evidence featured forensic pathology, the location of gunshot residue, and the interpretation of the death-site blood-spatter patterns. Los Angeles deputy coroner Dr. Louis Pena visited the scene and conducted the autopsy. He found bruises on the victim's right arm and wrist that suggested a struggle. A missing fingernail on Clarkson's right hand also indicated some kind of violence prior to the shooting. Her bruised tongue led Dr. Pena to conclude that the gun had been forced into her mouth. Its recoil had shattered her front teeth. The victim's purse was slung over her right shoulder. Since she was right-handed and would have used that hand to hold the gun, the doctor questioned suicide as the manner of death. Based on his crime-scene examination and the autopsy, Dr. Pena ruled Lana Clarkson's death a criminal homicide. The police arrested Spector, who retained his freedom by posting the $1 million bail.

Blood-spatter experts from the sheriff's office concluded that after the shooting, Spector had pressed the victim's right hand around the gun handle, put the revolver temporarily into his pants pocket, later wiped it clean of his fingerprints, then laid it near her body. From the bloodstains on his jacket, the government experts concluded he had been standing within two feet of the victim when the gun went off. The absence of her blood spray on a nearby wall led the spatter analysts to believe that Spector had been standing between the victim and the unstained surface when he fired the bullet into her mouth. Gunshot-residue experts found traces of gunpowder on the victim's hands and tongue. They also found traces of gunpowder on Spector's hands.

The forensic work performed by the Los Angeles Coroner's Office and the sheriff's department had not been flawless. A dental evidence technician had lost one of the victim's teeth; a criminalist had used lift-off tape to retrieve trace evidence from the victim's dress, which had interfered with the serology analysis; and the corpse had been moved at the scene, causing unnatural, postmortem blood flow from her mouth, which compromised that aspect of the blood-spatter analysis.

At nine o'clock at night on February 4, 2003, after the law enforcement team had completed processing the crime scene, two forensic scientists working for the defense—Dr. Henry Lee and the renowned forensic pathologist Dr. Michael Baden—entered the house. Following their on-site examination of the death scene as well as their later evaluations of the physical evidence, Drs. Lee and Baden concluded that Lana Clarkson had shot herself in the mouth. They based this conclusion on, among other things, the blood-spatter patterns, the gunshot residue, DNA evidence, and the bullet's trajectory, which suggested a shooter taller than the five foot seven defendant. According to Dr. Lee, Spector could have been standing as far as six feet from the victim when the gun was fired and still have been stained by her blood. His prosecution counterparts considered that impossible. The forensic battle lines in the Spector case had been drawn.

The Spector murder trial got under way in May 2007. But a month before the televised proceeding began, a man named Gregory Diamond called the Los Angeles District Attorney's Office with a startling accusation. In 2003 Diamond had been a law student working as an intern in Robert Shapiro's law office. Shapiro was, at the time, Spector's lead defense attorney. According to Diamond, he had seen Shapiro's associate, attorney Sara Caplan, pick up a white item at the crime scene that looked like a tooth and hand it to Dr. Michael Baden. This piece of physical evidence, according to Diamond, had never been turned over to the prosecution as prescribed by law. The district attorney's office immediately forwarded this information to Larry Fidler, the judge who would be presiding over the Spector trial. Judge Fidler assigned law school professor Laurie Levenson to look into the allegation.

As a result of her investigation, Levenson determined that in 2004 a former Los Angeles sheriff's deputy working as a private investigator for the Spector defense had reported to the prosecution that he had seen Dr. Henry Lee take something from the crime scene that looked like a fingernail. The defense had called this accusation, made by Stanley White, ridiculous. Nothing came of the allegation and the matter appeared to have been forgotten.

The Spector trial began in May as scheduled, but two weeks into the proceeding, Judge Fidler held a three-day hearing, without the jury present, to get to the bottom of the resurfaced accusations. Gregory Diamond took the stand and, albeit reluctantly, repeated his accusation under oath. Dr. Baden gave testimony denying that he had handled a tooth at the Spector mansion. Attorney Sara Caplan confirmed Dr. Baden's denial by testifying that she had not picked up a death-scene tooth. She did say, however, that she had seen Dr. Lee pick up a small white object from the carpet of the foyer and place it into a clear evidence vial. The item, she said, was about the size of a fingernail.

Stanley White, the former defense private investigator, testified that he had heard Dr. Lee announce that he had in his possession a death-scene fingernail that he had placed in a handkerchief. White said he had looked at the object and told Dr. Lee that what he saw didn't look like a fingernail. White's observation led to a brief argument with the famed forensic scientist.

Following the testimony of Diamond, White, Caplan, and Dr. Baden, Judge Fidler said he wanted to hear directly from Dr. Lee, who was at the time traveling in China and not expected back in the United States for two weeks. "I believe," the judge said, "that the court has an obligation when there is any inference or allegation that any evidence has been tampered with, to make an inquiry. The judge also said, "I'm a long way from making a determination."

The allegation that Dr. Lee had found of piece of Lana Clarkson's fingernail at the Spector house and kept it from the district attorney fit the prosecution's theory that the victim's fingernail had broken off when Spector forced the gun into her mouth and pulled the trigger. The judge, aware of the implications, was taking the allegation seriously. In mid-May, at a second hearing on the matter, Dr. Lee denied any wrongdoing at the Spector scene. He said he was astonished and insulted by the claims of the former defense investigator and attorney Sara Caplan. Dr. Lee wondered if perhaps Caplan had seen him put a cotton swab into a vial, mistaking it for a fingernail. "I think my reputation is seriously damaged," he was reported to have said.

On May 23, 2007, Judge Fidler made his finding of fact, ruling that it was his belief that Dr. Lee had either hidden or destroyed physical evidence from the death scene. The judge said: "I find the following: Dr. Lee did recover an item. It is flat, white, with rough edges. I cannot say it is a fingernail. It has never been presented by the prosecution" Although he implied it, the judge did not make a finding that Dr. Lee had committed perjury. But he did say this: "If I had to choose between the two [Dr. Lee and Sara Caplan], I am going to chose Ms. Caplan. [She] is more credible than Dr. Lee. Dr. Lee has a lot to lose if this turns out to be true." The judge also disclosed that notwithstanding his finding, Dr. Lee could testify for the defense when the time came, but the prosecution would be allowed to put attorney Caplan on the stand to impeach his credibility. Following the hearing, one of Spector's attorneys said that Dr. Lee would still be taking the stand on behalf of the defense.

Dr. Lee, on May 30, issued a written statement denying hiding or destroying shooting-scene evidence. He said the prosecution was trying to undermine his upcoming testimony for the Spector defense. The only physical evidence he had collected from the scene was "two small bundles of white, thread-like fibers embedded in blood."

Sara Caplan, on June 6, told Judge Fidler that if she were called to the stand by the prosecution to testify as an impeachment witness, she would refuse to testify. She would refuse on grounds of attorney-client privilege. Although she no longer represented Phil Spector, the privilege, in her opinion, still applied. Judge Fidler did not agree. "My sense of justice is not only to Mr. Spector . . . but also to Miss Clarkson and justice herself. She [attorney Caplan] went to a scene and said she saw someone manipulate, conceal, or destroy evidence. There is no privilege not to testify."

Ten days later, after Sara Caplan refused to answer a series of questions put to her by a deputy district attorney regarding Dr. Lee's death-scene activities, Judge Fidler held her in contempt of court. But instead of sending Caplan straight to jail for the duration of the Spector trial, the judge gave her attorney time to appeal his contempt decision. On June 23, an appellate court stayed Caplan's sentence until the appeals court could decide if her refusal to testify fell within the attorney-client privilege.

On June 26, after the government rested its case, the Spector defense led off with Dr. Vincent Di Maio, the former chief medical examiner of Bexar County, Texas. Dr. Di Maio, considered one of the leading experts on the subject of gunshot wounds, testified that he disagreed with the prosecution's experts who had asserted that blood spatter can travel only three feet from a person struck by a bullet. Dr. Di Maio said blood can travel more than six feet if a gun is fired into a person's mouth, explaining that when a revolver is discharged inside a person's mouth, the pressure from the muzzle gas that is trapped in the oral cavity creates a violent explosion. "The gas," he said, "is like a whirlwind, it ejects out of the mouth, out of the nose." Stating that, because 99 percent of intra-oral gunshot deaths are suicides, he believed that Lana Clarkson had killed herself, Dr. Di Maio said that in his thirty-five years as a medical examiner, he had seen only "three homicides that were intra-oral."

In an aggressive cross-examination by the deputy district attorney, Dr. Di Maio was asked how much he had been paid for his work on the case. The former medical examiner said that his bill was $46,000, which did not include his trial testimony. Courtroom spectators laughed when Dr. Di Maio told his cross-examiner that the longer he kept him on the stand, the more it would cost the defendant.

On July 10, 2007, Dr. Lee told an Associated Press reporter in a telephone interview that he was flying to China and didn't expect to be testifying at the Spector trial. "It's up to the defense and my schedule whether I come back to testify. But it will cost a lot for them to bring me back." Dr. Lee said he would be spending the next two weeks traveling to seven provinces teaching courses and helping the local authorities work on some of their cold cases. "I can live with my conscience," he said. "I didn't even know a fingernail was broken and I didn't find a fingernail. . . . When they can't destroy the science, they try to destroy your reputation."

Two days later, with Dr. Lee in China, attorney Sara Caplan took the stand at Spector's trial to avoid jail time for contempt and testified before the jury that she had seen Dr. Lee pick up something white from the death scene. Under sharp cross-examination by one of Spector's attorneys, Caplan said she did not know what the object in question was, or what had happened to it. At the conclusion of her testimony, Judge Fidler told the jury he would put her testimony into context at a later time.

The Spector case will not be good for Dr. Lee's career or for the general cause of forensic science. When celebrity experts get involved in high-profile murder investigations and trials, the stakes for them are extremely high. On September 18, 2007, the jury, following a week of deliberation, announced they were deadlocked seven to five. Two days later Judge Fidler sent them back to the jury room with a new set of instructions on how to determine reasonable doubt. In the Spector trial, the celebrity experts for the defense did more than just muddy the water by pointing out mistakes and erroneous conclusions by the government's experts. They offered a conflicting scenario backed by forensic science. In circumstantial trials like this, deadlocked juries are to be expected. The hung jury is what Phil Spector paid for, and it's what he got.

DR. HENRY LEE'S participation at various levels in so many cases involving such a variety of evidence and analysis is unusual for a forensic scientist. In the field, he is almost a one-of-a-kind practitioner. He has created for himself a role that borders on superhuman. At the core of his expertise, he is a forensic serologist, one who examines crime-scene biological stains to determine their identity and origin. Using a variety of chemicals such as luminal, and crime-site techniques such as special lighting, a forensic serologist can determine, generally and presumptively, if a stain—say, on a bedsheet—is blood, semen, sweat, urine, or saliva. In the crime lab, biological evidence can be individualized through DNA analysis.

As a crime-scene reconstruction expert, one who determines what happened at the crime site by taking into consideration all the physical clues, Dr. Lee is also a blood-spatter analyst. For example, the shape and size of a drop of blood might determine how far it traveled before landing on the wall, ceiling, floor, or table top; its direction of flight; and its velocity or speed. The higher the velocity, the smaller the drop. A gunshot wound to the head will produce tiny dots of blood; a hammer blow, larger, medium-velocity stains; and a punch to the nose, much larger droplets. Blood-spatter experts also analyze a variety of other crime-scene blood patterns such as pools, splashes, cast-offs, swipes, impressions, and prints.

As one who studies physical evidence to figure out, after the fact, what occurred at the scene of the crime, that is, as a criminalist (a term often confused with criminologist, one who studies the sociology of crime), Dr. Lee analyzes all kinds of physical evidence, including hair follicles, fibers, bite marks, bone fragments, brain matter, tissue, gunshot powder residue, soil, dust, pollen, and other forms of trace evidence. Dr. Lee also studies latent footwear and fingerprint patterns and analyzes bullet trajectories. He's a generalist in a field of narrowly defined specialists. This is his appeal and why he has been able to insert himself in so many cases. It may also be his weakness, because his expertise and knowledge cover so much forensic territory, they are spread thin. One man can only know so much.

Because the field of forensic science has become so specialized, limiting the role of its experts to a narrow set of issues, the forensic generalist may soon be a relic of the past. But as long as there is crime television, there will be forensic science celebrities fighting each other for public attention and the big expert-witness fees. While this puts a popular and glamorous face on forensic science, it may not, in the final analysis, be good for the profession. Because science and ego are a bad mix, forensic science is best conducted by behind-the-scenes people who are not worried about living up to their press clippings.

Conclusion

The nature of science itself, and the fact that forensic science is a service mainly delivered by the government, makes solving its problems a real challenge. Science is complex, constantly in flux, and often subject to disagreement. Government is slow, resistant to change, and difficult to hold accountable. The general difficulty with government is exacerbated by the convoluted structure of our criminal justice system and the adversarial nature of the trial process, where winning, not justice, has become the principal goal. Most problems in forensic science can be placed into one of three categories: personnel, jurisprudence (courts and law), and science itself. Many of these problems—governmental funding, the quality of law enforcement personnel, and what legislators and judges do or don't do—are beyond the control of forensic scientists.

DNA Backlogs

There are shortages of qualified practitioners in all the forensic science fields, but the need for more DNA analysts is a critical shortfall that creates enormous frustration and reveals the gap between public expectation and forensic science reality. With the growing awareness among criminal investigators of the identification value of DNA, crime laboratories all over the country have been overwhelmed with submissions of bodily fluid and tissue evidence.

The recent story of a violent robbery in a Bethesda, Maryland, parking garage illustrates how the crime-solving and crime-prevention potential of DNA profiling is seriously weakened by DNA backlogs in the nation's crime labs. Six months after DNA evidence from the garage where the victim was shot seven times (the man survived), the Montgomery County crime laboratory had still not processed this evidence for submission to the FBI's CODIS, a step that might have immediately identified the assailant. A spokesperson for the Montgomery County lab, in explaining the delay, said that six months is not an unusually long time to wait in cases like this. "The demand for analysis exceeds what our lab can do," she said. In the meantime, instead of relying on cutting-edge forensic science, the police were trying to solve the case by distributing poorly rendered composite drawings of the shooter, a technique never used on the television series *CSI*. The robber is still at large.

In 2005, the director of the state crime lab in Missoula, Montana, had to shut down his DNA section for seven months after three of his four analysts left the lab for better-paying positions. It took eight months to fully train their replacements, which created a major DNA backlog. In December 2006, a Montana judge had to postpone a Cascade County murder trial for five months to give the crime lab time to process DNA evidence central to the case. If the evidence did not get analyzed within five months, the judge would have no choice but to dismiss the case and release the murder defendant due to unreasonable delay. Concerned that the laboratory would not handle the assignment even within this period, the judge suggested that the blood-spatter evidence be sent to an out-of-state facility.

In Wisconsin, despite a new state crime lab and an expanded staff of DNA analysts, the facilities in Madison and Milwaukee had a total of 1,775 cases waiting for DNA results at the close of 2006. At the end of 2003, there were only 478 cases waiting, but the Wisconsin backlog has been growing every year, even after a $12 million renovation of the Madison lab in 2004. In Massachusetts, there have been problems with DNA backlogs and identification matches in the state crime lab at Sudbury. In 2002, after the police arrested a suspect in a Cape Cod murder case, it took more than a year to process his DNA, which did link him to the crime scene. In January 2007, the veteran forensic chemist in charge of a Massachusetts crime lab's DNA unit was suspended with pay for delaying the notification of DNA matches in several cases so long that the statute of limitations ran out. This meant that rapists who had not been prosecuted within the time limit had to be released. In some cases, the DNA administrator had sat on the DNA identification information seven or eight months before passing it on to prosecutors.

Misidentification has plagued Massachusetts DNA analyses as well. In 2006, a Massachusetts crime lab chemist identified traces of semen on the body of a murdered woman who, as it turned out, had not been sexually assaulted. The misidentification had been caused by evidence contamination in the lab. In another lab, state DNA auditors uncovered DNA misidentifications in at least ten cases.

The president of the labor union representing the suspended Massachussets administrator blamed the problem on the state's unwillingness to properly staff the crime lab. In four years, the lab had grown from four to thirty employees but had continued to fall behind because of its overload of cases. Funding had been increased from $6.2 million in fiscal 2005 to $16.2 in 2007. The Massachusetts State Police had lent the crime lab two employees, but they had no training in forensic science. The superintendent of the state police said he would ask the FBI to conduct an independent audit of the state's DNA operation.

In Minnesota, a new attorney general who took office in January 2007 promised to eliminate the state's twenty-month DNA backlog, a situation he called "monstrous," according to an Associated Press report by David Weber. The governor had promised to hire fifteen new DNA analysts at a two-year cost of $2.5 million, but according to the attorney general, that would not be enough to eliminate the backlog. To ensure that DNA results would be available to investigators and prosecutors no later than three weeks after submission, the state would have to outsource the work to private laboratories at a cost of $13 million a year. This was not about to happen.

Forensic Pathology

Personnel shortages in the DNA field are not the only staffing problem in forensic science. There are only 400 board-certified forensic pathologists working full time in the nation's medical examiner and coroner offices. An additional 200 medicolegal pathologists perform autopsies part-time and 300 or so pathologists are inactive. According to studies by the National Association of Medical Examiners, at least 850 full-time forensic pathologists are needed to handle the country's autopsy workload. Until the number of forensic pathologists is doubled, many of these cause- and manner-of-death specialists will be performing more than 350 autopsies a year. Studies by the National Association of Medical Examiners have shown that forensic pathologists who perform that many autopsies a year are prone to flagrant mistakes and serious errors in judgment. Exhaustion and overwork, more than incompetence and corruption, are the reasons behind bungled autopsies.

The main reason for the forensic pathology shortage is money. Of the forty forensic pathologists trained every year, a third go into hospital pathology. In too many jurisdictions the chief medical examiner earns less than $150,000 a year. Lack of public money also explains why 40 percent of the medical examiner's offices in the country do not have an in-house toxicology lab. In addition to the salary of the toxicologist, it takes $300,000 to equip one laboratory. In jurisdictions without proprietary toxicological services, specimens have to be sent to state crime labs for analysis. These laboratories are already overwhelmed with submissions, and the added work extends the existing backlogs.

In Marion County, Indiana, the shortage of forensic pathologists combined with the political nature of the coroner's system to make a problematic situation worse. Cause- and manner-of-death rulings in Indianapolis and the surrounding county had been made by an elected coroner acting on the advice of private forensic pathologists working for the county on a contract basis. In 2004, the man who had been the coroner for the past eight years couldn't run for reelection due to term limits. That opened the door for Dr. Kenneth Ackles, a seventy-year-old chiropractor who ran for the position on the Democratic ticket. Although his Republican opponent had a medical degree, Dr. Ackles, backed by the local Democratic Party, won the election and became the new coroner of Marion County.

Dr. Ackles took over a medicolegal operation that, while overworked and under the strain of a thousand autopsies a year, had managed to keep its head above water. The system was operated by Dr. Stephen Radentz, a board-certified forensic pathologist who performed the autopsies with the help of three other highly qualified forensic pathologists. Affiliated with Indiana University School of Medicine, the forensic pathology team worked for the county on a contract basis. While not employed directly by the county, they had worked well with the previous coroner. Following the election of Dr. Ackles, a man some of the contract forensic pathologists considered unqualified for the job, the county and the medical school pathology professors parted ways.

In December 2005, the coroner, on behalf of the county, entered into a five-year contract with a new private pathology group headed by Dr. Radentz, Forensic Pathology Associates of Indiana (FPA). The firm consisted of three full-time board-certified forensic pathologists, a part-time pathologist, and a forensic pathologist-in-training. In June 2006, Dr. Ackles cancelled the FPA contract effective the end of the year to save the county money. He planned to hire instead an in-house staff of forensic pathologists. Since proprietary operations in government and the private sector almost always cost more than corresponding contract services, there were many in the county who questioned the wisdom of this decision. Dr. Stephen Radentz was among those who considered Dr. Ackles's decision a terrible mistake.

In a letter to the mayor of Indianapolis, Dr. Radentz expressed his concerns and laid out the reasons he felt this way:

> Having been in administrative positions for the past seven years, I have experienced the longstanding shortage and difficulty of recruiting qualified board-certified forensic pathologists. For multiple reasons, this chronic deficit has recently culminated in a severe national shortage of forensic pathologists. Several weeks ago I attended the annual National Association of Medical Examiners meeting and every forensic pathologist I spoke to was already aware of the situation, and concerned about "that coroner in Indianapolis." . . .
>
> The re-establishment of a nationally accredited and effective Forensic Pathology Office . . . will, conservatively, take years. There will be no one readily available to testify regarding the majority of current and future homicide trials. . . . The estimated expenses to re-furnish, re-equip, and re-establish the current operating system will be $80,000 to $120,000, and cost several years of valuable lost time.

On December 20, 2006, the day after the $850,000-per-year FPA contract with Marion County expired, Dr. Ackles hired an in-house forensic pathologist from another state. Until he could fulfill his promise to bring in four more board-certified forensic pathologists, most of the county's autopsies had to be farmed out to neighboring jurisdictions. The new pathologist, Dr. Joyce Carter, had signed a five-year contract that would pay her $175,000 the first year with 3 percent increases the following years. She had inherited 150 pending homicide cases, and the transition was expected to be slow and difficult. Dr. Ackles's critics believed he had changed from a contract system to an in-house operation to give himself direct control over the operations of the coroner's office. While this was good for Dr. Ackles, homicide detectives and county prosecutors were not optimistic about the future of the county's medicolegal operation. The new coroner had replaced a medicolegal service that at least had been adequate to the challenge with one that seemed doomed to failure. In "Coroner Defends the Job He's Doing," published in the *Indianapolis Star* December 3, 2006, Dr. Ackles responded to his critics. An ex-boxer, he told reporter John Murray: "I look very easy, but once you rough me up, it's a different thing. I work about eighteen hours a day, and the people that say I'm incompetent . . . in the olden days I'd have put them in a ring, and we'd figure it out."

Coroner versus Medical Examiner Systems

Twenty-two states have replaced the elected coroner system like the one in Marion County with statewide medical examiner's offices. In these jurisdictions, appointed chief medical examiners oversee regional offices staffed by deputy medical examiners. However, in many of these states, the personnel shortage means that not all the medical examiners are forensic pathologists. In some states they are not even physicians; in others, the law requires that the chief medical examiner is a board-certified forensic pathologist and the deputies have medical degrees. Only a handful of states have medical examiner systems completely staffed by board-certified forensic pathologists. Still, any form of medical examiner operation is usually more reliable than its coroner counterpart, because the personnel are usually better qualified and do not have to run for political office. Under the best of circumstances a lot of pressure is brought to bear on those responsible for making manner-of-death rulings. Keeping politics and politicians out of the equation helps relieve some of that pressure and contributes to the credibility of these vital determinations.

In Indiana and the thirty-seven other states that still have counties operating under coroner's offices, there are no statutory qualifications for this elected position. Traditionally, the vast majority of coroners have been morticians or people in the funeral-home business. In the state of Washington, due to conflict-of-interest concerns, morticians are prohibited from running for the job. In four states—Minnesota, Kansas, Louisianna, and Ohio—coroners must at least be physicians. In Wisconsin's seventy-two counties, coroners tend to be nurses, EMTs, physicians, and LPNs, but not morticians. In Pennsylvania, three of the sixty-seven counties have medical examiner systems. Half the coroners in the remaining counties are morticians, with 10 percent EMTs or paramedics. In New York State, each county can set qualifications and decide whether to employ a medical examiner or a coroner. One county prohibits morticians from holding either position.

The Proviano Case

In Ohio, where coroners have to be physicans—a sort of middle ground between the traditional coroner's operation and the typical statewide medical examiner's system—problems can still occur when it comes to determining manner-of-death in cases where there should be no debate. An example is the Anthony Proviano shooting case. On December 28, 1997, near St. Clairsville on the eastern edge of the state across the line from the West Virginia panhandle, the body of a University of Cincinnati medical student was found along a back road not far from the I-70 motel he had checked into for the night. Anthony Proviano had left Cincinnati five days earlier en route to his home in Baldwin, Pennsylvania, where he planned to celebrate Christmas with his family. Olen Martin, chief deputy of the Belmont County Sheriff's Office, was in charge of the investigation.

Proviano had been shot once in the chest by a .25-caliber pistol found a hundred feet from his body. The handgun, registered to him, contained an unfired round. His pullover sweater was balled up, exposing his bloodstained T-shirt, which contained powder stains and the bullet hole. Near his body, crime-scene investigators found a spent shell casing and items of his clothing, including his hat, winter coat, right shoe, and leather gloves.

Dr. Manuel Villaverde, the Belmont County coroner, arrived at the scene and stunned the investigating officers by declaring that Anthony Proviano had committed suicide. According to Dr. Villaverde, who was not a forensic pathologist, Proviano had lifted up his sweater and shot himself in the chest, dropped the gun, then walked a hundred feet, collapsed, and died.

Chief Deputy Martin was absolutely convinced that Proviano had been murdered. The ground around his body had been disturbed and there were several fresh abrasions on his face, elbows, and knees. Later that day, at a funeral home in Bellaire, Dr. Villaverde removed the bullet from the victim's chest, a procedure that should have been performed by a forensic pathologist conducting an autopsy. Chief Deputy Martin, having on four occasions begged the coroner to send Proviano's body to Franklin County for an autopsy, openly questioned the coroner's professionalism. According to *St. Clair Times Leader* reporter Eric Ayres, in justifying his actions, Dr. Villaverde said, "Maybe we'll save the county some money."

In Ohio the county prosecutor, not the coroner, is the ultimate authority in determining, in sudden death cases, if foul play is a possibility. As a result, notwithstanding the coroner's official ruling of suicide, the investigation went forward. However, the fact the coroner didn't agree with the prosecutor's manner-of-death assessment would be a serious impediment if the case ever went to trial. Moreover, there had been no autopsy.

Dr. Villaverde's manner-of-death ruling had infuriated the police and the victim's family, who hired a private forensic pathologist, Dr. Leon Razin, to perform an autopsy. Dr. Razin, the former chief of forensic pathology with the Allegheny County, Pennsylvania, coroner's office, concluded that Anthony Proviano had been murdered.

On October 23, 1998, Dr. Villaverde, under pressure from the victim's family, two local congressmen, and the Belmont County prosecutor, changed the manner of death on Proviano's death certificate to "undetermined." Dr. Villaverde made it quite clear, however, that he still believed that the twenty-nine-year-old had taken his own life; according to the coroner, he had changed the manner-of-death designation simply to make the family feel better.

Dr. Villaverde ran for reelection in November 2000 and lost by a wide margin to Dr. Gene Kennedy, who had campaigned on the incumbent's role in the Proviano case. In August 2001, the new coroner changed Anthony Proviano's manner of death to "homicide." On February 21, 2006, a Belmont County jury found a fifty-year-old drug dealer/posititute named Marlene "Slim" Smith guilty of murdering Proviano with his own gun. The judge sentenced her to life. As it turned out, Dr. Villaverde's attempt to save the county the cost of an autopsy had not, thanks to Chief Deputy Martin and others, allowed a killer to get away with murder.

Backlogs in Fingerprint Identification

Personnel shortages exist even in the fingerprint identification field. In June 2006, the fingerprint unit in the Oakland Police Department Crime Laboratory, consisting of three examiners, had to be closed due to lack of money and staff. In Oakland, about 113 crime scenes a month produce latent fingerprint evidence. Many of these crime sites involve criminal homicides, an offense committed up to 150 times a year in the city. Following the closing, Oakland's crime-scene latents were sent to the sheriff's office laboratory in Contra Costa County. Seven months after the shutdown, with a backlog of 162 cases, many investigators in Oakland weren't bothering to submit fingerprint evidence. The head of the Oakland lab predicted in December of 2006 that the fingerprint unit wouldn't be reopened until the summer of 2007. By then, this vital unit would have been closed for a year. Kenneth Moses, the head of a private crime lab in San Francisco, called the inability to fund and staff the Oakland fingerprint unit outrageous, noting that the latent fingerprint is the most commonly gathered form of crime-scene evidence.

While television shows like *CSI* have popularized forensic science as a career, resulting in hundreds of new academic majors in the nation's colleges and universities, limited resources have prohibited or slowed the needed growth of this service within the public sector. As a governmental priority, forensic science falls well below basic law enforcement, the war on drugs, and homeland security.

Quality Control

Limited resources have affected the quality of forensic science training, supervision, and quality control. Maintaining a high-quality government service has never been easy. Independent advisory boards are a good idea but are not enough to ensure that crime labs are performing at the highest professional levels. Over the years, had it not been for the persistent cries of government whistleblowers, the public would not have been alerted to serious flaws in the nation's crime labs. Journalistic inquiries such as those conducted by the *Chicago Tribune*, the *Seattle Post-Intelligencer*, and *USA Today* have exposed problems in forensic science. The ACLU and private sector attorneys like DNA specialist Barry Scheck and innocence projects modeled after his have provided a measure of after-the-fact quality control. The lesson here is simple: the government can't police itself.

Quality control is especially important in the fields of DNA analysis and latent fingerprint identification. The fact that DNA analysts like Fred Salem Zain and Joyce Gilchrist operated so long without detection is reason for concern. The news that most of the nation's fingerprint examiners have not taken or passed proficiency tests is also alarming. Moreover, of all the forensic sciences, the latent fingerprint examiner is the least educated. Most don't have bachelor's degrees and even fewer have college degrees in science. Only a few police agencies and crime laboratories require minimum qualifications for the job beyond a high school degree and a felony-free background. If latent fingerprint identification is a science, why are its practitioners so poorly educated? In the post-*Daubert*, Judge Pollak era, fingerprint examiners without science backgrounds are vulnerable to challenge in court.

Even the most qualified fingerprint examiners, handwriting experts, and footwear identification specialists make honest mistakes. Particularly in fields of subjective identification, bias has a way of creeping into the analysis. A series of studies and experiments involving fingerprint examiners in England by a pair of cognitive psychologists has shown that "biasing contextual information" can lead to mistaken conclusions. For example, when fingerprint examiners were told a suspect had confessed, these experts made identifications in cases where, without this knowledge, they had previously declared the same set of prints a mismatch. These studies, conducted at the University of Southampton, suggest that latent fingerprint work and, by implication, handwriting identification and footwear impression comparison are more subjective than previously believed. In light of these findings, the less these forensic experts know about the crime in question, the better. Within the fingerprint field, erecting a wall between the examiner and the criminal investigation is much more difficult when the expert is employed directly by the law enforcement agency.

In April 2007, a latent-fingerprint examiner with the Seminole County Sheriff's Office in central Florida complained to her boss that Donna Birks, an eleven-year veteran of the identification unit, was biased and unethical. After fingerprint examiners with the Florida Department of Law Enforcement determined that Birks had made false identifications in six cases, she resigned. The state reviewers found that another member of the sheriff's fingerprint bureau had made one misidentification. The sheriff demoted this employee to dispatcher. According to the fingerprint auditors, none of these misidentifications were intentional. The scandal caused the closing of the county fingerprint operation until qualified personnel could be hired.

Scientific Objectivity

In order to maintain scientific objectivity, forensic science practitioners have to rise above the adversarial nature of the trial process. They have to be true to their science. This is especially difficult when their conclusions conflict with the law enforcement view of the case. Staying at arm's length from law enforcement is much easier for experts in the private sector. Crime lab employees who get too involved in the overall crime investigation are more vulnerable to prosecutorial pressure and influence. Keeping a firewall between forensic science and criminal prosecution is vital but difficult. It's easy to understand, for example, how a forensic pathologist in a medical examiner's office might feel as if he or she is part of a law enforcement team, particularly in emotional cases, such as those involving suspected infanticide and child abuse.

Pseudoscience

Criminal trial judges have to protect juries from pseudoscience, unqualified expert witnesses, and prosecutors who draw inappropriately incriminating inferences from valid scientific testimony. For example, there should be no such thing as forensic footprint morphology. People like Dr. Louise Robbins, Dr. Ralph Erdmann, Dr. Michael West, and Fred Salem Zain have no place in a court of law. Beyond that, prosecutors should not be allowed to assure jurors that a crime-scene hair follicle (without a root that would allow for DNA analysis), bite-mark wound, or questioned handwriting came from the defendant to the exclusion of all others. No person should be convicted solely on the basis of these forms of identification.

Dueling Experts

The increasing presence of dueling expert witnesses, given the procedural nature of the U.S. trial, is a problem without a satisfying remedy. The blowing smoke phenomenon is particularly vexing. If there is an answer to the blowing smoke, muddying the waters problem, it will have to come from within the forensic science community in the form of a tighter code of ethics. Regarding battling experts, judges could help by imposing stricter standards in the area of expert qualification. This would keep out the phonies and reduce the opportunity for opposing testimony, particularly in the field of questioned document examination.

A Final Word

As I have said, many of the difficulties affecting the quality of forensic science have nothing to do with the profession itself. As a result, there is little administrators and practitioners in the profession can do to solve these problems. The inherent difficulty of crime-scene investigation—the fragile nature of some physical evidence, weather conditions (heat, rain, snow, etc.), and problems associated with protecting the scene from unauthorized visitors—attenuates the effectiveness and application of forensic science. Most of the people involved in forensic science—police officers and detectives—are not scientists and do not work under laboratory conditions. Once a crime scene has been bungled or ignored, there is not much the forensic scientist can do to correct the problem.

The sheer volume of crime and the dwindling number of competent police investigators, a problem created in part by the escalating wars on drugs and terrorism, has detracted from the crime-fighting potential of forensic science. Also diminished is the role of forensic science in the protection of the innocent. This aspect of forensic science has been undermined by prosecutorial misconduct and weak defense advocacy. Jury nullification, as illustrated in the O. J. Simpson case, can be a problem as well.

There is no question that the U.S. criminal justice system has benefited greatly from forensic science. The recent wave of public interest in, and knowledge of, forensic science has created high expectations for the profession. Perhaps rather than try to lower these reasonable expectations, prosecutors should demand more from police officers and detectives. As long as convictions continue to be based on the testimony of eyewitnesses, questionable confessions, and jailhouse informants, the great potential of forensic science won't be realized. In forensic science and law enforcement, it would be a good idea if life started imitating art.

NOTES

CHAPTER 1 FORENSIC PATHOLOGISTS FROM HELL

20 For Wood's autopsy report, "Man Freed Amid Questions Raised over Autopsy," Associated Press, November 21, 2002.

CHAPTER 2 A QUESTION OF CREDIBILITY

33 For Berkland quotes, "Female Employee Found Dead in Scarborough's District Office," Associated Press, July 21, 2001.

CHAPTER 3 THE SUDDEN INFANT DEATH DEBATE

52–55 All Marybeth Tinning quotes are from "Marybeth Tinning Killed Her Nine Babies," Court TV Crime Library, *www.crimelibrary.com.*

57 For Williams quote, "Baby Deaths Doctor 'Breached Duty to Be Fair,'" Press Association, *Society Guardian* (UK), May 31, 2005.

59 For Southall quote, Sandra Laville, "Professor Admits Murder Claim: Expert Accuses Father of Killing Babies on Basis of Documentary," *Guardian* (UK), June 8, 2004.

60 For Drucker quote, Tony Leather, "An Expert Manipulation of Justice," *www.nhsexposed.com,* August 21, 2003.

CHAPTER 6 FINGERPRINT IDENTIFICATION

118 For Foilb quote, Jonathan Saltzman, "SJC to Hear Arguments on Banning Fingerprint Evidence," *Boston Globe,* September 5, 2005.

122 For Cole quote, Jonathan Saltzman, "SJC Bars a Type of Prints at Trial," *Boston Globe,* December 28, 2005.

CHAPTER 7 FINGERPRINTS NEVER LIE

126 For Judge Johnston's quote, "Inquiring into Fingerprint Evidence," BBC News, *www.bbc.co.uk,* February 7, 2000.

129 Wertheim's message is quoted in "Finger of Suspicion," *Frontline Scotland,* BBC-Scotland, transcript of broadcast January 18, 2000.

129 For Wertheim quote, "False Impression," *Frontline Scotland,* BBC- Scotland, transcript of broadcast May 16, 2000.

129 For Wallace quote, "Fingerprints to Be 'Double-Checked,'" BBC News, *www.bbc.co.uk,* June 22, 2000.

131 For Bayle quote, "Experts Seek Fingerprint Inquiry," BBC News, *www.bbc.co.uk,* September 19, 2002.

131 For McKie quote, "'Cover-Up' Claims over Prints," BBC News, *www.bbc.co.uk,* October 9, 2003.

131 For Jamieson quote, "Fingerprint Service Defended," BBC News, *www.bbc.co.uk,* November 25, 2003.

132 For Iain McKie quote, "Anger at Fingerprint Case Claim," BBC News, *www.bbc.co.uk,* April 7, 2004.

133 For Scottish ministry quote, "'Relief' over Fingerprint Verdict," BBC News, *www.bbc.co.uk,* February 7, 2006.

134 For McBride quote, Liam McDougall, "Revealed: Detective in Charge of Murder Case Wanted Shirley McKie 'Gagged' by Executive," *Herald* (UK), February 25, 2006.

134 For Zeelenberg quote, "Father's Fingerprint Error Anger," BBC News, *www.bbc.co.uk,* February 8, 2006.

138 For Marquise quote, Eben Harrell, "Pressure Mounts for McKie Inquiry as U.S. Announces FBI Investigation," *Scotsman,* March 22, 2006.

CHAPTER 8 SHOE-PRINT IDENTIFICATION AND FOOT MORPHOLOGY

148 For Riddle quote, Flynn McRoberts, Steve Mills, and Maurice Possley, "Unproven Techniques Sway Courts, Erode Justice," *Chicago Tribune,* October 17, 2004.

151 For prosecutor James Yeatts quote, Phoebe Zerwick, "Mixed Results; Forensics, Right or Wrong, Often Impresses Jurors," *www.journalnow.com,* August 29, 2005.

CHAPTER 9 BITE-MARK IDENTIFICATION

156 For Maryland odontologist's quote, Flynn McRoberts, and Steve Mills, "From the Start, a Faulty Science," *Chicago Tribune,* October 19, 2004.

CHAPTER 10 EAR-MARK IDENTIFICATION

170–174 For all quotes in the section on the David Kunze case, *State v. David Wayne Kunze,* Court of Appeals of Washington, Division 2, 97 Wash. App. 832, 988 P. 2d 977 1999.

176 For Judge Mitchell quote, Andre A. Moenssens, "DNA Evidence Proves Ear ID Wrong," *www.forensic-evidence.com,* 2004.

CHAPTER 11 EXPERT VERSUS EXPERT

183 For Wecht quote, Clay Evans, "Coroner Apologizes for Wecht's Remarks," *Boulder Daily Camera,* March 20, 1997.

188–189 For Hoffman letter, "Darnay Hoffman. The Hoffman Files," *www.acandyrose.com,* accessed 2001.

190 For district court judge's quote, Matt Sebastian, "Judge Rejects Suit against D.A," *Boulder Daily Camera,* January 21, 1998.

192–193 For Kane letters to Liebman and Wong, Michael J. Kane letter to Cina L. Wong, January 20, 1999, "The Hoffman Files," *www.acandyrose.com.*

194 For Owens quote, "Detective Lou Smit, Ret." *www.lousmit.com.*

196–201 For Epstein quotes, Gideon Epstein, Deposition, *Wolf v. Ramsey,* May 17, 2002, pp. 178, 179.

CHAPTER 12 JOHN MARK KARR

210 For Karr's e-mails, Helen Kennedy, "Twisted Suspect Obsessed with Kids," *New York Daily News,* August 18, 2006.

216 For Owens quote, Kevin Simpson, "Tests Rule Out Karr as Source of DNA, *New York Times,* August 29, 2006.

216 For Lacy quote, "Boulder D.A. Defends Decision to Arrest Karr," Associated Press, August 29, 2006.

CHAPTER 13 HAIR AND FIBER IDENTIFICATION

219 For Blake quote, Mark Wrolstad, "Hair-Testimony Flawed as a Forensic Issue: DNA Testing Reveals Dozens of Wrongful Verdicts Nationwide," *Dallas Morning News,* March 31, 2001.

CHAPTER 15 BULLET IDENTIFICATION, FBI STYLE

247 For Peters quote, *State of New Jersey v. Michael S. Behn,* Superior Court of New Jersey, Appellate Division.

CHAPTER 16 THE CELEBRITY EXPERT

258–259 For CNN transcript, "Defense to Examine Laci Peterson Remains," CNN TV, *www.transcripts.cnn.com,* transcript of broadcast August 8, 2003.

263 For Peterson quotes, John Springer, "Woman's Death in Germany 18 Years Ago Enters Novelist's Murder Trial," *www.courttv.com,* August 19, 2003.

268 For chauffeur's quote, Linda Deutsch, "Criminalist Says Spector's DNA Wasn't Detected on Gun," Associated Press, June 9, 2007.

271 For judge's first quote at first hearing, Harriet Ryan, "Spector Insiders Call into Question the Conduct of Famed Criminalist Henry Lee at Death Scene," *www.courttv.com.* May 4, 2007; for judge's second quote, Peter Y. Hong, "Spector Trial Claim: Bombshell or Sideshow," *Los Angeles Times,* May 7, 2007.

271 For Lee quote, Harriet Ryan, "Misstep in Spector Case Could Haunt Renowned Forensic Scientist for Cases to Come," *www.courttv.com,* May 25, 2007.

271 For first quote from judge's finding of fact, Linda Deutsch, "Renowned Forensic Scientist Henry Lee's Credibility Challenged in Phil Spector Murder Trial," Associated Press, May 25, 2007; for second quote, Harriet Ryan, "Judge in Spector Trial Rules Defense Expert Dr. Henry Lee Hid or Destroyed Evidence," *www.courttv.com,* May 23, 2007.

272 For Lee quote, Linda Deutsch, "Dr. Lee Says He's Being Slandered to Undermine Spector Testimony," Associated Press, May 31, 2007.

272 For Di Maio quote, Peter Y. Hong, "Spector Trial Expert Backs Suicide Theory," *Los Angeles Times,* June 27, 2007.

273 For Lee quote, Linda Deutsch, "Expert Says He's Going to China, May Not Testify in Spector Trial," Associated Press, July 10, 2007.

CONCLUSION

276 For spokesperson quote, Stephanie Siegel. "Family Will Not Sleep While Gunman Free," *Gazette*.net (Montgomery County, Md.), December 6, 2006.

279 For Radentz letter, "Public Announcement: Marion County Coroner's Office Going Out of Business," *www.advanceindiana.blogspot.com,* November 8, 2006.

SOURCES

CHAPTER 1 FORENSIC PATHOLOGISTS FROM HELL

Dr. Ralph Erdmann

Bragg, Roy. "Autopsy Record of Pathologist Who Quit Raises Many Eyebrows." *Houston Chronicle.* March 8, 1992.

———. "Botched Autopsy Causes Judge to Reduce Charges." *Houston Chronicle.* April 4, 1992.

Brown, Chip. "Pathologist Accused of Falsifying Autopsies." *Los Angeles Times.* April 12, 1992.

Campbell, Geoffrey A. "Erdmann Faces New Legal Woes." *ABA Journal.* November 1995.

Dutton, Rex. "Agencies Seen As the Real Bad Guys in '93 Death Case." *San Diego Union-Tribune.* March 30, 1998.

Fricker, Richard L. "Pathologist's Plea Adds to Turmoil." *ABA Journal.* March 1993.

Rubin, Frank. "The Homicide That Almost Slipped Away." *National Law Journal.* September 28, 1992.

Suro, Roberto. "Ripples of a Pathologist's Misconduct in Graves and Courts of West Texas." *New York Times.* November 22, 1992.

Van Story, Jay B. "Lubbock County D.A. Travis Ware's Reign of Error and Terror." February 10, 2005. *www.innocentinmates.org/vanstory/terror.html.*

Dr. Joan E. Wood/McPherson Case

"Another Man Jailed Because of Medical Examiner's Mistakes." *ABC Action News,* Tampa–St. Petersburg. November 23, 2002.

Davis, Cary. "Former Medical Examiner Wood Snubs Subpoenas." *St. Petersburg Times.* May 6, 2001.

Farley, Robert. "Scientologists Settle Death Suit." *St. Petersburg Times.* May 29, 2004.

Felony Indictment. November 13, 1998. Circuit Court of the Sixth Judicial Circuit of Florida, Pinellas County.

"Inmate Gets 30-Day Stay of Execution." Associated Press. December 3, 2002.

Levesque, William R. "Coroner: Old Case Won't be Reopened." *St. Petersburg Times.* January 20, 2003.

———. "Ex-Medical Examiner Unable to Testify." *St. Petersburg Times.* August 23, 2002.

———. "Pinellas-Pasco Medical Examiner Retires." *St. Petersburg Times.* June 29, 2000.

David Long's Ordeal

"Man Freed Amid Questions Raised over Autopsy." Associated Press. November 21, 2002.

"Medical Examiner's Apparent Mistakes Put Man in Jail." *ABC Action News,* Tampa–St. Petersburg. November 21, 2002.

Freeing John Peel

"Another Man Jailed Because of Medical Examiner's Mistake." *ABC Action News,* Tampa–St. Petersburg. November 23, 2002.

Martin, Susan Taylor. "Scientology Case Takes Its Toll." *St. Petersburg Times.* July 9, 2005.

Dr. Richard O. Eicher

Meinhardt, Jane. "151 Autopsies Rechecked: The Mistakes Have Resulted in an Exhumation and a New Investigation by Police." *St. Petersburg Times.* June 24, 2000.

Tobin, Thomas C. "Questions Put 150 Autopsies in Doubt." *St. Petersburg Times.* April 13, 2000.

Dr. Angelo Ozoa/Galbraith Case

Hinton, William Dean. "Skeletons in the Coroner's Office." *Silicon Valley Metro.* June 16, 22, 2004.

Levey, Noam. "Claims of Inadequate Procedure May Give Lift to Family's Lawsuit." *San Jose Mercury News.* July 27, 2002.

Lynem, Julie N. "Palo Alto Man Sues Former County Coroner; Botched Autopsy of Wife Led to His Arrest, Suit Says." *San Francisco Chronicle.* September 3, 1999.

Nelson Galbraith v. County of Santa Clara and Ngelo Ozoa, M.D. Filed October 9, 2002, No. 00–17369, D.C. No. CV-99–20887-SW. Appeal from the U. S. District of California.

Stein, Loren. "Digging Up Buried Bones." *www.truthinjustice.org.* 2002.

Woolfolk, John. "Santa Clara County's Suspended Coroner Quits." *San Jose Mercury News.* November 7, 2003.

Zapler, Mike. "Decade-Old Autopsy Casts Long Shadows." *San Jose Mercury News.* January 23, 2005.

Dr. Charles Harlan

"Botched Autopsies, Odd Behavior Catch Up to Medical Examiner." Associated Press. May 1, 2005.

Clancy, Chris. "Medical Panel Rejects Coroner's Request." *Nashville City Paper.* November 19, 2003.

"Findings against Medical Examiner Dr. Charles Harlan." *Nashville Tennessean.* April 22, 2005.

Johnson, Rob. "Officials: Remains in Former Harlan Home May Be Pet." *Nashville Tennessean.* January 19, 2005.

———. "Witness Says Harlan Rushed Man's Autopsy." *Nashville Tennessean.* May 22, 2003.

Loggins, Kirk. "Harlan's Problems Could Cause Court Cases to be Re-Examined." *Nashville Tennessean.* January 15, 2002.

"New State Medical Examiner Appointed." *www.forensicmed.com.* March 19, 1998.

"Public Autopsies in Private Hands." *www.forensicmed.com.* August 1, 1999.

"State Files Suit against Dr. Charles Harlan." *News release.* Division of Consumer Affairs, Tennessee. December 12, 1997.

"State of Tennessee Department of Health in the Matter of Charles Harlan, M.D. Respondent, Nashville, Tennessee License Number 7778; Before the Board of Medical Examiners." Docket No. 17.18–022307A, Final Order.

Wardle, Amanda. "Harlan Sits in Hot Seat." *Nashville City Paper.* May 20, 2003.

CHAPTER 2 A QUESTION OF CREDIBILITY

Michael Berkland Debacle

Berkland/Missouri

George, Chris, and Denis Wright. "Unwrapped, Part 2: Dr. Michael Berkland in Missouri." *www.onlinejournal.com.* November 21, 2001.

Berkland/Klausutis Case

"An Untidy Wrap-Up for Klausutis Case." Editorial. *Northwest Florida Daily News.* January 9, 2002.

"Associate Medical Examiner Suspended in Licensing Flap." *http://www.beachbrowser.com/Archives/Local-News/August-99/Associate-medical-examiner-suspended-in-licensing-flap.htm.*

Baker, Tamara, and Celeste Harrison. "Big Wheel Keeps on Turning." *American Politics Journal.* July 21, 2003.

"Female Employee Found Dead in Scarborough's District Office." Associated Press. July 21, 2001.

George, Chris, and Denis Wright. "Never Bound by the Truth." *American Politics Journal.* September 2, 2001.

McLaughlin, Tom. "Examiner: Klausutis' Death Was Accidental." *Northwest Florida Daily News.* August 7, 2001.

"NFDN Defends Press' Right to Ask Questions." *Northwest Florida Daily News.* August 5, 2001.

Pivnick, Derek. "Heart Problem Contributed to Aide's Death; Medical Examiner Says Woman Hit Her Head during Fall, Killing Her." *Pensacola News-Journal.* August 7, 2001.

"Police Chief Says No Indication of Foul Play in Klausutis Death." *Northwest Florida Daily News.* July 21, 2001.

Van Bergen, Jennifer. "Part 1: Congressional Aide Found Dead in Congressman's Office." *www.truthout.org.* January 4, 2002.

Wright, Denis, and Chris George. "A Death in the Congressman's Office: Does Anybody in the Press Care about Lori Klausutis?" *American Politics Journal.* August 8, 2001.

———. "Unwrapped, Part 1: A Strange Way to Die." *www.onlinejournal.com.* November 10, 2001.

Dr. Charles Siebert/Martin Anderson Boot Camp Case

Bender, Michael C. "Lawmakers to Continue Push on Medical Examiner's Removal in Boot Camp Death Case." *www.palmbeachpost.com.* March 25, 2006.

Burlew, Jeff. "Complaints Filed against Bay County Medical Examiner." *Tallahassee Democrat.* May 26, 2006.

CBS/Associated Press, "Boot-Camp Body May Be Exhumed." *www.cbsnews.com*. February 25, 2006.

Dorsey, Sara. "Family of Boy Who Died after Beating at Florida Boot Camp . . ." Tampa Bay's 10 News. March 13, 2006.

"Families File Complaints against Coroner in Boot Camp Case." WESH 2 (Orlando). May 25, 2006. Accessed at *http://caica.org/ NEWS%20DEATHS%20MARTIN%20coroner %20coverup%202.htm*.

"Florida Attorney General Wants Probe of Coroner." Associated Press. April 26, 2006.

Goodnough, Abby. "2nd Autopsy in Youth Boot Camp Death Fails to End Questions." *New York Times*. March 18, 2006.

Green, Robert. "Florida Boot Camp Death Not from Sickle Cell: Doctor." *www.reuters.com*. March 14, 2006.

Greene, Lisa. "A Medical Problem Stirs a Societal Debate." *St. Petersburg Times*. October 7, 2006.

Hiaasen, Carl. "Autopsy: At Least Doc. Got Gender Right." *Miami Herald*. March 15, 2006.

Kaczor, Bill. "Medical Examiner Defends Boot Camp Finding." Associated Press. March 18, 2006.

Kallestad, Brent. "Black Caucus and NAACP Want Juvenile Boot Camp Closed." *Lakeland (Fla.) Ledger, www.theledger.com*. February 15, 2006.

———. "Mother Says Boot Camp Guards 'Murdered' Her Son." *Lakeland (Fla.) Ledger, www.theledger.com*. February 16, 2006.

Leary, Alex, and Joni Jones. "Another Mom Assails Autopsies." *St. Petersburg Times*. February 21, 2006.

"Medical Examiner Defends Boot Camp Finding." Associated Press. *Houston Chronicle, www.chron.com*. March 16, 2006.

Miller, Carol Marbin. "E-Mails Put Florida Investigator in Hot Seat." *Miami Herald*. March 18, 2006.

———. "Prosecutor Has Options Despite Tests." *Miami Herald*. March 16, 2006.

Miller, Carol Marbin, and Jacob Goldstein. "Examiner: Prompt Aid Might Have Saved Teen." *Miami Herald*. April 17, 2006.

Miller, Carol Marbin, and Marc Caputo. "Autopsy Uproar in Teen's Boot Camp Death." *Miami Herald*. February 21, 2006.

———. "E-mails Lead to FDLE's Ouster in Case." *Miami Herald*. March 31, 2006.

Miller, Kimberly. "Almost No Place Is Safe in Tornado, Town Learns." *Palm Beach Post*. September 18, 2004.

Nelson, Melissa. "Boot Camp Medical Examiner Plans Appeal in Other Cases." Associated Press. September 13, 2006.

———. "Boy Beaten in Boot Camp Died." *www.heraldtribune.com*. February 17, 2006.

———. "NAACP: Boot Camp Withheld Details of Teen's Encounter with Guards." *Bradenton (Fla.) Herald, www.bradenton.com*. March 19, 2006.

———. "Tape Shows Kneeing, Hitting." *Lakeland (Fla.) Ledger. www.theledger.com*. February 18, 2006.

———."Teen's Boot Camp Death 'Natural'; Medical Examiner: Boy Died from Internal Bleeding Brought on by Sickle Cell Trait." *Lakeland (Fla.) Ledger, www.theledger.com*. February 17, 2006.

"Our Position: It's Important to Revise Autopsy Report in Boot-Camp Death." Editorial. *www.orlandosentinel.com*. March 17, 2006.

Price, Stephen. "State Panel Puts Siebert on Probation." *www.tallahassee.com*. August 9, 2006.

Skoloff, Brian. "Supervised Probation Advised for Boot Camp Medical Examiner." *www.gainsville.com*. August 9, 2006.

Stacy, Mitch. "Expert, 2nd Autopsy Indicates Teen Didn't Die of Natural Causes." *Bradenton (Fla.) Herald, www.bradenton.com*. March 14, 2006.

———. "New Autopsy Finds Florida Teen Was Suffocated." Associated Press. May 6, 2006.

"Statement by Bay County Medical Examiner on Autopsy." *St. Petersburg Times*. March 17, 2006.

Vansickle, Abbie, and Alex Leary. "Youth Suffocated, Report Says." *St. Petersburg Times*. May 6, 2006.

CHAPTER 3 THE SUDDEN INFANT DEATH DEBATE

SIDS in General

Bittner, Emily. "Baby-Death Cases Revived by Improved Technology." *Arizona Republic*. December 4, 2004.

Condon, Deborah. "Brain Abnormalities Linked to SIDS." *www.irishhealth.com*. November 1, 2006.

Hall, Seth. "Study Links Genetic Factors to Cot Deaths." *Guardian* (UK). November 1, 2006.

"New Virus Suspected in Two SIDS Cases." Associated Press. *www.msnbc.com*. September 2, 2004.

Norman, Ken. "The Causes of Cot Death." *www.portia.org*. June 2000.

Rao, Vasanti, M.D. "Sudden Infant Death Syndrome." Cayuga Medical Center at Ithaca. *www.cayugamed.org*. 2000.

Stein, Rob. "Study Finds Natural Causes in Multiple SIDS Cases." *Washington Post*. January 1, 2005.

Teichroeb, Ruth. "SIDS Often Used as a 'Catchall' Category." *Seattle Post-Intelligencer*. October 31, 2002.

Rise of Dr. Roy Meadow

Cohen, David. "Professor's Obsession with Child Deaths Has Robbed Me of My Little Girl Too." *Guardian* (U.K.). June 14, 2003.

Leather, Tony. "An Expert Manipulation of Justice." *www.nhsexposed.com*. August 21, 2003.

Meadow, Roy. "Munchausen Syndrome by Proxy: The Hinterlands of Child Abuse." *Lancet*. August 13, 1977.

"Munchausen's Syndrome by Proxy." BBC News. *www.bbc.co.uk*. March 3, 2004.

"Sir Roy Meadow, Paediatrician in Leeds England." *www.whonamedit.com*. 2004.

Verkaik, Robert. "Six Women to Have Verdicts Reviewed." Mothers Against MSBP Allegations. *www.MSBP.com*. 2004.

Tinning Case

"Marybeth Tinning." Bedford Hills Correctional Facility. *www.schenectadyny.info*. 2004.

"Marybeth Tinning Killed Her Nine Babies." Court TV Crime Library. *www.crimelibrary.com*. 2004.

Clark Case and Repeal of Meadow's Law

Batt, John. "Doctors on Trial: Time to Put Experts to the Test." BBC News. *www.bbc.co.uk*. July 11, 2006.

Byard, Roger W. "Unexpected Infant Death: Lessons from the Sally Clark Case." *Medical Journal Australia*. August 2004.

Cohen, David. "Scourge of the Child Snatchers." *London Evening Standard*. February 24, 2003.

Crook, Amanda. "Riddle as Sally Clark Dies." *Manchester Evening News*. March 17, 2007.

"Disappointed and Disheartened." BBC News. *www.bbc.co.uk*. February 17, 2006.

Doward, Jamie. "Parents Demand Gag on Cot Death Doctor's Lectures: Outrage at International Acclaim for Meadow." *London Observer*. January 16, 2005.

Driscoll, Margarette. "Up to 5,000 Cases That Had Been through the Family Courts Might Need to be Looked at Again." *London Sunday Times*. February 1, 2004.

"Experts Buoyed by Meadow Ruling." BBC News. *www.bbc.co.uk*. February 18, 2006.

Marsh, Paula. "Doubts over Dead Baby Cases." *Birmingham Evening Mail*. January 30, 2003.

"Minister Intervenes over Meadow." BBC News. *www.bbc.co.uk*. May 20, 2006.

"Paediatrician Attracts Complaints." BBC News. *www.bbc.co.uk*. April 17, 2006.

"Profile: Sir Roy Meadow." BBC News. *www.bbc.co.uk*. July 15, 2005.

Reid, Sue. "The Children Snatcher in Chief." *London Daily Mail*. June 30, 2003.

———. "Mother Is Banned by Law from Caring for Her Infant Son; Lost a Baby to Cot." *London Daily Mail*. July 7, 2003.

Swingler, Steve. "New Hope for Julie." *Birmingham Evening Mail*. January 30, 2003.

Williams, Margaret. "The Rise and Fall of a Medical 'Expert.'" *www.meactionuk.org*. June 22, 2003.

Dr. David Southall/Clark Case

Laville, Sandra. "Professor Admits Murder Claim: Expert Accuses Father of Killing Babies on Basis of Documentary." *Guardian* (UK). June 8, 2004.

"Murder Claim Doctor Guilty of Misconduct." *Guardian* (UK). August 6, 2004.

"Sally Clark's Husband Thought Murder Claim Was 'Sick Joke,'" PA News, *www.injustice-busters.com*. June 8, 2004.

Dr. Alan Williams/Clark Case

"Baby Deaths Doctor 'Breached Duty to Be Fair,'" Press Association. *Society Guardian* (UK). May 31, 2005.

Dr. Roy Meadow/Patel Case

Cohen, David. "Medical Expert in the Firing Line." *London Evening Standard*. June 12, 2003.

Jenkins, Simon. "Trupti Patel and the Rotten Courts of Salem." *London Times*. June 13, 2003.

McCall, Felicity. "Trupti Patel." *www.portia.org*. 2004.

Yates, Nathan. "GMC Probing Trupti Expert." *London Daily Mirror*. June 13, 2003.

Meadow's Law in the United States

"Bed-Sharing Raises Cot Death Risk." *Daily Mail*. August 7, 2005.

Buchanan, Katy. "Understanding Why Some Infants Don't Gasp." *Pittsburgh Post-Gazette*. January 25, 2006.

"Child Taken into Custody after Mother Arrested in Possible Abuse." Associated Press. September 23, 2005.

Geddes, Linda. "Brain Defects May Cause 50% of Cot Deaths." *www.newscientist.com.* October 31, 2006.

"Scientists Identify Genetic Underpinnings of Some Sudden Infant Deaths." *www.scientificamerican.com.* July 20, 2004.

Talaga, Tanya. "Pacifiers Cut Risk of SIDS: Study." *Toronto Star.* December 9, 2005.

MSBP in Great Britain

Gornall, Jonathan. "Was Message of Sudden Infant Death Misleading?" *British Medical Journal.* December 2, 2006.

CHAPTER 4 INFANTS WHO CAN'T BREATHE

Dr. Alfred Steinschneider/H. Case

Busch, Frederick. "A Mother on Trial." *New York Times.* September 14, 1997.

"Cot Death Error 'Costs Lives.'" BBC News. *www.newsbbc.co.uk.* February 25, 1999.

MacKeen, Dawn. "New Controversy over Sudden Infant Death Syndrome." *www.salon.com,* September 1997.

Steinberg, Jacques. "Mother Goes on Trial in 5 Deaths." *New York Times.* March 31, 1995.

Gedzius Case

Tanner, Lindsey. "One Mother, Three Fathers, Six Cases of SIDS?" Associated Press. *South Coast Today.* December 30, 1997.

Wetsch, Elisabeth. "Deborah Gedzius." Serial Killer True Crime Library. *www.crimezzz.net.* 2005.

"Woman Killed in Crash Once Investigated." *Las Vegas Review-Journal.* July 13, 2002.

Dr. Thomas Truman

Knox, Richard A. "Suffolk D.A. May Revisit Infant Deaths to Seek Evidence of Homicide." *Boston Globe.* September 10, 1997.

———. "Suspicions Surface in Cases Termed 'Sudden Infant Death.'" *Boston Globe.* September 9, 1997.

MacKeen, Dawn. "An Interview with Dr. Thomas Truman." *www.salon.com.* September 19, 1997.

CHAPTER 5 SWOLLEN BRAINS AND BROKEN BONES

Dr. Holmes Morton/Glick Case

Morton, Holmes, M.D. "A Modern Miracle in the Most Traditional of Communities." *www.affymetrix.com.* October 2004.

Norton, Ken. "The Causes of Cot Death." *www.portia.org.* June 12, 2000.

Ruder, Kate. "Genomics in Amish Country." *www.genomenewsnetwork.org.* July 23, 2004.

Brandy Briggs Case

"Court Overturns Mother's Conviction 2nd Time in 3 Days." *www.2houston.com.* December 16, 2005.

Fergus, Mary Ann. "Testimony Aids Woman Convicted in Son's Death; Medical Examiner Doubts Homicide." *Houston Chronicle.* July 10, 2004.

Lezon, Dale. "Man's Conviction Thrown Out on Baby Death." *Houston Chronicle.* October 9, 2004.

Tilghman, Andrew. "Autopsies by Former Examiner Reviewed; Several Cases Got a Second Look after Questions about Neutrality." *Houston Chronicle.* July 22, 2004.

———. "Judge: New Autopsy Evidence Insufficient to Reverse Guilty Plea." *Houston Chronicle.* October 7, 2004.

Dr. Colin R. Paterson

Blum, Andrew. "Defective Bones, Not Child Abuse." *Forensic Echo. echo.forensicpanel.com.* December 5, 2000.

"Brittle Truths." *Frontline Scotland.* BBC-Scotland. Transcript of broadcast November 14, 2000.

Bullock, Blake, Charles J. Schubert, Patrick D. Brophy, Neil Johnson, Martin H. Reed, and Robert A. Shapiro. "Cause and Clinical Characteristics of Rib Fractures in Infants." *Pediatrics.* April 2000.

"Doctor Accused of Flawed Evidence." *Dundee (Scotland) Courier.* November 6, 2003.

Dyer, Clare. "Inexpert Witness." *Guardian* (UK). April 6, 2004.

Dyer, Owen. "Doctor Accused of Misrepresenting Evidence in Child Abuse Cases." *British Medical Journal.* January 24, 2004.

———. "GMC Strikes Off Proponent of Temporary Brittle Bone Disease." *British Medical Journal.* March 13, 2004.

General Medical Council Professional Conduct Committee Hearing. Transcript. November 5–14, 16–27, 2003. Respondent doctor: Colin Ralston Paterson. *www.gmc-uk.org/concerns/decions/search_database/pcc_paterson_20040304.asp.*

Glueck, Michael Arnold, M.D., and Robert J. Cihak, M.D. "Temporary Brittle Bone Disease and Infant Fractures." *www.newsmax.com.* October 6, 2005.

Innis, Michael D. "Re: Dr. Paterson and the GMC." *British Medical Journal.* March 8, 2004.

McDoughall, Liam. "Scores of Scottish Child Abuse Cases Reopened after Expert Witness Struck Off." *Sunday Herald* (Scotland). May 9, 2004.

Moreno, Joelle Anne. "A Courtroom Diagnosis: Countering the Defense of Temporary Brittle Bone Disease as Mild OI." National District Attorneys Association. *www.ndaa-april.org.* November 8, 2004.

Paterson, Colin, M.D. "Corruption and Miscarriages of Justice in Child Care Cases." Lecture at NCHR Symposium. *www.nkmr.org.* June 5, 2004.

State v. Talmadge. 196 Ariz. 436, 999 P.2d 192, 197 (2000).

Struthers, Mark. "Re: Time for an Overhaul of Child Protection." *British Medical Journal.* March 20, 2004.

von Kaehne, Peter M. R., M.D. "Time for an Overhaul of Child Protection." *British Medical Journal.* March 14, 2004.

Womersley, Tara. "Brittle Bones 'Are Being Diagnosed as Child Abuse.'" *London Daily Telegraph.* October 18, 2000.

CHAPTER 6 FINGERPRINT IDENTIFICATION

Jennings Case

Carleton v. People. 150 Ill. 181, 37 N.E. 244 (1908).

People v. Jennings. 252 Ill. 534, 96 N.E. 1077 (1911).

Fingerprint Identification in General

Coghlan, Andy, and James Randerson. "How Far Should Fingerprints Be Trusted?" *www.NewScientist.com*. September 17, 2005.

Cole, Simon A. "Is Fingerprint Identification Valid? Rhetoric of Reliability in Fingerprint Proponent's Discourse." *Law and Policy*. January 2006.

Epstein, Robert. "Fingerprints Meet *Daubert*: The Myth of Fingerprint 'Science' Is Revealed." *Southern California Law Review*. October 2002.

Mnookin, Jennifer L. "Fingerprints: Not a Gold Standard." *Issues in Science and Technology*. September 22, 2003.

Monaghan, Peter. "Scholars Challenge the Infallibility of Fingerprints." *Chronicle of Higher Learning*. November 13, 2006.

Newman, Andy. "Fingerprinting's Reliability Draws Growing Court Challenges." *New York Times*. April 7, 2001.

Mitchell Case

Wayman, James L., M.D. "When Bad Science Leads to Good Law: The Disturbing Irony of the *Daubert* Hearing in the Case of *U.S. v. Bryon C. Mitchell*." *Biometrics Publications*. San Jose State University. www.engr.sjsu.edu/biometrics/publications_daubert.html. February 2, 2000.

Harding Case

Mark Alan Prentice v. State of New York. Case No. 2004–009–01, Claim No. 91731, March 30, 2004, Syracuse.

"Police Investigation Supervisor Admits Faking Fingerprints." *New York Times*. July 30, 1993.

Taylor, Gary. "Fake Evidence Becomes Real Problem." *National Law Journal*. October 9, 1995.

Yan, Ellen. "Probe of State Cops Shows Tampering." *New York Newsday*. February 4, 1997.

Douglas Case

Crogan, Jim. "LAPD Fiction: How a Detective Made Up Evidence and Got Away with It." *LA Weekly*. March 1–7, 2002.

Jackson Case

"Fingerprints: Infallible Evidence?" CBS News. *www.cbsnews.com*. July 20, 2003.

Judge Louis H. Pollak

Berry, Steve. "Judge Reverses Fingerprint Decision." *Los Angeles Times*. March 14, 2002.

Loviglie, Joann. "Judge Mulls Testimony of Fingerprints." *Los Angeles Times*. February 28, 2002.

Maier,Timothy W. "Federal Judge Slams Fingerprint 'Science.'" *Insight in the News*. February 22, 2002.

Mnookin, Jennifer L. "Questioning the Credibility of Fingerprint Evidence." *Issues in Science and Technology*. Fall 2003.

Neufeld, Peter, and Barry Scheck. "Will Fingerprinting Stand Up in Court?" *New York Times*. March 9, 2002.

Cowans Case

Barter, Andrea R. "Fingerprint Identification and Its Aura of Infallibility." *Lawyers Journal* (Massachusetts Bar Association). March 1, 2006.

McRoberts, Flynn, and Steve Mills. "U.S. Seeks Review of Fingerprint Techniques." *Chicago Tribune.* February 21, 2005.

Richardson, Franci. "Fake Fingerprint Probe." *Boston Herald.* April 25, 2004.

Richardson, Franci, and Maggie Mulvchell. "Stephen Cowens: 'This Boy Did Six Years in the Joint for Nothing.'" *Boston Herald.* May 7, 2004.

Smalley, Suzanne. "Boston Police Shutting Down, Revamping Fingerprint Unit; Identification Error, Critical Report Cited." *Boston Globe.* October 14, 2004.

Smith, Ron, and Associates. "Boston Police Department Latent Print Unit Evaluation Report." September 4, 2004.

Patterson Case

Barter, Andrea R. "Fingerprint Identification and Its Aura of Infallibility." *Lawyers Journal* (Massachusetts Bar Association). March 1, 2006.

Commonwealth v. Terry L. Patterson. SJC-09478. Transcript. September 7–December 17, 2005.

Ostrowski, Steve. "*Daubert* Challenge to Simultaneous Impressions in Massachusetts." *Weekly Detail.* www.clpex.com/TheDetail.htm. September 19, 2005.

Saltzman, Jonathan. "SJC Bars a Type of Prints at Trial." *Boston Globe.* December 28, 2005.

———. "SJC to Hear Arguments on Banning Fingerprint Evidence." *Boston Globe.* September 5, 2005.

Madrid Bombing Case

Feige, David. "Printing Problems: The Inexact Science of Fingerprint Analysis." *www.slate.msn.com.* May 27, 2004.

Frieden, Terry, and Harry Schuster. "Report: Sloppy FBI Work Led to Wrong Man." CNN. January 6, 2006.

Heath, David. "Bungled Fingerprints Expose Problems at FBI." *Seattle Times.* June 7, 2004.

Kramer, Andrew. "Court Dismisses Case against Mayfield." Associated Press. May 25, 2004.

McRoberts, Flynn, Steve Mills, and Maurice Possley. "Forensics under the Microscope." *Chicago Tribune.* October 17, 2004.

Mnookin, Jennifer L. "The Achilles' Heel of Fingerprints." *Washington Post.* May 29, 2004.

Ryan, Jason. "Report: FBI Problems Led to Wrongful Terror Arrest." *ABC News.* January 6, 2006.

Saker, Anne. "Basic Flaws Led FBI to Bungle Mayfield Case, Report Finds." *Oregonian.* January 7, 2006.

Stacey, Robert B. "Report on the Erroneous Fingerprint Individualization in the Madrid Train Bombing Case." *Forensic Science Communications. www.fbi.gov.* January 2005.

Stout, David. "Report Faults FBI's Fingerprint Scrutiny in Arrest of Lawyer." *New York Times.* November 17, 2004.

Verhovek, Sam Howe. "2 Million, Apology Settle FBI Fingerprint Case." *Los Angeles Times.* November 30, 2006.

CHAPTER 7 FINGERPRINTS NEVER LIE

Adams, Lucy. "Plan to Change Prints System Angers Experts." *Herald* (UK). January 4, 2006.

"Anger at Fingerprint Case Claim." BBC News. *www.bbc.co.uk.* April 7, 2004.

"Appeal for McKie 'Fighting Fund.'" BBC News. *www.bbc.co.uk.* March 9, 2006.

Bargeton, Steve. "McKie Case: Public Inquiry Call Rejected." BBC News. *www.bbc.co.uk.* February 17, 2006.

" 'Cover-Up' Claims over Prints." BBC News. *www.bbc.co.uk.* October 9, 2003.

"Crime Office Faces Shake-Up." BBC News. *www.bbc.co.uk.* September 14, 2000.

"Crown Office Rules Out Fingerprint Action." BBC News. *www.bbc.co.uk.* September 7, 2001.

"Expert Criticises Print 'Mindset.' " BBC News. *www.bbc.co.uk.* June 24, 2006.

"Experts Question Fingerprint Evidence." BBC News. *www.bbc.co.uk.* March 21, 2000.

"Experts Seek Fingerprint Inquiry." BBC News, *www.bbc.co.uk.* September 19, 2002.

"The Fall-Out: Suspicion, Experts at War and a Question Mark over Thousands of Convictions All Over the World." *Herald* (UK). February 19, 2006.

"False Impression." *Frontline Scotland.* BBC-Scotland. . Transcript of broadcast May 16, 2000.

"Father's Fingerprint Error Anger." BBC News. *www.bbc.co.uk.* February 8, 2006.

"Finger of Suspicion." *Frontline Scotland.* BBC-Scotland. Transcript of broadcast January 18, 2000.

"Fingerprint Battle Takes New Turn," BBC News. *www.bbc.co.uk.* June 28, 2005.

"Fingerprint Detective Claim Rejected." BBC News. *www.bbc.co.uk.* February 14, 2002.

"Fingerprint Experts Suspended." BBC News. *www.bbc.co.uk.* August 3, 2000.

"Fingerprint Officer Fights On." BBC News. *www.bbc.co.uk.* March 21, 2002.

"Fingerprint Perjury Inquiry Launched." BBC News. *www.bbc.co.uk.* July 6, 2000.

"Fingerprint Service Defended." BBC News. *www.bbc.co.uk.* November 25, 2003.

"Fingerprints to Be 'Double-Checked.'" BBC News. *www.bbc.co.huk.* June 22, 2000.

"Fresh Allegations in McKie Case." BBC News. *www.bbc.co.uk.* February 19, 2006.

Gray, Louise. "McKie 'Should Not Have Been Put on Trial.'" *Scotsman.* June 22, 2006.

Harrell, Eben. "FBI Ordered McKie Case 'Swept under Carpet.'" *Scotsman.* March 8, 2006.

———. "Pressure Mounts for McKie Inquiry as U.S. Announces FBI Investigation." *Scotsman.* March 22, 2006.

———. "Prints in McKie Case 'Had Been Manipulated.'" *Scotsman.* March 9, 2006.

Howie, Michael. "50,000 Is Pledged to Shirley McKie Fighting Fund." *Scotsman.* February 25, 2006.

———. "Parliament Committee Wins Right to McKie Report." *Scotsman.* July 13, 2006.

"Inquiring into Fingerprint Evidence." BBC News. *www.bbc.co.uk.* February 7, 2000.

"Internet Makes Mark on Fingerprint Case. BBC News. *www.bbc.co.uk.* May 14, 1999.

"Judge Rejects Fingerprint Defense." BBC News. *www.bbc.co.uk.* December 24, 2004.

Macaskill, Mark. "McConnell 'Lied' over McKie Case." *Scotland Sunday Times.* February 12, 2006.

McDougall, Liam. "Fingerprint Experts Boycott Conference over McKie Affair." *Herald* (UK). March 12, 2006.

———. "Revealed: Detective in Charge of Murder Case Wanted Shirley McKie 'Gagged' by Executive." *Herald* (UK). February 25, 2006.

———. "Shirley McKie: Was It Really an Honest Mistake?" *Sunday Herald* (UK), February 12, 2006.

"McKie Campaigners Launch Cash Fund," Press Association. *Scotsman.* March 9, 2006.
"McKie Case Inquiry Calls Rejected." BBC News. *www.bbc.co.uk.* February 22, 2006.
"Murder Appeal after Print Error." BBC News. *www.bbc.co.uk.* August 17, 2000.
"Murder Conviction Is Quashed." BBC News. *www.bbc.co.uk.* August 14, 2002.
O'Hare, Paul. "25 Years of 'Truly Evil' Murder of Widow, 91." *Scotsman.* March 2, 2005.
"Praise for Criminal Records Agency." BBC News. *www.bbc.co.uk.* May 24, 2001.
"Print Case 'Shames Scots Justice.'" BBC News, *www.bbc.co.uk.* September 20, 2004.
"'Relief' over Fingerprint Verdict." BBC News. *www.bbc.co.uk.* February 7, 2006.
"Shirley McKie Case: Inquiry Urged." Press Association. *Scotsman.* February 25, 2006.
Spector, Michael. "Do Fingerprints Lie? The Gold Standard of Forensic Evidence Is Now Being Challenged." *New Yorker.* May 27, 2002.

CHAPTER 8 SHOE-PRINT IDENTIFICATION AND FOOT MORPHOLOGY

Dr. Louise Robbins

Barnum, Art. "Footprints Key in 1984 Murder Retrial, Court Told." *Chicago Tribune.* June 8, 1989.
Gibson, Bob. "When Justice Hides Its Face." *Journal of the DuPage County (Ill.) Bar Association.* May 2005.
Gibson, Ray. "Courted Expert Steps on Toes with Footprints." *Chicago Tribune.* April 6, 1986.
Gregory, Ted, and James Kimberly. "After 2 Decades, Longtime Suspect Dugan Is Charged." *Chicago Tribune.* November 30, 2005.
Hanson, Marc. "Believe It or Not." *ABA Journal.* June 1993.
Mahany, Barbara, and John Schmeltzer. "Nicarico Death Suspect Freed, Expert Too Ill to Testify, Buckley Charges Dropped." *Chicago Tribune.* March 6, 1987.
Zerwick, Phoebe. "Mixed Results; Forensics, Right or Wrong, Often Impresses Jurors." *www.journalnow.com.* August 29, 2005.

Sergeant Robert B. Kennedy

McRoberts, Flynn, Steve Mills, and Maurice Possley. "Unproven Techniques Sway Courts, Erode Justice." *Chicago Tribune.* October 17, 2004.
Rupert, Jake. "Courts Trample Life's Work." *Ottawa Citizen.* March 22, 2004.

Penland Case

Whitmire, Tim. "New Trial Ordered for Death Row Inmate." *Greensboro (N.C.) News-Record.* www.news-record.com. July 25, 2005.
Zerwick, Phoebe, "Inmate to Get Second Chance." *Winston-Salem Journal.* July 26, 2005.
———. "Mixed Results; Forensics, Right or Wrong, Often Impresses Jurors." *www.journalnow.com.* August 29, 2005.

Shoe-Print Identification in General

William B. Ratliff v. State of Alaska. Court of Appeals No. A-8651, April 15, 2005. 110 P.3rd 982 (2005).

CHAPTER 9 BITE-MARK IDENTIFICATION

Brown Case

Elliott-Engel, Amaris. "Memories Clash in Brown Case." *Auburn (N.Y.) Citizen.* February 25, 2007.

Santos, Fernanda. "Inmate Finds Vindication in His Quest for a Killer." *New York Times.* February 25, 2007.

Smith, John. "Ex-DA Denies Withholding Evidence in Brown Case." *Syracuse Post-Standard.* January 31, 2007.

Dr. Michael West

60 Minutes. CBS. Transcript of broadcast February 17, 2002.

Coyle, Marcia. "'Expert' Science under Fire in Capital Cases; *Daubert v. Frye.*" *National Law Journal.* July 11, 1994.

"Ex-Death Row Inmate Seeks New Prosecution for Rape-Murder Retrial." Associated Press. *Mississippi Clarion-Ledger.* March 12, 2006.

Hanson, Mark. "Out of the Blue." *ABA Journal.* February 1996.

Murr, Andrew. "A Dentist Takes the Stand." *Newsweek.* August 20, 2001.

Sotos, James G. "Expert Witness to Face Malicious Prosecution Suit." *Chicago Daily Law Bulletin.* January 16, 2003.

Dr. Raymond D. Rawson/ Krone Case

Bommersback, Jana. "Arizona Sent an Innocent Man to Death Row." *Phoenix Magazine.* July 2004.

"Death Is Different." Editorial. *New York Times.* April 10, 2002.

"Man Convicted on Erroneous Bite Mark Identification Evidence Finally Free; Served 10 years for Crime He Didn't Commit." *www.forensic-evidence.com.* 2002.

Nelson, Robert. "About Face." *Phoenix News Times.* April 21, 2005.

Potter, Chris. "Live Man Walking." *www.pittsburghcitypaper.com.* October 2002.

Sebelius, Steve. "Democracy in Peril,." *Las Vegas Mercury.* December 2, 2004.

Sherrer, Hans. "Twice Wrongfully Convicted of Murder; Ray Krone Is Set Free after 10 Years." *Justice Denied Magazine.* September 2002.

Dr. Raymond D. Rawson/Tankersley Case

McRoberts, Flynn. "Bite Mark Verdict Faces New Scrutiny." *Chicago Tribune.* November 29, 2004.

———. "Judge Questions Bite-Mark Evidence He Applied in 1994." *Chicago Tribune.* December 24, 2004.

State of Arizona v. Bobby Lee Tankersley. Supreme Court of Arizona En Banc, No. CR-94–0168-AP. Yuma County.

Dr. John Kenney/Young-Hill Case

McRoberts, Flynn, and Steve Mills. "From the Start, a Faulty Science." *Chicago Tribune.* October 19, 2004.

Mills, Steve, and Jeff Coen. "Men Exonerated in 1990 Murder." *Chicago Tribune.* January 31, 2005.

Possley, Maurice. "New Tests Requested in Victim's Bite Marks." *Chicago Tribune.* July 25, 2003.

Dr. Lowell Levine/Edmund Burke Case

Burns, Brian. "Judge Recommends Dismissal of Burke's Claims." *Walpole (Mass.) Times.* October 16, 2003.

Glynn, Tom. "State Had Exculpatory Evidence." *Walpole (Mass.) Times.* August 14, 2003.

———. "Questionable Procedure Propped Up Failing Case." *Walpole (Mass.) Times.* September 25, 2003.

———. "What the Dentist Said." *Walpole (Mass.) Times.* September 4, 2003.

Nonscience of Bite-Mark Identification

Bowers, C. Michael. "DNA and Bite Mark Analysis." Expert Law Web site. *www.expertlaw.com.* Winter 1996.

———. A Statement Why Court Opinions on Bite Mark Analysis Should Be Limited," Expert Law Web site. *www.expertlaw.com.* December 1996.

Levine, Lowell J. "The Forensic Odontologist." In *Murder Ink: The Mystery Reader's Companion,* ed. Dylis Winn. New York: Workman, 1977.

Santos, Fernanda. "Evidence from Bite Marks, It Turns Out, Is Not So Elementary." *New York Times.* January 28, 2007.

Whittaker, D. K. "Bite Marks: The Criminal's Calling Cards." *British Dental Journal.* Fall 2003.

CHAPTER 10 EAR-MARK IDENTIFICATION

Kunze Case

"Charges Dropped in Earprint Case." Associated Press. March 23, 2001.

State v. David Wayne Kunze. Court of Appeals of Washington, Division 2. 97 Wash. App. 832, 988 P. 2d 977 1999.

Dallagher Case

Moenssens, Andre A. "Another Ear Print Conviction Reversed." *www.forensic-evidence.com.* 2004.

———. "DNA Evidence Proves Ear ID Wrong." *www.forensic-evidence.com.* 2004.

Woffinden, Bob. "Earprint Landed Innocent Man in Jail for Murder." Gu*ardian (UK).* January 23, 2004.

Ear-Mark Identification in General

Egan, Toby. "Are Dutch Ears Different From American Ears? A Comparison of Evidence Standards." *www.forensic-evidence.com.* 1996.

———. "Netherlands Court of Appeal at Amsterdam Rejects Ear Reliability." www.forensic-evidence.com. November 1, 2000.

"Sci/Tech: Police Play It by Ear." BBC News. *www.bbc.co.uk.* January 2, 1999.

CHAPTER 11 EXPERT VERSUS EXPERT

The Crime

Brennan, Charlie. "Complete Ransom Note Unveiled; Magazine Somehow Gets Copy." *Rocky Mountain News.* September 3, 1997.

———. "Patsy Ramsey Handwriting Probed: Sources Say Sample Contains Characters with Some Features Seen in Ransom Note." *Rocky Mountain News.* March 19, 1997.

———. "Ramsey Lawyers Want to Publish Ransom Note." *Rocky Mountain News.* August 8, 1997.

Evans, Clay. "Coroner Apologizes for Wecht's Remarks." *Boulder (Colo.) Daily Camera.* March 20, 1997.

———. "JonBenét's Dad Didn't Write Note." *Boulder (Colo.) Daily Camera.* March 15, 1997.

"Ex-FBI Experts: Ramsey Note a Deception." *Boulder (Colo.) Daily Camera.* September 8, 1997.

Giannelli, Paul C. "Expert Qualifications: Who Are These Guys?" *American Bar Association.* Spring 2002.

Keene-Osborn, Sherry, and Daniel Glick "A Case Forever Unraveling. *Newsweek.* September 15, 1997.

McCullen, Kevin. "Writing Samples Taken in Michigan; Detectives Sought Unrehearsed Examples from Patsy Ramsey." *Rocky Mountain News.* November 22, 1997.

"*Newsweek* Publishes Copy of Ransom Note." *Rocky Mountain News.* September 8, 1997.

The Nongraphologist Document Examiners

Chet W. Ubowski

Anderson, Christopher. "CBI Expert Testifies for 2nd Day at Hearing." *Boulder (Colo.) Daily Camera.* October 28, 1998.

Brennan, Charlie. "Handwriting Test Fails to Clear Patsy Ramsey; Lawyers Say Results Mean JonBenét's Mom Wasn't Author of Note." *Rocky Mountain News.* July 9, 1997.

"CBI Expert Seen with Prosecutor; Handwriting Analyst Apparently Testified Before Ramsey Panel." *Boulder (Colo.) Daily Camera.* October 16, 1998.

Howard C. Rile Jr.

"Re: *Wolf v. Ramsey.*" March 22, 2002. Document Report. www.acandyrose.com.

Rile, Howard.

———. Ramsey Case Presentation. August 22, 2004, American Society of Questioned Document Examiners Convention, Memphis.

Lloyd W. Cunningham

Rile, Howard. Interviews by the author. August 24, 2003, American Society of Questioned Document Examiners Convention, Baltimore; August 22, 2004, American Society of Questioned Document Examiners Convention, Memphis.

Darnay Hoffman

"Darnay Hoffman, The Hoffman Files." *www.acandyrose.com.* 2001.

"Hoffman's Folly." Editorial. *Boulder (Colo.) Daily Camera.* November 24, 1997.

Hunter, Alex. Brief in Support of Motion to Dismiss, District Court, County of Boulder, State of Colorado, Case No. 97 CV 1732 Div. 2, December 9, 1997.

McCullen, Kevin. "Lawyer Files Suit to Force Changes in Ramsey Case." *Rocky Mountain News.* November 20, 1997.

"Patsy Ramsey Asked If She Wrote Ransom Note; Lawyer Ends Ramsey Deposition Early." Denver's ABC 7. *www.thedenverchannel.com.* December 21, 2001.

Sebastian, Matt. "Judge Rejects Suit against D.A." *Boulder (Colo.) Daily Camera.* January 21, 1998.

"What Refusal to Prosecute?" Editorial. *Boulder (Colo.) Daily Camera.* January 25, 1998.

Thomas C. Miller

Abbott, Karen. “Judge Says He Can't Stop Editor's Indictment.” *Rocky Mountain News.* October 29, 1999.

“Boulder Officers Are Subpoenaed in Tabloid Case.” Associated Press. *Boulder (Colo.) Daily Camera.* February 18, 2000.

Colorado v. Craig Allen Lewis. December 20, 1999. Charged with commercial bribery and extortion. *www.acandyrose.com.*

Colorado v. Thomas C. Miller. August 20, 1999. Charged with commercial bribery. *www.acandyrose.com.*

Good, Owen S. “Handwriting Expert Says He Didn't Know Reporter [Who] Sought Ramsey Note.” *Rocky Mountain News.* June 14, 2001.

McPhee, Mike. “Judge Halts Probe in Rare Ruling.” *Denver Post.* October 9, 1999.

Miller, Thomas C. Affidavit. *www.thewebsafe.tripod.com.* November 12, 1997.

———. “Questioned Document Examiner Report.” *www.thewebsafe.tripod.com.* November 3, 1997.

Pankratz, Howard, and Marilyn Robinson. “Lawyer Faces Bribery Charge.” *Denver Post.* August 21, 1999.

Sebastian, Matt. “Globe Decision Delayed; a Decision on Whether Prosecutors Can Indict Editor Craig Lewis Is Expected Next Week.” *Boulder (Colo.) Daily Camera.* October 19, 1999.

“Tabloid Editor Arrested over Ramsey.” Associated Press. *www.crimelynx.com.* December 20, 1996.

David Liebman

Liebman, David. Letter to D.A. Hunter requesting grand jury appearance, September 28, 1998, “The Hoffman Files.” *www.acandyrose.com.* 2001.

———. Ramsey Handwriting Report, November 26, 1997. *www.acandyrose.com.* 2001.

Wickham, Melissa. “The Hidden Truth.” *Barbados Daily Nation.* March 6, 2005.

Ted Widmer

“Hey, Ken—Tell Us about Your Girlfriend.” *National Enquirer.* August 25, 1998.

Cina L. Wong

Kane, Michael J. Letter to Cina L. Wong, January 20, 1999, “The Hoffman Files.” *www.acandyrose.com.*

Wong, Cina L. Affidavit. U.S. District Court, Northern District of Georgia, Atlanta Division, August 28, 2002, *Wolf v. Ramsey*, “The Hoffman Files.” *www.acandyrose.com.*

———. Deposition. U.S. Court, Northern District of Georgia, Atlanta Division, May 13, 2002, *Wolf v. Ramsey.* “The Hoffman Files.” *www.acandyrose.com.*

———. Handwriting report to Darnay Hoffman, November 14, 1997. “The Hoffman Files.” *www.acandyrose.com.*

———. Letter to Alex Hunter, September 20, 1998. “The Hoffman Files,” www.acandyrose.com.

Lou Smit

Anderson, Christopher. “Reactions Follow Book Claims.” *Boulder (Colo.) Daily Camera.* March 16, 2000.

Brennan, Charlie. “Investigator Now Helping Ramseys.” *Rocky Mountain News.* December 29, 1999.

"Detective Lou Smit, Ret." *www.lousmit.com.*

"Ex-Prosecutor Investigator Supports Intruder Theory in JonBenét Ramsey Case." Associated Press. *www.longmontfyi.com.* May 3, 2001.

Hartman, Todd. "Evidence Man: Persistence, Keen Eye Have Helped Sleuth Solve Other Killings." *Rocky Mountain News.* May 5, 2001.

Henry, Travis. "Keenan Brings Investigator's 'Fresh Set of Eyes' to JonBenét Ramsey Case." *Northern Colorado Daily Times-Call.* June 10, 2003.

McCullen, Kevin. "DAs Tried to Block Testimony by Smit." *Rocky Mountain News.* March 15, 2000.

"Ramsey Case Update #81, Chief Mark Beckner Responds to Media Questions about Lou Smit's Television Appearance." *www.ci.boulder.co.us.com.* April 30, 2001.

"Smit Letter: Ramseys Innocent." Associated Press. *Boulder (Colo.) Daily Camera.* September 20, 1998.

Wolf v. Ramsey and the Carnes Decision

Brennan, Charlie. "U.S. Judge's Ramsey Ruling Questioned." *Rocky Mountain News.* April 25, 2003.

Epstein, Gideon. Affidavit. U.S. District Court, Northern District of Georgia, Atlanta Division, August 27, 2002, *Wolf v. Ramsey.* "The Hoffman Files." *www.acandyrose.com.*

———. Deposition. U.S. District Court, Northern District of Georgia, Atlanta Division, May 17, 2002, *Wolf v. Ramsey.* "The Hoffman Files." *www.acandyrose.com.*

———. Ramsey Case Presentation. American Society of Questioned Document Examiners Conference, Memphis. August 22, 2004.

"Gideon Epstein, Forensic Examiner of Questioned Documents." *www.gideonepstein.com.*

Henry, Travis. "If Not Patsy Ramsey, Who? The Unusual Suspects." *Northern Colorado Daily Times-Call.* April 14, 2003.

Mocine-McQueen, Marcos, and Paula Woodward. "JonBenét Evidence Points to Intruder, Judge Rules." *Denver Post.* June 12, 2003.

Wolf v. Ramsey. Civil case, 1-CV-1187-JEC, U.S. District Court, Northern District of Georgia, Atlanta Division, Carnes Order, March 31, 2003. *www.acandyrose.com.*

CHAPTER 12 JOHN MARK KARR

A Murder Suspect Emerges

Cocco, Marie. "Return of Cable TV Experts." www.truthdig.com. August 21, 2006.

Geller, Adam. "Suspect's Life Story Proves to Be Tangled Tale." *Chicago Tribune.* August 18, 2006.

Goldstein, Amy, and Anne Hull. "JonBenét Mentioned in '01 Probe of Suspect." *Washington Post.* August 19, 2006.

Karr, John Mark. Transcript of press conference. MSNBC. www.msnbc.msn.com. August 17, 2006.

Kennedy, Helen. "Twisted Suspect Obsessed with Kids." *New York Daily News.* August 18, 2006.

The Arrest and Confession

"John Mark Karr: In His Words." *People.* August 24, 2006.

Meyer, Jeremy P. "Thai Police Detail JonBenét Suspect Investigation." *Denver Post.* August 23, 2006.

Mitchell, Kirk, and Howard Pankratz. "Case against Karr Still Being Built." *Denver Post*. August 24, 2006.

Moore, Oliver. "Man Arrested in Thailand Linked to Ramsey Slaying." *Globe and Mail*. August 17, 2006.

Sarche, Jon, and Chase Squires. "John Mark Karr's Attorney Wants Handwritten Application Sealed." Associated Press. August 25, 2006.

Squires, Chase. "DNA a Battlefield in JonBenét Slaying." Associated Press. August 26, 2006.

Willing, Richard. "Some Say Karr Links Always Looked Flimsy." *USA Today*. August 30, 2006.

Wohlsen, Marcus. "Family: JonBenet Suspect Researched Case." Associated Press. August 17, 2006.

The Graphologists Surface

David Liebman

Valenty, Richard. "A Karr Trip to Boulder?" *Colorado Daily*. August 21, 2006.

Wiggins, Stacy. "Local Connection to JonBenet Ramsey Case." News Channel 3. Norfolk, Virginia. August 19, 2006.

Don Lehew

Kilzer, Lou. "Handwriting Expert Holds to Opinion." *Rocky Mountain News*. August 29, 2006.

———. "Handwriting Expert Points Finger at Karr." *Rocky Mountain News*. August 22, 2006.

Lehew, Don. Curriculum vitae. *www.handwritingexpert.com*. Accessed 2005.

Nancy Grace. CNN. Transcript of broadcast August 21, 2006.

Young, Kelly. "Local Handwriting Expert Says He Believes Karr Wrote Ransom Note." *Jacksonville (Tex.) Daily Progress*. August 19, 2006.

Curtis Baggett

Baggett, Curtis. Web site. *www.expertdocumentexaminer.com*.

Brown v. State. 1999 Tex. App. Dallas, Texas, 1999.

Kilzer, Lou. "Handwriting Analyst Was Disqualified." *Rocky Mountain News*. August 23, 2006.

———. "Handwriting Expert Holds to Opinion." *Rocky Mountain News*. August 29, 2006.

———. "Handwriting Expert Points Finger at Karr." *Rocky Mountain News*. August 22, 2006.

Mitchell, Kirk. "Ransom Note's Link to Karr May be Shaky, Experts Say." *Denver Post*. August 24, 2006.

United States v. Bourgois. 950 F.2nd. 980, 1992.

Wael Abdin v. Delores Abdin. Pulaski County Circuit Court, CA 05–169, January 18, 2006.

Wheeler v. Olympia Sports Center, Inc. U.S. District Court, District of Maine, 03–265-P-H, October 12, 2004.

DNA Trumps Confession

"Boulder D.A. Defends Decision to Arrest Karr." Associated Press. August 29, 2006.

Brennon, Charlie. "Miss Steps; Experts: Resolution of Ramsey Case Unlikely." *Rocky Mountain News*. December 23, 2006.

Simpson, Kevin. "Tests Rule Out Karr as Source of DNA." *New York Times*. August 29, 2006.

CHAPTER 13 HAIR AND FIBER IDENTIFICATION

Hair Identification in General

Haggo, Regina, and Douglas Haggo. "Somestimes Splitting Hairs Isn't Good Enough." *Hamilton (Ont.) Spectator.* June 9, 2007.

Rivera Case

Devise, Daniel. "Hair Strands Were Not Young Victim's." *Miami Herald.* March 21, 2003.
Grimm, Fred. "By a Hair, Death Row Turns into Doubt Row." *Miami Herald.* March 25, 2003.
McMahon, Paula. "Crime Lab Botches Murder Inquiry." *South Florida Sun-Sentinel.* June 24, 2003.
Norman, Bob. "A Single Hair." *New Times Broward-Palm Beach.* June 28, 2001.

Fain Case

Bonner, Raymond. "Death Row Inmate Is Freed after DNA Test Clears Him." *New York Times.* August 24, 2001.
Freedberg, Sydney P. "Good Cop, Bad Cop." *St. Petersburg Times.* March 4, 2001.
"Snitches, Junk Science, and Big Business, Part 6." September 1, 2001. *www.unquiet-mind.com.*

The Career of Michael P. Malone

Buchanan, Edna. "Did FBI Help Send Wrong Man to Death Row?" *Miami Herald.* May 31, 2003.
Carnell, Brian. "Man's Conviction Overturned after FBI Agent's False Testimony Revealed." Associated Press. July 10, 2003.
Cohen, Laurie P. "Strand of Evidence: FBI Crime-Lab Work Emerges as New Issue in Famed Murder Case." *Wall Street Journal.* April 16, 1997.
Freedberg, Sydney P. "Good Cop, Bad Cop." *St. Petersburg Times.* March 4, 2001.
———. "Sloppy Lab Work Casts Doubts on Some Florida Cases." *St. Petersburg Times.* March 5, 2001.
Soloman, John. "Conviction Tossed on FBI Lab Misconduct." Associated Press. *www.truthinjustice.com.* 2003.
Stanfield, Frank. "Officer's New Trial May Hang by a Hair." *Orlando Sentinel.* October 24, 1998.

Arnold Melnikoff/Bromgard Case

Bohrer, Becky. "More Cases Questioned in Crime Lab Case." Associated Press. *www.law-forensics.com.* May 29, 2003.
Gillis, Charles. "Scandal in Forensic Labs: Hundreds of Cases Undergoing Review in Montana." *National Post* (Canada). February 1, 2003.
Liptak, Adam. "DNA Will Let a Montana Man Put Prison behind Him, but Questions Linger." *New York Times.* October 1, 2002.
Olsen, Lise. "Crime Lab Work Failed to Qualify to Test Hair Samples; Case Reviews Ongoing in Two States after DNA Analysis Clears Montana Inmates." *Seattle Post-Intelligencer.* January 2, 2003.
Smith, Ericka Schenck. "Bad Testimony? State Finds Five More Persons Convicted Partially on Evidence from Controversial Forensic Scientist." *Missoulian.* May 30, 2003.

Teichroeb, Ruth. “Counties to Be Told of Crime Lab Flaws.” *Seattle Post-Intelligencer.* March 17, 2004.

———. “First Sentence to Be Vacated after Errors by State Crime Lab.” *Seattle Post-Intelligencer.* September 30, 2004.

———. “State Patrol Fires Crime Lab Scientist.” *Seattle Post-Intelligencer.* March 24, 2004.

The Career of Charles A. Linch

Green, Frank. “Hair Analysis Use Faulted.” *Richmond (Va.) Times-Dispatch.* October 19, 2002.

Wrolstad, Mark. “Hair-Matching Flawed as a Forensic Science: DNA Testing Reveals Dozens of Wrongful Verdicts Nationwide.” *Dallas Morning News.* March 31, 2001.

Charles A. Linch/Routier Case

Geringer, Joseph. “Darlie Routier: Doting Mother/Deadly Mother.” Crime Library. *www.crimelibrary.com.*

“Justice for Darlie.” *www.justicefordarlie.net* Web site.

Charles A. Linch/Blair Case

Hixson, Josh. “DNA Evidence Points to New Suspects in Blair Case.” *Star Community Newspapers* (Tex.). November 3, 2006.

Housewright, Ed. “Michael Blair Sits on Death Row,” *Dallas Morning News.* October 29, 2006.

“Michael Blair.” Texas Moratorium Network and *Dallas Morning News. www.texasmoratorium.com.* October 4, 2002.

CHAPTER 14 DNA ANALYSIS

DNA Analysis in General

Axtman, Kris. “Growing Introspection in Death-Penalty Capital.” *Christian Science Monitor.* November 16, 2004.

Eaton, Sara. “Mold Found on Lab’s DNA Evidence.” *Ft. Wayne (Ind.) Journal Gazette.* October 19, 2005.

Lozano, Juan A. “Latest Report Continues Criticism of Houston Crime Lab.” Associated Press. January 4, 2006.

McMahon, Paula. “Crime Lab Botches Murder Inquiry: Prosecutors Must Drop Charges after DNA Evidence Is Contaminated.” *South Florida Sun-Sentinel.* June 24, 2003.

McVicker, Steve. “More DPS Labs Flawed; DNA Testing Woes across State Threaten Thousands of Cases.” *Houston Chronicle.* March 27, 2004.

Orlov, Rick. “Lab Used by LAPD Falsified DNA Data.” *Los Angeles Daily News.* November 18, 2004.

Possley, Maurice, Steve Mills, and Flynn McRoberts. “Scandal Touches Even Elite Forensics Labs across U.S.” *Kansas City Star.* November 3, 2004.

Rondeaux, Candace. “Felons’ DNA Missing from Va. Database.” *Washington Post.* December 21, 2006.

Roth, Mark. “DNA Evidence: Debate on Who Gets DNA Tests.” *Pittsburgh Post-Gazette.* December 20, 2005.

Teichroeb, Ruth. “Rare Look inside State Crime Labs Reveals Recurring DNA Test Problems.” *Seattle Post-Intelligencer.* July 22, 2004.

“Watchdog Commission States Find DNA Labs Aren’t Infallible.” *Journal News.* April 3, 2006.

Willing, Richard. "Many DNA Matches Aren't Acted On." *USA Today*. November 21, 2006.

Wilson, Ring. "DNA Backlog Undercuts Crime-Solving Value." *Washington Post*. April 27, 2005.

Karla Carmichael

Boyd, Deanna. "Crime Lab Subject of Criminal Inquiry." *Ft. Worth Star-Telegram*. April 13, 2003

———. "Doubts Increase about Crime Lab." *Ft. Worth Star-Telegram*. August 26, 2003.

———. "Lab Inquiry Finds Flaws but No Injustices." *Ft. Worth Star-Telegram*. November 27, 2005.

———. "Scientist at Crime Lab Is Fired." *Ft. Worth Star-Telegram*. April 22, 2003.

McDonald, Melody. "DNA Test Sways Prosecutors." *Ft. Worth Star-Telegram*. October 10, 2002.

Indianapolis–Marion County/Kuppareddi Balamurugan

Horne, Terry. "Crime Lab Boss Placed on Leave; Mayor Removes Longtime Director amid Allegations He Helped Cover Up Wrongdoing." *Indianapolis Star*. January 25, 2004.

Ryckaert, Vic. "Judge Asked to Halt DNA Tests; Crime Lab Less Than Candid about Cases under Review, Attorney Says." *Indianapolis Star*. August 13, 2003.

Fred Salem Zain

Chan, Sau. "Scores of Convictions Reviewed as Chemist Faces Perjury Accusations." *Los Angeles Times*. August 21, 1994.

"Death of Lying Chemist Fred Zain." *www.talkleft.com*. December 8, 2002.

McAdams, Dona. "The Face of Innocence." DNA and Human Rights Center. *www.hrcberkeley.org*. 2002.

Nyden, Paul J. "DNA Advocates Get No Hearing in W.Va." *Charleston (W. Va.) Gazette*. May 8, 2006.

Rosen, Edward D. "When Experts Lie." *www.truthinjustice.com*. 2002.

Staff. "Crime Lab: More Zain Fallout." *Charleston (W. Va.) Gazette*. May 10, 2006.

Joyce Gilchrist

"Chemist Fraud." *www.talkleft.com*. April 20, 2004.

Cooper, Scott. "Judge Pleases McCarty, Rips Former Chemist." *Oklahoma Gazette*. May 11, 2007

Hastings, Deborah. Associated Press. "Testimony Doubted in Execution Case." *www.truthinjustice.org*. August 29, 2001.

"In the Matter of Renewed Investigation of State Police Crime Laboratory." Summary of Recently Issued West Virginia Supremem Court Decisions. *www.state.wv.us/wvsca. June 16, 2006.*

Jeffrey Todd Pierce v. Joyce Gilchrist and Robert H. Macy. United States Court of Appeals, Tenth Circuit, March 2, 2004. Appeal from the United States District Court for the Western District of Oklahoma D.C. NO. CIV-02–509-C; denial of defendants' motion to dismiss affirmed.

Luscombe, Belinda. "When the Evidence Lies." *Time*. May 13, 2001.

Romano, Lois. "Police Chemist's Missteps Cause Okla. Scandal." *Washington Post*. November 26, 2001.

"Three Death Row Cases Due Review." *Shawnee (Okla.) News-Star.* November 10, 2004.

Yardley, Jim. "Inquiry Focuses on Scientist Employed by Prosecutors." *New York Times.* May 2, 2001.

Dr. Pamela Fish

Fisk, Margaret Cronin. "Lawyer Frees Chicago Two after Retesting of Lab Sample; Technician Had Testified in Earlier Bad Conviction." *National Law Journal.* December 17, 2001.

Mills, Steve, Flynn McRoberts, and Maurice Possley. "Forensics under the Microscope; When Labs Falter, Defendants Pay." *Chicago Tribune.* October 20, 2004.

Mills, Steve, and Maurice Possley. "Report Alleges Crime Lab Fraud Scientist Is Accused of Providing False Testimony." *Chicago Tribune.* January 14, 2001.

Possley, Maurice, and Ken Armstrong. "Lab Tech in Botched Case Promoted; Testimony Helped Wrongfully Convict Man of Rape." *Chicago Tribune.* March 19, 1999.

Possley, Maurice, and Steve Mills. "Crime Lab Analyst Hit as 3 Seek New Trials; Testimony Disputed in '86 Slaying Case." *Chicago Tribune.* January 27, 2001.

CHAPTER 15 BULLET IDENTIFICATION, FBI STYLE

Bullet Identification in General

Peters, Charles A. "The Basis for Compositional Bullet Lead Comparisons." *Forensic Science Communications.* July 2002.

United States v. Cornell Winfrei McClure. U.S. District Court for the District of Maryland, *Daubert* Hearing, November 29, 2004.

Earhart Case

Graczyk, Michael. "Junkman Pays for Killing Girl." *Amarillo Globe News.* August 12, 1999.

Halperin, Rich. "Texas Execution." *Death Penalty News.* August 11, 1999.

Behn Case/Charles A. Peters

Gallagher, Mary P. "Shooting Holes in Bullet Analysis." *www.whistleblowers.org.* September 26, 2005.

Gramza, Joyce. "FBI Bullet Lead." *ScienCentral Video News.* November 21, 2003.

Piller, Charles, and Robin Mejia. "Science Casts Doubts on FBI's Bullet Evidence Method." *Los Angeles Times.* February 3, 2003.

"Reasonable Doubt: Can Crime Labs be Trusted?" *CNN Presents.* Transcript of broadcast March 13, 2006.

Sniffen, Michael J. "FBI Drops Controversial Bullet Tests." *www.washingtonpost.com.* September 1, 2005.

State of New Jersey v. Michael S. Behn. Superior Court of New Jersey, Appellate Division, Argued October 12, 2004, Decided March 7, 2005.

William Tobin/Erik Randich

"FBI Bullet-Matching Tests under Fire." Associated Press. *www.msnbc.com.* November 21, 2003.

"FBI Laboratory Announces Discontinuation of Bullet Lead Examinations." Press release. *www.fbi.gov.* September 1, 2005.

Saltzman, Jonathan. "New Trial Sought in Couple's Slaying; Former FBI Agent Challenges

Evidence." *Boston Globe.* February 10, 2005.

Scheck, Barry C. "Statement Calling for Reopening of Cases in Light of National Academies Report on FBI Bullet Lead Analysis." *www.truthinjustice.org.* February 10, 2004.

Solomon, John. "FBI Bullet Analysis Flawed, Imprecise." Associated Press. November 21, 2004.

Ragland Case/Kathleen Lundy

Ortiz, Brandon. "Courts Sift for 'Junk' Science; Ragland Case Shows How Evolving Theories Made More Work for Judges." *Lexington (Ky.) Herald-Leader.* March 25, 2006.

———. "Loss of Evidence Played Down; Prosecutor Talks about Ragland Trial Bullets." *Lexington (Ky.) Herald-Leader.* March 26, 2006.

———. "State's High Court Orders New Trial in Ragland." *Lexington (Ky.) Herald-Leader.* March 23, 2006.

Taylor, Louise. "Ragland Case Lie Sparks Call for FBI Review: Expert Admitted to Perjury about Bullet-Lead Tests." *Lexington (Ky.) Herald-Leader.* July 20, 2002.

CHAPTER 16 THE CELEBRITY EXPERT

Dr. Henry Lee in General

Shaffrey, Ted. "Doctor Lee Will Catch You, Fool!" *New Haven Advocate.* October 6, 2005.

Testimony in Smith, Foster, and Simpson Cases

"O. J. Simpson: Week-by-Week." Week 31, August 21–25, 1995. *www.courttv.com.*

Ruddy, Christopher. "Controversial Lee to Issue Report on Foster's Death." *www.newsmax.com.* November 24, 1995.

———. "'New' Evidence in Foster Case Would Stain Credibility." *www.newsmax.com.* October 26, 1995.

Ramsey and Scott Peterson Cases

Brennan, Charles. "JonBenét Case Adrift after Mom's Death." *Rocky Mountain News.* June 26, 2006.

"Defense to Examine Laci Peterson Remains." CNN TV. *www.transcripts.cnn.com.* Transcript of broadcast August 8, 2003.

Michael Peterson Case

Barnes, Larry A. Elizabeth A. Ratliff Autopsy Report. Ninety-seventh General Hospital, Frankfurt am Main, Germany. November 25, 1985.

Bean, Matt. "Expert Maintains Death of Novelist's Wife Was an Accident." *www.courttv.com.* September 16, 2003.

———. "Forensic Scientist: Blood Evidence 'Inconsistent' with Beating." *www.courttv.com.* September 15, 2003.

———. "Novelist Convicted of First-Degree Murder in Wife's Staircase Death." *www.courttv.com.* October 10, 2003.

Lee, Demaris. "Expert Leads Off the Defense." *Raleigh News and Observer.* September 9, 2003.

North Carolina v. Michael Iver Peterson. Appeals Brief, Fourteenth District, North Carolina Court of Appeals, No. CoA 05–973, *www.vanceholmes.com.* 2006.

Radisch, Deborah L. Elizabeth A. Ratliff Autopsy Report. North Carolina Medical Examiner's Office, Chapel Hill. April 16, 2003.

———. Kathleen Peterson Autopsy Report. North Carolina Medical Examiner's Office, Chapel Hill. December 9, 2001.

Rockman, Bonnie. "Peterson Lawyer Seeks New Trial." *Raleigh News and Observer.* April 19, 2006.

Springer, John. "1985 Death Continues to Dominate Novelist's Murder Trial." *www.courttv.com.* August 26, 2003.

———. "Blood Spatter Expert Spends Sixth, Grueling Day on Stand at Novelist's Trial." *www.courttv.com.* August 20, 2003.

———. "Defense Continues to Grill Evidence Technician." *www.courttv.com.* July 23, 2003.

———. "Expert Gives Jury a Tutorial in Blood Spatter." *www.courttv.com.* August 13, 2003.

———. "Medical Examiner Sticks to Analysis: Peterson's Death a Homicide." *www.courttv.com.* September 5, 2003.

———. "Neuropathologist Says Peterson's Wife Bled for Several Hours." *www.courttv.com.* September 2, 2003.

———. "Novelist's Murder Trial Opens in North Carolina." *www.courttv.com.* July 1, 2003.

———. "Pathologist: Someone Tried to Strangle Kathleen Peterson," *www.courttv.com.* September 24, 2003.

———. "Testimony in 1985 Death in Germany Allowed into Novelist's Murder Trial," *www.courttv.com.* August 22, 2003.

———. "Witness Suggests Blood Spatter Links Novelist to Wife's Death." *www.courttv.com.* August 18, 2003.

———. "Woman's Death in Germany 18 Years Ago Enters Novelist's Murder Trial." *www.courttv.com.* August 19, 2003.

Phil Spector Case

Deutsch, Linda. "Criminalist Says Spector's DNA Wasn't Detected on Gun." Associated Press. June 9, 2007.

———. "Dr. Lee Says He's Being Slandered to Undermine Spector Testimony." Associated Press. May 31, 2007.

———. "Expert Says He's Going to China, May Not Testify in Spector Trial." Associated Press. July 10, 2007.

———. "Ex-Spector Lawyer Goes before Judge." Associated Press. July 12, 2007.

———. "Renowned Forensic Scientist Henry Lee's Credibility Challenged in Phil Spector Murder Trial." Associated Press. May 25, 2007.

Hong, Peter Y. "Spector Trial Claim: Bombshell or Sideshow." *Los Angeles Times.* May 7, 2007.

———. "Spector Trial Expert Backs Suicide Theory." *Los Angeles Times.* June 27, 2007.

Mikulan, Steven. "Phil Noir: Doth the Doc Protest Too Much?" *LA Weekly.* May 31, 2007.

Muskal, Michael. "Ex-Spector Lawyer Relents, Testifies on Possible Evidence." *Los Angeles Times.* July 12, 2007.

Ryan, Harriet. "Criminalist to Spector Jury: Don't Believe Tarantino or Henry Lee When It Comes to Blood." *www.courttv.com.* June 19, 2007.

———. "Judge in Spector Trial Rules Defense Expert Dr. Henry Lee Hid or Destroyed Evidence." *www.courttv.com.* May 23, 2007.

———. "Judge Threatens to Hold Spector's Former Attorney in Contempt of Court for Refusing to Testify." *www.courttv.com*. June 7, 2007.

———. "Misstep in Spector Case Could Haunt Renowned Forensic Scientist for Cases to Come." *www.courttv.com*. May 25, 2007.

———. "Spector Insiders Call into Question the Conduct of Famed Criminalist Henry Lee at Death Scene." *www.courttv.com*. May 4, 2007.

Spano, John. "Ex-Spector Lawyer Avoids Jail—For Now." *Los Angeles Times*. June 23, 2007.

CONCLUSION

General

Giannelli, Paul C. "Crime Labs Need Improvement: The Quality of the Lab Is Criminal; Government Must Invest in Personnel and Facilities." *Issues in Science and Technology*. September 22, 2003.

Oklahoma District Attorneys Council. "Oklahoma's State Plan for the Improvement of Forensic and Medical Examiner Services." State of Oklahoma. February 2006.

Risinger, D. Michael, and Michael J. Saks. "A House with No Foundation: Forensic Science Needs to Build a Base of Rigorous Research to Establish Its Reliability." *Issues in Science and Technology*. September 22, 2003.

Stone, Kit R., and Dan Morrison. "The CSI Effect." *usnews.com*. April 25, 2005.

DNA Backlogs

"New Attorney General Decries 'Monstrous' DNA Backlog." Associated Press. *St. Paul Pioneer Press*. January 14, 2007.

Richmond, Todd. "Labs' Backlog of DNA Testing Grows." *St. Paul Pioneer Press*. December 20, 2006.

Saltzman, Jonathan, and John R. Ellement. "Mass. DNA Lab's Lapse Draw Beacon Hill Inquiry." *Boston Globe*. January 17, 2007.

Siegel, Stephanie. "Family Will Not Sleep While Gunman Free." *Gazette*.net (Montgomery County, Md.). December 6, 2006.

Tracy, Jan. "Union Defends Suspended DNA Database Administrator." *Boston Globe*. January 14, 2007.

"Trial Delayed Because State Crime Lab Hasn't Processed Evidence." Associated Press. *Helena Independent Record*. December 15, 2006.

Weber, David. "Withholding DNA Evidence." Associated Press. January 14, 2007.

Forensic Pathology

Murray, John. "Coroner Defends the Job He's Doing." *Indianapolis Star*. December 3, 2006.

———. "Marion County Coroner Signs Doctor to Contract." *Indianapolis Star*. December 16, 2006.

O'Shaughnessy, Brendan. "Indy Autopsies May Be Farmed Out." *Indianapolis Star*. December 5, 2006.

"Public Announcement: Marion County Coroner's Office Going Out of Business." *www.advanceindiana.blogspot.com*. November 8, 2006.

Standing Bear, Zug G. "Conflict of Interest in U.S. Coroner Systems." www.funeralethics.org/summary.htm. 2004.

Proviano Case

Ayres, Eric. "Villaverde: Proviano Committed Suicide." *St. Clairsville, Ohio, Times Leader.* February 16, 2006.

Gibben, Mark. "A Not Quite Cold Case." *The Malefactor's Register.* markgibben.com. February 27, 2006.

Levin, Steve. "Testimony Ties Woman to Slaying of Medical Student." *Pittsburgh Post-Gazette.* February 16, 2006.

———."Woman Convicted in '97 Killing of Medical Student." *Pittsburgh Post-Gazette.* February 22, 2006.

Fingerprint Identification

Dror, Itiel E., and David Charton. "Why Experts Make Errors." *Journal of Forensic Identification.* April 2006.

Matier, Phillip, and Andrew Ross. "Police Department Dusting Dilemma." *www.sfgate.com.* December 11, 2006.

Monroe, Willie. "Oakland Police without Fingerprint Crime Unit." *www.abclocal.gocom.* December 11, 2006.

Stutzman, Rene. "Analyst in Fingerprint Scandal Quits before Being Fired." *Orlando Sentinel.* June 6, 2007.

———. "Fingerprint Scandal Costs Analyst Her Job." *Orlando Sentinel.* June 7, 2007.

INDEX

ABOUT THE AUTHOR

A graduate of Westminster College (Pennsylvania) and Vanderbilt University Law School, Jim Fisher conducted criminal investigations as a special agent for the FBI from 1966 to 1972. He spent the next thirty years teaching criminal investigation, criminal law, and forensic science at Edinboro University of Pennsylvania. During this period, he investigated a series of crimes that led to the publication of six books in the crime and law genre. Two of these books have been nominated for Edgar Allan Poe Awards in the Best Fact Crime category. His criminal investigations have been featured three times on *NBC Dateline,* as well as on *Current Affair,* the Learning Channel, *Inside Edition,* PBS, CBS, and Fox News. He has spoken at national law enforcement and forensic science conferences and has guest lectured at numerous colleges and universities on the subjects of criminal investigation and forensic science. You can reach Jim Fisher at *jfisher@edinboro.edu* and keep up with developments in forensic science through his Web site, *jimfisher.edinboro.edu.*